# Lecture Notes in Mathematics

Edited by A. Dold, B. Eckmann and F. Takens

W0245839

1428

Karin Erdmann

# Blocks of Tame Representation Type and Related Algebras

Springer-Verlag
Berlin Heidelberg New York London Paris Tokyo Hong Kong

**Author**
Karin Erdmann
Mathematical Institute, Oxford University,
24-29 St. Giles, Oxford OX1 3LB, England

Mathematics Subject Classification (1980): 20C20, 20C05, 20C15, 16A46, 16A48, 16A64

ISBN 3-540-52709-5 Springer-Verlag Berlin Heidelberg New York
ISBN 0-387-52709-5 Springer-Verlag New York Berlin Heidelberg



Printing and binding: Druckhaus Beltz, Hemsbach/Bergstr.
2146/3140-543210 – Printed on acid-free paper

# Introduction

In these notes we shall study algebras which are associated to blocks of tame representation type. These are the 2-blocks whose defect groups are dihedral or semidihedral or (generalized) quaternion. Over the last few years, a range of new results on a class of algebras including such blocks have been obtained. The algebras are essentially defined in terms of their stable Auslander-Reiten quivers (which we shall describe later), and it has been proved that any such algebra is Morita equivalent to one of the algebras in a small list which is explicitly given by generators and relations. In particular, this describes tame blocks; and allows one to extend classical results on the arithmetic properties of these blocks.

The aim here is to provide a comprehensive account of these developments. We include new results, also some new proofs of known results; and the original work has been revised. We also present some general theory on algebras, including study of particular classes of algebras, which we think is important to understand the subject.

Suppose G is a finite group and K is a field of characteristic p; we assume that K is algebraically closed. The group algebra KG is a direct sum of indecomposable algebras, $KG = B_1 \oplus \ldots \oplus B_n$, and the $B_i$ are the blocks of KG. Equivalently, the identity of KG is a sum of orthogonal centrally primitive idempotents $e_i$, and $B_i = e_iKG$.

One main topic in modular representation theory is the study of such blocks, as algebras, and their module categories. It is known that a block is a symmetric algebra and is in particular self-injective. When the prime p divides the group order, then the blocks of KG are usually not semisimple.

An analogue of the role played for KG by a Sylow p-subgroup of G is the *defect group of a block*. In our context, we define a defect group of a block B to be a minimal subgroup D of G such that every B-module is D-projective. (A module M is D-projective if M is isomorphic to a direct summand of $W \otimes_{KD} KG$ for some KD-module W.) The defect groups of a block form one conjugacy class of p-subgroups of G.

Considering a block as an algebra, it has a representation type. In general, an algebra is of *finite representation type* if it has finitely many isomorphism classes of indecomposable modules; otherwise it is *of infinite type*. An algebra of infinite type is *tame* if, roughly speaking, there is a good parametrization of the indecomposable modules; and it is *wild* if its module category is comparable with that of the free algebra in two generators, see I.4.2 and I.4.4. The representation type of blocks is characterized as follows, summarizing a number of results ([BD], [Hi], [$Br_1$]):

THEOREM *Consider the group algebra KG of a finite group G over a field K of characteristic p; or a block B of KG. Suppose D is a Sylow p-subgroup of G; or a defect group of B. Then the representation type of KG; or of B, is*

*(i) finite if D is cyclic;*

*(ii) tame if p = 2 and D is dihedral or semidihedral or generalized quaternion;*

*(iii) wild, otherwise.*

(For comparison, Maschke's theorem on semisimplicity corresponds to the case D = 1.) Blocks of finite type, from the module theoretic point of view, are now well-understood, thanks to [J], [K], but also [$Gr_1$], [$M_2$] and more recently [GR] and many others. Our interest lies in the study of tame blocks and their module cateories.

Historically, block theory started off with the functional approach, and this was developed by R. Brauer. There the object of study are characters, Brauer characters, numerical data such as k(B) (the number of irreducible characters of the block B) and l(B) (the number or irreducible Braer characters of B, that is, the simple B-modules). Much work has been done by Brauer and others; and one of the original motivations seems to have been classification problems for finite simple groups. (In fact, there are powerful results, for example [BS], [GW], [ABG]). Experience shows that these arithmetic data are to a large extent locally determined, that is, depend on p-local subgroups of the given group. In fact the defect groups of a block were discovered first in that context, but in a different form (determined by character values reduced modulo p). This suggests the following approach : Fix a p-group D and study arithmetic properties of arbitrary blocks with defect groups D.

There were results for particular Sylow p-groups of G, or defect groups of a block: An early paper of this type [$B_2$] deals with defect groups of order p, and these results have been generalized by Dade to arbitrary blocks with cyclic defect groups [$Da_1$]. Later, Brauer studied blocks whose defect groups are Klein 4-groups and dihedral groups, and Olsson obtained analogous results for semidihedral and quaternion defect groups. For cyclic defect groups, these arithmetic results include more or less complete information about decomposition numbers and Cartan matrices. For other types of defect groups, however, there were only partial results obtained by functional methods; even with hypotheses on the groups, as in $[L_1]$.

Returning to the module approach, originally the methods used were based on Green correspondence, which exploits module categories of p-local subgroups. This works well for cyclic defect groups and in particular cases but is not as powerful in general.

Since then new discoveries on representations of finite dimensional algebras more generally have had an impact. For tame representation type, Ringel gave a classification of tame local algebras $[R_1]$. Going further in this direction, Donovan classified symmetric tame algebras with two simple modules [Do]; and using Brauer's arithmetical results on dihedral blocks, he obtained the following:

If B is a block with dihedral defect groups then the basic algebra of B belongs to a small list of algebras explicitly given by generators and relations (at least modulo the socle).

The hypotheses he was using, namely "tame, symmetric, elementary divisors of the Cartan matrix" seemed to be too weak to deal with larger numbers of simple modules.

In the meantime, Auslander and Reiten had discovered almost split sequences $[AR_{1,2}]$ and they were used with great success in classification problems.

Concerning blocks, Gabriel and Riedtmann noticed that the graph structure of the stable Auslander-Reiten quiver is the same for all blocks with cyclic defect groups; it is a graph of the form $\mathbb{Z}A_m/e$ (here m is the order of the defect group, and e is the number of projective modules). In [GR] they classified all symmetric algebras, up to Morita equivalence whose stable Auslander-Reiten quiver is of the form $\mathbb{Z}A_n/k$, for arbitrary n and k; this includes an explicit description of the basic algebras by generators and relations. The list contains all blocks of finite type. (As in [Do], this work is independent of groups.)

Later, Riedtmann proved a general theorem which describes the graph structure of stable Auslander-Reiten components (and she used this result as an order principle to classify self-injective algebras of finite type). In the special case of group algebras, the graph structure of Auslander-Reiten components is even more restricted, by [W] (and [HPR]). This encourages one to try a strategy similar to [GR] also for tame blocks; and in fact, this approach is successful.

The algebras we study are defined as follows: Suppose $\Lambda$ is a finite-dimensional algebra over a field K which is algebraically closed, of arbitrary characteristic. We say that $\Lambda$ is of *dihedral (semidihedral, quaternion) type* if it satisfies the following conditions:

(1) $\Lambda$ is tame, symmetric and indecomposable.

(2) The Cartan matrix of $\Lambda$ is non-singular.

(3) the stable Auslander-Reiten quiver of $\Lambda$ has the following components:

| | dihedral type | semidihedral type | quaternion type |
|---|---|---|---|
| tubes | rank 1 and 3 | rank $\leq 3$ | rank $\leq 2$ |
| | at most two 3-tubes | at most one 3-tube | |
| others | $\mathbb{Z}A_{\infty}^{\infty}/\Pi$ | $\mathbb{Z}A_{\infty}^{\infty}$ and $\mathbb{Z}D_{\infty}$ | |

(For the dihedral type, we do not need the hypothesis that $\Lambda$ is tame; this follows from the classification.) If B is a block with dihedral (semidihedral, quaternion) defect groups then B is an algebra of dihedral (semidihedral, quaternion) type, see (VI.1, VII.1, VIII.1).

The aim is to determine these algebras up to Morita equivalence, that is, to determine their basic algebras by generators and relations.

Comparing finite type and tame type, the difficulties are rather different. Given an algebra with stable Auslander-Reiten quiver $\mathbb{Z}A_m/k$; for fixed m and k it is possible with some combinatorics to determine all symmetric basic algebras explicitly by hand. The problem is to give a unified classification for arbitrary m, k. This was solved elegantly by Riedtmann with help of covering theory. On the other hand, for tame type any component of the stable Auslander-Reiten quiver is infinite; and the number of components of a given type is almost always infinite which means that one needs new methods again (and which makes regularity conditions, especially (2), necessary). The crucial observations are that for tame blocks the number of 3-tubes in the stable Auslander-Reiten quiver is finite, and that particular modules must lie at ends of components.

Condition (2) and the restriction to symmetric algebras is motivated by modular representation theory; in fact, symmetric algebras have been first studied by Brauer and Nesbitt [BN]. The importance in our context is that symmetric algebras are self-injective, and, moreover, that for any indecomposable projective module, the simple quotient and the socle are isomorphic. The other main property we use is that for symmetric algebras, $\Omega$-periodicity coincides with $\tau$-periodicity, where $\tau$ is the Auslander translation.

We will now describe the content of these notes in more detail. The first chapter is a general introduction into some representation theory of algebras. It includes a survey on Auslander-Reiten theory; and we also give an outline of some covering theory, in simple special cases, and we apply it to recognize wild algebras. Covering theory for representations of algebras has been developed and used extensively during the last years, especially for finite type (see for example [$Ri_{1,2,3}$], [$G_3$]); but also for tame type ([DS], [Sk]) and in general ([BG], [MP]).

In Chapter II we study special biserial algebras and also local "semidihedral algebras" and their representation theory. Special biserial algebras form an important class of algebras of tame or finite type. Well-known examples are blocks of group algebras with cyclic defect groups (see [Ku], [GR], [SW]), but also algebras appearing in the Gelfand-Ponomarev classification of Harish-Chandra modules over the Lorentz group [GP]. In fact, our results provide another instance: Blocks of group algebras with dihedral defect groups are also special biserial (modulo the socle); this property characterizes them amongst the blocks of infinite type.

The classification of indecomposable modules for special biserial algebras is essentially due to Gelfand-Ponomarev. In [GP] they have considered a special case, however the proof generalizes, see [$R_2$] (also [$DF_3$] and others). There are a number results on Auslander-Reiten theory for special biserial algebras, such as [BS]. More recently Butler and Ringel showed that the Gelfand-Ponamarev technique can be exploited to obtain certain irreducible maps [BR], and our account is based on this. There is also a new and elegant way to deal with the representation theory of special biserial algebras with covering theory, due to [DS].

For a long time, no explicit classification for the modules of the local semidihedral algebras $\Lambda_m$ (see II.9) was known; but recently B. Crowley-Boevey has obtained a good parametrization [$CB_{2,3}$]. In fact, he has found a natural way to extend the idea of functorial filtration in [GP], and his work is of great importance in our context. We use his results to prove that the non-periodic stable Auslander-Reiten components of local semidihedral algebras are of tree class $A_\infty^\infty$ or $D_\infty$. This is new; and it is essential later for determining the stable Auslander-Reiten quiver for arbitrary blocks with semidihedral defect groups.

In Chapter III we study tame local symmetric algebras in general. This is a revision of Ringel's work [$R_1$], restricted to the special case of symmetric algebras. We give a list of these algebras, by generators and relations. One reason for doing this is the observation that they are all either of dihedral type, or of semidihedral type, or else the simple module has $\Omega$-period 4 (in which case, the algebra is very likely of quaternion type). Moreover, we use local tame algebras later to determine relations for arbitrary algebras of quaternion type. Another application is the structure of the centre of these algebras. We find that a local tame symmetric algebra which is not commutative must be of dimension 4k and its centre has dimension k+3. The interest of centres in modular representation theory arises from the fact that the dimension of the centre of a block B is the same as k(B); see also [CK].

In Chapter IV we collect general observations on modules with particular (small) $\Omega$-period; we also determine all quivers for arbitrary tame local symmetric algebras with at most three simple modules. Then we come to a central part, namely to principles how to exploit the graph structure of the stable Auslander-Reiten quiver to information on the structure of projective modules. (A short illustration may be found in IV.3.8 where we determine all indecomposable symmetric algebras whose stable Auslander-Reiten quiver contains a component $\cong \mathbb{Z}\tilde{A}_{1,2}$.) The most important results here are methods to show that certain modules lie at ends of components; these go back to [$E_4$]. In IV.4 we give the proof of a theorem due to Butler and Ringel [BR], for arbitrary algebras; and in IV.5 we prove a generalization of (2.8),[$E_4$], valid for self-injective algebras. Then we study whether these modules lie at "a-ends" (components with tree class $A_\infty$ or $A_n$), or at "d-ends" (components with tree class $D_\infty$ or $D_n$ or $\tilde{D}_n$). Our motivation is the study of algebras of dihedral (and semidihedral) type; here the only ends occur at tubes (and also at $\mathbb{Z}D_\infty$-components). Actually, since the results of these sections use only the local structure of an Auslander-Reiten component, they can also be used for self-injective algebras of finite type, of type $A_n$ or $D_n$.

Chapter V deals mostly with group representations; the main aim is to determine the structure of the stable Auslander-Reiten quiver $\Gamma_s(B)$ for an arbitrary tame block B. We start with a survey on the relevant modular representation theory; then we determine the Morita equivalence classes for a few tame blocks of 2-local groups by hand; here we use a number of rather small general algebras, such as skew group rings and exploit the fact that the structure of these small algebras is rather restricted. (We note that these results also follow from more general theorems in [Ku], [KP], [$Pu_{1,2}$].) In the next section we summarize more generally methods for relating stable Auslander-Reiten components of group algebras for groups with those for p-local subgroups. This includes new results due to Kawata [$K_{1,2}$]. We use these techniques and earlier results to determine $\Gamma_s(B)$ for arbitrary tame blocks. In the last section we apply the theory of [AB] to determine the number of irreducible characters of tame blocks; the results are due to [$B_{2,3}$] and [O].

In Chapter VI we classify algebras of dihedral type. The proofs we give here are new; and we also include the algebras with two simple modules which have not been studied in [$E_4$]; and therefore we have a new approach (and improvement) of Donovan's results. Moreover, we have now simplified and generalized the definition, compared with [$E_4$] (where we assumed that each 3-tube is fixed by $\Omega$). Therefore, we obtain a number of new algebras. To classify them, creates no problem.

In Chapter VII, we deal with algebras of quaternion type. We have a small list of algebras explicitly given by generators and relations, and we show that the basic algebra of an algebra of quaternion type belongs to this list. It is convenient for this type of algebras to work directly with the basic algebra since the hypotheses give immediately information about $J^2$ (modulo $J^3$). The treatment here is streamlined and shortened compared with [$E_6$]; we also include algebras which were omitted there; and the relations were not always correctly stated.

In Chapter VIII we study algebras of semidihedral type. We show that any such algebra which is basic belongs to a small list of algebras given explicitly by generators and relations. This is a different approach from that in [$E_5$]: we do not make assumptions about the number of simple modules. As an order principle we use various possible positions of simple modules in the stable Auslander-Reiten quiver and as for the dihedral type, modules at ends of components; however, the ends of $\mathbb{Z}D_\infty$-components increase the complexity of the problem considerably. We obtain as a corollary that the number of simple modules of these algebras is bounded by three; this is a new result. The work for this type of algebra is rather more involved than for the others; in fact, for part of the work one is close to algebras of finite type (or algebras with Euclidean components); this recalls how much more work the self-injective algebras of type $D_n$ required, compared with $A_n$ (see [$Ri_{2,3}$]). Parallel to this, one might compare the relative lengths of [ABG] and [GW].

The main aim of Chapter IX is to determine which of these algebras can occur as blocks, and to prove functional results. The proofs here are new (and simplified), continuing [$E_9$]; they do not assume results from [$B_{2,3}$], [O]. Instead, we use only formulae for dim $Z(\Lambda)$ from our list of algebras and k(B), together with a few general principles. The results are a complete list of possible Cartan matrices for tame blocks, and for each of them the decomposition matrix (which is unique). We note that finding the decomposition matrix is very elementary; it also does not depend on defect groups. The method also determines without extra work generalized decomposition matrices $D^j$ where j is a central involution of the defect group.

In Chapter X, we give a proof of the theorem by Brauer and Suzuki. As a further application we determine the Morita equivalence classes for a family of blocks for alternating groups. Moreover, we include some comments about misprints and errors in earlier work.

The families of algebras obtained in VI to VIII are listed at the end. We have adopted the following labelling: D, SD, Q denotes the type of algebra, and then we name all possible quivers; in most cases there is not more than one family per type of algebra with a given quiver, sometimes there are two (or three). Algebras which are isomorphic modulo the socle usually belong to the same family. In the tables, we do not give details on socle scalars, except for dihedral type; although in Chapter VI, VII we included some scalar transformations. The second part of the tables contains conditions to be satisfied for blocks, and the decomposition matrices, together with examples for blocks.

When determing relations, we tacitly use some standard arguments. Namely, there are automatic zero-relations satisfied by symmetric basic algebras, due to the fact that $e\Lambda f \cap \mathrm{soc}\,\Lambda = 0$ for orthogonal idempotents e, f. Moreover, to show that the given relations are sufficient, one verifies that dim $KQ/I$ is the same as dim $\Lambda$ where $\Lambda$ is the algebra considered; and to show that an algebra $KQ/I$ is symmetric one easily defines a symmetrizing form.

A number of problems remain open. Concerning the algebras, it is not proved whether the algebras in the list are tame. Possibly a new result by Bautista [B] could help. Going further, one would like to have a good parametrization for the indecomposable representations; perhaps there are refinements of the methods in [$CB_{2,3}$]. However, for quaternion type, it is still open at present how to deal with the local algebra. As for blocks, we ask whether there are examples in the remaining cases where the arithmetic conditions are satisfied. It may not be possible to answer this by general theory; bearing in mind that the similar question for blocks of finite type is also not answered.

There is a conjecture by Brauer, namely: Given a p-group D, there are only finitely many Cartan matrices which can occur for p-blocks with defect group D. A stronger question was asked by P. Donovan: Given a p-group D, is the number of Morita equivalence classes of p-blocks with defect group D then finite?

For tame blocks, we confirm Brauer's conjecture; on the other hand, we are not able to answer Donovan's question. For some families of blocks of quaternion type, there are infinitely many non-isomorphic algebras which are isomorphic modulo the socle.

The mathematical work for these notes and the writing was mostly done whilst I was visiting the University of Essen; I am grateful to Professor Michler for inviting me, and to the DFG for financial support.

The very first attempts started with the group theoretic approach via Green correspondence. However, J. Alperin pointed out to me that my work on Klein 4-groups contained an error (and I am indebted to him for this); so I had to revise my strategy, and it was possible to find a correct proof for the result using Auslander-Reiten theory. Thanks to the suggestion by Professor Michler, I continued, first with the dihedral case and then with the other types, and this was a rather interesting project. I greatly appreciate various discussions with members of the groups in Essen and in Bielefeld and with B. Crowley-Boevey, L. Kovacs, M. Schaps and with A. Skowronski. Also, I should like to thank the Universities of Essen (and Oxford) for hospitality and for providing technical facilities.

# TABLE OF CONTENTS

# I. *Algebras, quivers, representation type, Auslander-Reiten theory, coverings*

## I.1 *Background*

We usually assume that K is an algebraically closed field of arbitrary characteristic. By an algebra A, we mean a finite dimensional K-algebra (associative, with 1). The radical of A is denoted by J or J(A) or rad A. Recall that J is nilpotent, and A/J is semisimple; in our context, A itself is not semisimple.

We consider only finite-dimensional modules; and most often we work with right modules, in the category mod A. Note that the Krull-Schmidt theorem applies; that is, every $M \in \text{mod } A$ is a direct sum of indecomposable A-modules, and such a decomposition is unique up to isomorphism.

Recall that a module M is semisimple if and only if $MJ = 0$. In general, soc M is the largest semisimple submodule of M; and top M is the largest semisimple factor module of M, that is, top $M = M/MJ$. Note also that $MJ \cong \text{rad } M$, the intersection of all maximal submodules of M.

The socle series of M is, by definition, the sequence of submodules

$$0 \subsetneq \text{soc}_1(M) \subsetneq \text{soc}_2(M) \subsetneq \dots \subsetneq \text{soc}_k(M) \subsetneq \dots \subsetneq M$$

with $\text{soc}_1(M) = \text{soc } M$ and $\text{soc}_k(M)/\text{soc}_{k-1}(M) = \text{soc}(M/\text{soc}_{k-1}(M))$. The socle length of M is the first integer r such that $\text{soc}_r(M) = \text{soc}_{r+1}(M)$.

The radical series of M is defined to be the sequence of submodules

$$0 \subsetneq \dots \subsetneq \text{rad}^k(M) \subsetneq \dots \subsetneq \text{rad}^2(M) \subsetneq \text{rad}(M) \subsetneq M$$

with $\text{rad}^k(M) = \text{rad}(\text{rad}^{k-1}(M))$ [ $\cong MJ^k$]. The radical length of M is the first integer l such that $\text{rad}^l(M) = \text{rad}^{l+1}(M)$.

We note that $M \neq 0$ implies soc $M \neq 0$ and top $M \neq 0$. In particular, if M has socle length r and radical length l then $\text{soc}_r(M) = 0$ and $\text{rad}^l(M) = 0$. Actually, $r = l$ and this number is called the Loewy length of M.

An indecomposable projective module P is an indecomposable summand of $A_A$. Any such P is of the form eA for some primitive idempotent e of A; recall that decomposing $A_A$ into a direct sum of indecomposable projectives is the same as expressing $1_A$ as a sum of orthogonal primitive idempotents. If $P = eA$ is indecomposable projective then

$eA/eJ$ is simple, and one has a bijection between the indecomposable projectives and the simple A-modules, via $eA \longleftrightarrow eJ$. If S is simple we sometimes write $S = S(e)$ when $S \cong eA/eJ$. Dually, any indecomposable injective A-module has a simple socle.

Suppose P is indecomposable projective and $S = P/PJ$. Let $D = \text{End } S$; then by Schur's Lemma, D is a division ring. If K is algebraically closed then $D \cong K$. Moreover, $\text{End } P/ \text{ rad End } P \cong D$, and for $n \geq 1$ more generally $\text{End}(nP)/\text{rad End}(nP)$ is isomorphic to the full matrix ring $M_n(D)$.

The Cartan matrix of the algebra A has entries $c_{ij} = \dim \text{Hom}(P_j, P_i)$ where $P_i$, $P_j$ run through the isomorphism types of indecomposable projective A-modules. Alternatively, $c_{ij} = \dim e_i A e_j$ where $e_i$ and $e_j$ run through a complete set of non-isomorphic primitive orthogonal idempotents. Note that $c_{ij}$ is also equal to the number of composition factors of $P_i$ which are isomorphic to $P_j/P_jJ$ when K is algebraically closed.

Usually we assume that the algebra A is self-injective, that is, $A_A$ is an injective module. Then indecomposable projectives and indecomposable injectives coincide; in particular they have simple socles and tops. If P is indecomposable projective we write $H(P) = \text{rad } P/\text{soc } P$, and we call this module the "heart" of P.

It is well-known that modules over arbitrary rings have an injective hull. In particular, for a module M of a finite-dimensional algebra A, there is an injective module $E = E(M)$ containing M and $\text{soc } E = \text{soc } M$.

Every module M of a finite-dimensional algebra A has a projective cover, that is, there is an epimorphism $\pi: P \to M$ where P is projective and $\text{top } P \cong \text{top } M$. The "Heller translate" of M, denoted by $\Omega M$, is by definition the kernel of $\pi$; it is unique up to isomorphism.

Suppose that A is self-injective. If M is indecomposable then so is $\Omega M$; and if M is not projective then $\Omega M$ is non-zero and not projective. Moreover, for $Z \in \text{mod } A$ one defines $\Omega^{-1}Z$ to be the cokernel of an injective hull $0 \to Z \to E(Z)$. As the notation suggests, for M indecomposable and non-projective we have that $M \cong \Omega^{-1}[\Omega M]$ and also $M \cong \Omega[\Omega^{-1}M]$. Moreover, $\text{top } M \cong \text{soc } \Omega M$ in case $\Lambda$ is symmetric.

A module is ($\Omega$-)periodic if $M \neq 0$ and for some $k \in \mathbb{N}$ we have $M \cong \Omega^k M$; if k is the smallest such integer then M has period k.

Let $M$ be an $A$-module, and suppose $\dots P_n \to P_{n-1} \to \dots \to P_1 \to P_0 \to M$ is a minimal projective resolution of $M$. The complexity of $M$, which describes the rate of growth of $\{ P_n \}$, is by definition the least integer $s \geq 0$ such that $\lim_{n \to \infty} (\dim P_n)/n^s = 0$, if such $s$ exists. A projective module has complexity zero, and the complexity of an $\Omega$-periodic module is $\leq 1$.

If $X$ and $Y$ are $A$-modules we write sometimes $(X, Y)$ for $\mathrm{Hom}_A(X, Y)$. Moreover $\mathcal{P}(X, Y)$ is the subspace of $(X, Y)$ consisting of those maps which factor through some projective module; and we write $\underline{\mathrm{Hom}}_A(X, Y)$ for the quotient $(X, Y)/\mathcal{P}(X, Y)$. Suppose $A$ is self-injective, then the following holds: If $S$ is simple and $X$ is a module which does not have projective summands then $\underline{\mathrm{Hom}}_A(S, X) \cong \mathrm{Hom}_A(S, X)$ and $\underline{\mathrm{Hom}}_A(X, S) \cong \mathrm{Hom}_A(X, S)$. Moreover, for arbitrary $X$ and $Y$, one has $\underline{\mathrm{Hom}}_A(X, Y) \cong \underline{\mathrm{Hom}}_A(\Omega X, \Omega Y)$.

It is well-known that for $A$ self-injective, we have that $\mathrm{soc}({}_A A) = \mathrm{soc}(A_A)$ (see [N]). This property has the consequence that for every $k \in \mathbb{N}$, the right annihilator and the left annihilator of $J^k$ coincide. We note also that a module $M$ has no projective summands if and only if $Mx = 0$ for all $x \in \mathrm{soc}\, A$.

Recall that the group algebra of a group $G$ over $K$, denoted by $KG$, is the algebra with $K$-basis $\{ \bar{g}: g \in G \}$, and where the multiplication is defined by $\bar{g}\,\bar{h} = \overline{gh}$, and linear extension. We usually identify $g$ and $\bar{g}$. We write $K\langle X, Y\rangle$ for the (non-commutative) free algebra in two generators. On the other hand, an algebra $K[X, Y, \dots ]$ is commutative.

Concerning further notation, we write uniserial modules as $\mathcal{U}(S_0, S_1, \dots, S_k)$ with simple modules $S_i$. If $M$ is any module, its dimension vector is denoted by $\underline{\dim}\, M$, and $|M|$ is the composition length of $M$. Moreover, we write ind $\Lambda$ for a set of representatives for the indecomposable $\Lambda$-modules.

## I.2 *Morita equivalence, basic algebras*

I.2.1 By definition, two rings R and S are *Morita equivalent* if they have equivalent module categories. We write $R \sim_M S$.

By Morita's Theorem, the following are equivalent:

(i) mod R is equivalent to mod S.

(ii) There is a projective generator $P_R$ in mod-R such that $S \cong End(P_R)$.

(iii) There is a projective generator ${}_RQ$ in R-mod such that $S \cong End({}_RQ)$.

(iv) R-mod is equivalent to S-mod.

Suppose (ii) holds, then an equivalence mod R ≅ mod S is obtained via the functors F: mod R → mod S and G: mod S → mod R defined by

$$F(M) = Hom_R(P, M) \quad \text{and} \quad G(W) = W \otimes_S P.$$

(see [M], or [AF] Chapter 22).

I.2.2 *The centre of a ring is Morita invariant.*

Proof: Let $\mathcal{A}$ be any abelian category, and let I be the identity functor. Then the set Nat(I) of natural transformations of I forms a ring, the "centre of $\mathcal{A}$".

Now assume that $\mathcal{A}$ = mod A for some ring A, and let C be the centre of A. There is a ring homomorphism $\mu$: C → Nat(I) where $\mu_c$ is right multiplication by c. This is an isomorphism.

I.2.3 *The lattice of right ideals is Morita invariant.*
(See for example [AF] 21.11).

For finite dimensional algebras, Morita equivalence classes are parametrized by basic algebras. The idea of a basic algebra goes back to Osima $[Os_{1,2}]$.

I.2.4 Let A be a finite-dimensional K-algebra where K is algebraically closed. We say that A is *basic* if all simple A-modules are 1-dimensional. More generally, if R is an arbitrary Artin ring then R is basic if R/J(R) is a direct sum of division rings.

I.2.5 LEMMA (a) *Assume that P is a projective module. Then End(P) is basic if and only if P is a direct sum of pairwise non-isomorphic indecomposables.*
(b) *Let A be an arbitrary algebra. Then A is basic if and only if $A_A$ is a direct sum of non-isomorphic indecomposables.*

Proof: (a) Fix a decomposition $P = \oplus P_{ij}$ where $P_{ij} \cong P_{kl}$ if and only if $i = k$. Let $E = \mathrm{End}(P)$. We may write $E = \{ [\phi_{\lambda\mu}]$ where $\phi_{\lambda\mu} \in \mathrm{Hom}(P_\mu, P_\lambda)\}$. Then $[\phi_{\lambda\mu}]$ belongs to J(E) if and only if $\phi_{\lambda\mu}$ has no right-or left inverse, for all $\lambda$, $\mu$. Hence $E/J(E) \cong \oplus M_{n_i}(D_i)$ where $D_i = \mathrm{End}(P_{i1})/\mathrm{rad}\,\mathrm{End}(P_{i1})$, and $n_i$ is the number of j with $P_{ij} \cong P_{i1}$. This is a direct sum of division rings if and only if $n_i = 1$ for all i. Part (b) follows from (a) since $A \cong \mathrm{End}(A_A)$.

I.2.6 LEMMA *Two basic algebras are Morita equivalent if and only if they are isomorphic.*

Proof: Suppose A and B are basic algebras, and $A \sim_M B$. By Morita's Theorem, there is a projective generator $P_A$ such that $B \cong \mathrm{End}(P)$. Since B is basic, we deduce from I.2.5 that P is multiplicity-free as a direct sum of A-modules. Moreover, since A is basic and P is a projective generator, it follows that $P \cong A_A$. Therefore End(P) = $\mathrm{End}(A_A) \cong A$. The other direction is trivial.

I.2.7 COROLLARY *If A is a finite-dimensional K-algebra then there is a unique basic algebra $A_0$ with $A_0 \sim_M A$.*

Proof: Take a full set of non-isomorphic indecomposable projective A-modules $P_1$, ..., $P_n$, say, let $P = \bigoplus_{i=1}^{n} P_i$ and $A_0$ = End P. Then P is a projective generator and End P is therefore Morita-equivalent to A. Moreover, $A_0$ is basic by I.2.5; and the uniqueness has been proved in I.2.6.

Let A be an arbitrary algebra. The algebra $A_0$ above is, by definition, *the basic algebra of A*. It may be constructed as follows: Write $1_A = \Sigma\ e_{ij}$ where the $e_{ij}$ are orthogonal primitive idempotents, such that $e_{ij}A \cong e_{kl}A$ if and only if i = k. Set $e = \Sigma\ e_{i1}$; and one can take $A_0 = eAe$.

I.2.8 *The Cartan matrix of an algebra is invariant under Morita equivalence.*

This is clear for an equivalence of module categories.

## I.3 *Symmetric Algebras*

Assume that A is a finite dimensional K-algebra.

I.3.1 The algebra A is *symmetric* if there is a K-linear map $\psi$: A → K such that $\psi(ab) = \psi(ba)$ for all a, b ∈ A and such that ker $\psi$ does not contain any non-zero right ideal of A. We say that $\psi$ is a symmetrizing form for A.

I.3.2 EXAMPLES (1) If A is the full matrix algebra then A is symmetric, here $\psi$ can be taken as the usual trace.

(2) Let A = KG where G is some finite group. Then A is symmetric, with symmetrizing form $\psi(\Sigma\ a_g g) = a_1$.

I.3.3 LEMMA *For an algebra A, the following are equivalent:*

*(i) A is symmetric.*

*(ii) For every idempotent e of A, the algebra eAe is symmetric.*

*(iii) The basic algebra of A is symmetric.*

Proof: (i) ⇒ (ii) Let f = 1-e, then e and f are orthogonal idempotents, and A = eAe ⊕ eAf ⊕ fAe ⊕ fAf. Suppose $\psi$ is a symmetrizing form for A; then the restriction $\psi_0$ of $\psi$ to eAe is a symmetric linear form. Suppose I is a right ideal of eAe and $\psi_0(I) = 0$. Then IA is a right ideal of A and I ⊆ IA; and IA = eIAe ⊕ eIAf = I ⊕ eIAf. We have that $\psi(eIAf) = \psi(feIAf) = 0$ and $\psi(eIAe) = \psi_0(I) = 0$. Hence IA = 0 and

$I = 0$.

(iii) is a special case of (ii).

(iii) ⇒ (i) (see also [Do]). Let $P_1, P_2, \ldots, P_n$ be a full set of indecomposable projective A-modules. We identify the basic algebra $A_0$ of A with $\mathrm{End}_A(\oplus P_i)$; and moreover, A with $\mathrm{End}_A(\oplus P_i^{d_i})$ for appropriate integers $d_i$. Now suppose $\psi_0$ is a symmetrizing form for $A_0$. We define $\psi: A \to K$ on elements $a_{ij} \in \mathrm{Hom}_A(P_i, P_j)$ by $\psi[a_{ij}] = \psi_0[a_{ij}]$ and extend this linearly. Then $\psi$ is a symmetric linear form since $\psi_0$ is symmetric. Suppose I is a right ideal of A and $\psi(I) = 0$. We have $A_0 = eAe$ for some idempotent e of A; and eIe is a right ideal of $A_0$ contained in ker $\psi_0$. Consequently $eIe = 0$; and using I.2.3 (or directly) one sees that $I = 0$.

In particular, we have the following:

I.3.3.1 *If A is symmetric and P is indecomposable projective then $\mathrm{End}_A(P)$ is also symmetric.*

It is important that symmetric algebras are self-injective. For basic algebras there is the following characterization, which was proved by Nakayama [N].

I.3.4 PROPOSITION *Let A be a finite-dimensional basic algebra. Then the following conditions are equivalent:*

*(i) A is self-injective.*

*(ii) There is an isomorphism $\Theta : A_A \to \mathrm{Hom}_K({}_AA, K)$ of A-modules.*

*(iii) There is a linear form $\psi: A \to K$ such that ker $\psi$ does not contain a non-zero right ideal.*

*(iv) If $A = \bigoplus_{i=1}^{q} Ae_i$ where the $Ae_i$ are indecomposable A-modules then there is a permutation $\nu$ of $\{1, 2, \ldots, q\}$ such that soc $Ae_i \cong$ top $Ae_{\nu(i)}$ and soc $e_{\nu(i)}A \cong$ top $e_iA$, $1 \le i \le q$.*

We see now that a basic algebra is symmetric if and only if it is self-injective and has a linear form $\psi$ satisfying (iii) which is in addition symmetric. If this is the case then the permutation $\nu$ in (iv) (the "Nakayama permutation") is the identity:

<u>I.3.5</u> LEMMA *Assume that A is symmetric, and let $S = soc(A_A)$. Then $eAf \cap S = 0$ if e, f are orthogonal idempotents of A.*

Proof: We may assume that A is basic and that e, f are primitive. Let $\psi$ be a symmetrizing form of A. Take $s \in eAf \cap S$. Then sK is a right ideal of A, and $\psi(s) = \psi(esf) = \psi(fes) = \psi(0) = 0$. It follows that $s = 0$.

<u>I.3.6</u> REMARK An algebra A is *weakly symmetric* if for any indecomposable projective A-module eA one has top eA $\cong$ soc eA. By I.3.4, a weakly symmetric algebra is self-injective. Moreover, by I.3.5 (and I.3.4) we see that symmetric implies weakly symmetric.

EXAMPLE Consider a 4-dimensional local algebra

$$A = K\langle X, Y\rangle/(X^2, Y^2, XY - rYX), \quad 0 \neq r \in K.$$

Then soc A is simple, and since A is local it must be weakly symmetric. In fact, one defines a non-degenerate linear form satisfying I.3.4(iii) by $\psi(XY) = 1$ and $\psi(X) = 0 = \psi(1) = \psi(Y)$.

Moreover, A is symmetric if and only if $r = 1$: Clearly, if $r = 1$ then $\psi$ is symmetric. Conversely, suppose A is symmetric, with symmetrizing form $\lambda$. Since soc A $= \langle YX \rangle$ we have that $0 \neq \lambda(YX) = \lambda(XY) = \lambda(rYX) = r\lambda(YX)$ and $r = 1$.

<u>I.3.7</u> LEMMA *Let A be a symmetric algebra. Then the Cartan matrix of A is symmetric.*

Proof: We may assume that A is basic. Let $\psi$ be a symmetrizing form of A. Fix two primitive idempotents e, f of A; we have to show that dim eAf = dim fAe. Define a bilinear map $eAf \times fAe \to K$ by $(w,z) \to \psi(wz)$; it suffices to show that this is non-singular. Let $w \in eAf$ such that $\psi(wz) = 0$ for all $z \in fAe$. Since $wA = wfA = wfAe + wfA(1-e)$ and $\psi[wfA(1-e)] = \psi[(1-e)wfA] = 0$, it follows that $\psi(wA) = 0$; and therefore $w = 0$.

<u>I.3.8</u> LEMMA *Let A be symmetric and indecomposable. Then soc $A \subseteq J^2$ unless A is local and $J^2 = 0$.*

Proof: Suppose soc $\Lambda \not\subseteq J^2$. Then there is a primitive idempotent e of $\Lambda$ such that soc $e\Lambda \not\subseteq eJ^2$. If $eJ^2$ were non-zero then $0 \neq$ soc $eJ^2 \subseteq$ soc $e\Lambda$, and since soc $e\Lambda$ is simple, soc $eJ^2$ = soc $e\Lambda \subseteq eJ^2$, a contradiction. Hence $eJ^2 = 0$ and $eJ \subseteq$ soc $e\Lambda$. In particular $e\Lambda = e\Lambda e$. Since $\Lambda$ is indecomposable, we deduce $\Lambda = e\Lambda e$ and $J^2 = 0$.

I.3.9 *Suppose the algebra $\Lambda$ is symmetric and basic. Then $e(soc_2\Lambda)e$ is contained in the centre of $\Lambda$, for any primitive idempotent e of $\Lambda$.*

Proof: Suppose $w \in e(soc_2\Lambda)e$, and let $\alpha \in J$ be an arrow; then $w\alpha - \alpha w$ spans an idieal contained in soc $\Lambda$. Let $\Psi$ be a symmetrizing form for $\Lambda$, then $\Psi[w\alpha - \alpha w] = 0$ and hence $w\alpha - \alpha w = 0$.

## I.4 *The representation type*

Let $\Lambda$ be a finite-dimensional K-algebra. Then $\Lambda$ is said to be *of finite representation type* (or "of finite type") provided there are finitely many indecomposable $\Lambda$-modules. Otherwise, $\Lambda$ is *of infinite type*.

I.4.1 EXAMPLE $\Lambda = K[X]/(X^n)$. A $\Lambda$-module is a vector space together with a linear transformation A satisfying $A^n = 0$. The module is indecomposable if and only if A is similar to a Jordan canonical form

$$\begin{bmatrix} 0 & 1 & 0 & \cdots & 0 \\ 0 & 0 & 1 & \cdots & 0 \\ \vdots & & \ddots & \ddots & \\ \vdots & & & \ddots & 1 \\ 0 & & \cdots & & 0 \end{bmatrix}$$

Hence there are precisely n indecomposable $\Lambda$-modules.

Amongst the algebras of infinite type, one distinguishes the *tame* algebras. This means, informally, for any $d \geq 1$, almost all indecomposable modules of dimension d belong to finitely many 1-parameter families. More precisely, following [D]:

I.4.2 The algebra $\Lambda$ is of *tame representation type* provided $\Lambda$ is not of finite type, whereas for any dimension $d > 0$, there are a finite number of $K[T]$-$\Lambda$-bimodules $M_i$ which are free as left $K[T]$-modules such that all but a finite number of indecomposable Λ-modules of dimension d are isomorphic to $N \otimes_{K[T]} M_i$ for some i and some simple K[T]-module N.

We note that an algebra is tame if and only if its basic algebra is tame.

I.4.3 EXAMPLE Let Λ be the 4-dimensional commutative algebra $\Lambda = K[X, Y]/(X^2, Y^2)$, the "Kronecker algebra". If char K = 2 then Λ is isomorphic to the group algebra of the Klein 4-group: If V = < a, b > then an isomorphism KV → Λ is given by a → (1 - X) and b → (1 - Y) and extension.

The indecomposable Λ-modules (see for example [Ba],[$C_1$]) Suppose M is a Λ-module which does not have projective summands. Then soc Λ, that is, $J^2$, annihilates M and $MJ \subseteq$ soc M. If we represent X and Y with respect to a basis of M which contains a basis of soc M then

$$X \to \begin{bmatrix} 0 & A \\ 0 & 0 \end{bmatrix} \quad \text{and} \quad Y \to \begin{bmatrix} 0 & B \\ 0 & 0 \end{bmatrix}.$$

Finding a canonical form for M is then equivalent to classifying pairs of matrices (A, B) with respect to the equivalence relation

$$(A, B) \sim (A', B') \iff \text{There are invertible matrices } P, Q \text{ such that } A' = PAQ \text{ and } B' = PBQ.$$

This has been solved by Kronecker at the end of the last century [Kr]. The list of indecomposables for K algebraically closed is as follows:

(a) K of dimension 1, and Λ.

(b) There are two indecomposable modules of dimension 2n + 1 for n = 1, 2, ... . The matrices A, B are

$$\text{(i)} \quad \begin{bmatrix} 1 & 0 & 0 & & & \\ & 1 & 0 & 0 & & \\ & & \ddots & & & \\ & & & & 0 & 0 \\ & & & & 1 & 0 \end{bmatrix} \quad \text{and} \quad \begin{bmatrix} 0 & 1 & 0 & & & \\ 0 & 0 & 1 & 0 & & \\ & & & \ddots & & \\ & & & & 1 & 0 \\ & & & & & 1 \end{bmatrix}$$

(ii) the transposes of the matrices in (i).

(These modules are isomorphic to $\Omega^{\pm n}(K)$, $n \geq 1$.)

(ii) For each $\lambda \in K \cup \{\infty\}$ and for each positive integer n, there is an indecomposable Λ-module $C_n(\lambda)$ of dimension 2n on which

$$A = I_n, \text{ and } B = J(n,\lambda) \text{ when } \lambda \in K, \text{ and}$$

$$A = J(n,0) \text{ and } B = I_n \text{ when } \lambda = \infty.$$

(Here $J(n,\lambda)$ is the indecomposable Jordan matrix of size n with eigenvalue $\lambda$.)

The algebra $\Lambda$ is tame: If d is odd then there are at most two indecomposable modules of dimension d, which we may ignore.

Now let $d = 2n$. Take $M = K[T]^{2n}$, and define a $\Lambda$-right action on M by setting

$$\alpha \to \begin{bmatrix} 0 & I_n \\ 0 & 0 \end{bmatrix} \quad \text{and} \quad \beta \to \begin{bmatrix} 0 & J(n,T) \\ 0 & 0 \end{bmatrix}.$$

Then M is a $K[T]$-$\Lambda$-bimodule. The simple $k[T]$-modules are the modules of the form $S_\lambda := K[T]/(T-\lambda)$, $\lambda \in K$. Then $S_\lambda \otimes_{K[T]} M \cong C_n(\lambda)$. This parametrizes the modules of dimension 2n, except for one (or two if $n = 2$).

I.4.3.1 *Let $\Lambda$ be a local algebra. Then $\Lambda$ is of finite type if and only if $\Lambda \cong K[X]/(X^m)$ for some $m \geq 0$.*

Proof: If $\Lambda$ is of finite type then its radical $J(\Lambda)$ must be cyclic: Otherwise $\Lambda$ would have a factor algebra $\bar{\Lambda} \cong K[X, Y]/(XY, X^2, Y^2)$. This algebra is of infinite type since all non-projective modules in I.4.3 may be considered as $\bar{\Lambda}$-modules. Hence $\Lambda \cong K[X]/I$ for some ideal $I \subseteq (X^2)$ ( or $\Lambda \cong K$ ), and then $I = (X^m)$. For the converse, see I.4.1.

Otherwise, there is the concept of *wild* representation type. This means, roughly speaking, that mod $\Lambda$ is comparable with mod K<X, Y>. It appears that the choice of name was inspired by the fact that for any finite-dimensional algebra R, mod R may be embedded into mod K<X, Y>; for a general discussion, see [$Br_2$] or [$G_1$]. Concerning a formal definition of wild, there is some variation in the literature. The following, due to Drozd [D] is now the prefered one.

I.4.4 We say that a finite-dimensional K-algebra $\Lambda$ is *wild* if there is a finitely generated K<X,Y>-$\Lambda$-bimodule B which is free as a left K<X,Y>-module such that the functor $- \otimes_{K<X,Y>} B$ from mod-K<X,Y> to mod-$\Lambda$ preserves indecomposability and reflects isomorphisms.

I.4.5 EXAMPLE Let $\Lambda$ be the 4-dimensional local algebra $\Lambda = K[X_0, X_1, X_2]/(X_iX_j)$.

If char K = 2 then $\Lambda \cong KD/J^2$ where D is the elementary abelian group $(\mathbb{Z}_2)^3$ of order 8. Let B be the free K<X, Y>-left module of rank 2, with basis $b_1$, $b_2$. On B, define a right $\Lambda$-action by

$$wb_1X_i = 0 \quad \text{and} \quad wb_2X_0 = wb_1, \quad wb_2X_1 = wXb_1 \ , \ wb_2X_2 = wYb_1$$

($w \in K\langle X, Y\rangle$). Then M is a K<X, Y>-$\Lambda$-bimodule.

Suppose W is a right K<X, Y>-module. Then W is a vector space together with a pair of linear transformations $\phi$ and $\psi$ representing X and Y respectively. Then $W \otimes_{K\langle X, Y\rangle} B$ has K-dimension 2dim W, and the action of $\Lambda$ is given by

$$X_0 \to \begin{bmatrix} 0 & 1 \\ 0 & 0 \end{bmatrix}, \quad X_1 \to \begin{bmatrix} 0 & \phi \\ 0 & 0 \end{bmatrix}, \quad X_2 \to \begin{bmatrix} 0 & \psi \\ 0 & 0 \end{bmatrix}.$$

It is visible that $- \otimes_{K\langle X,Y\rangle} B$ preserves indecomposability and reflects isomorphisms.

With this choice of definitions the following deep result has been established:

I.4.6 TAME-WILD-THEOREM [D, $CB_1$] *Suppose $\Lambda$ is a finite-dimensional algebra of infinite type. Then $\Lambda$ is either tame or wild.*

We shall frequently use the following elementary facts.

I.4.7 *Suppose $\Lambda$ is a finite-dimensional algebra. Then*

*(a) If I is some ideal of $\Lambda$ such that $\Lambda/I$ is wild then so is $\Lambda$.*

*(b) If e is an idempotent of $\Lambda$ such that $e\Lambda e$ is wild then so is $\Lambda$.*

*(c) $\Lambda$ is wild if and only if its basic algebra is wild.*

*(d) If $\Lambda$ is wild then so is $\Lambda^{op}$.*

*The same statements hold with wild replaced by "of infinite type". Moreover, (c) and (d) are true with tame instead of wild, and*

*(b') If $\Lambda$ is tame and e is an idempotent of $\Lambda$ then $e\Lambda e$ is tame or of finite type.*

For (b), (b'), see [Bo].

## I.5 *Algebras and quivers*

In order to work with basic algebras (and coverings) it is convenient to use the concept of quivers and relations.

<u>I.5.1</u> A *quiver* $Q$ is a directed graph $Q = (Q_0, Q_1, s, e)$ where $Q_0$ is the set of vertices and $Q_1$ is the set of arrows, and $s$, $e$ are maps $Q_1 \to Q_0$.
Given an arrow $\alpha \in Q_1$, we say it starts at vertex $s(\alpha)$ and terminates at $e(\alpha)$. The quiver is said to be finite provided both $Q_0$ and $Q_1$ are finite sets.

<u>I.5.2</u> REPRESENTATIONS OF QUIVERS Suppose $Q$ is a quiver; and $K$ is a fixed field.
A *representation* $\underline{V}$ of the quiver $Q$ over $K$ is given by $(V_i, \varphi_\alpha)$ where for any vertex $i \in Q_0$ we have a vector space $V_i$, and for any arrow $i \xrightarrow{\alpha} j$, there is a linear transformation $\varphi_\alpha: V_i \to V_j$. If $\underline{V} = (V_i, \varphi_\alpha)$ and $\underline{V}' = (V_i', \varphi_\alpha')$ are representations of $Q$ over $K$ then a *map* $\eta: \underline{V} \to \underline{V}'$ is defined to be $\eta = (\eta_i)$ where $\eta_i: V_i \to V_i'$ is a linear transformation such that for any arrow $i \xrightarrow{\alpha} j$ there is a commutative diagram

$$\begin{array}{ccc} V_i & \xrightarrow{\varphi_\alpha} & V_j \\ \eta_i \downarrow & & \downarrow \eta_j \\ V_i' & \xrightarrow{\varphi'_\alpha} & V_j' \end{array}$$

that is, $\varphi_\alpha \eta_j = \eta_i \varphi_\alpha'$.
Denote the category of representations of $Q$ over $K$ by $\mathfrak{R}(Q)$.

<u>I.5.3</u> THE PATH ALGEBRA OF A QUIVER

(a) Given $e, f \in Q_0$; then a *path* of length $l \geq 1$ from $e$ to $f$ is of the form $(e|\alpha_1, \ldots, \alpha_l|f)$ with arrows $\alpha_i$ satisfying $e(\alpha_i) = s(\alpha_{i+1})$ for all $i$, $1 \leq i \leq l$, such that $e$ is the starting point of $\alpha_1$, and $f$ is the end point of $\alpha_l$. In addition, we also define for any vertex $e$ of $Q$ a path of length zero (from $e$ to itself), denoted by $e$ also (or $(e||e)$ ).

(b) The path algebra $KQ$ of $Q$ is defined to be the $K$-vector space with basis the set of all paths in $Q$. The product of two paths is taken to be the composition if it exists, and zero otherwise. In this way, we obtain an associative K-algebra which has an identity if and only if $Q_0$ is finite ( then the identity is given by $\sum_{e\in Q_0} e$ ). Note that the path algebra is finite-dimensional if and only if $Q$ is finite, and there is no cyclic path in $Q$.

We denote by $KQ^+$ the ideal of $KQ$ generated by all arrows. Then $(KQ^+)^n$ is the ideal generated by all paths of length $\geq n$.

I.5.4 *The categories $\mathcal{R}(Q)$ and mod $KQ$ are equivalent:*

(i) Given $\underline{V} = (V_i, \varphi_\alpha)$ in $\mathcal{R}(Q)$, define the $KQ$-module $M_V$ with underlying vector space $\oplus V_i$ with action of the algebra as follows: Let $v \in V_i$, then

$$ve_i = v \text{ and } ve_j = 0 \text{ for } i \neq j,$$

$$v\alpha = \varphi_\alpha(v) \text{ if } \alpha \text{ starts at } i \text{ and } v\alpha = 0 \text{ otherwise.}$$

(ii) Suppose $M$ is a $KQ$-module, define $\underline{V} = (V_i, \varphi_\alpha)$ as follows:
If $i \in Q_0$ then take $V_i = Me_i$, and if $i \xrightarrow{\alpha} j$ is an arrow then $\varphi_\alpha$ is the linear transformation $Me_i \to Me_j$ which is given by right multiplication with $\alpha$.

(iii) If $\eta = (\eta_i)$ is a map $\underline{V} \to \underline{V}'$ and $M = M_V$ and $M' = M_{V'}$ then $\eta$ induces in an obvious way a $KQ$-homomorphism which we also denote by $\eta$. Any $\Lambda$-homomorphism arises from a map $\underline{V} \to \underline{V}'$.

In particular $\mathcal{R}(Q)$ is an abelian category.

(If one prefers to work with left modules then one may identify $\mathcal{R}(Q)$ with $KQ^{op}$-mod. A path of $Q^{op}$ is of the form $\beta_1\beta_2 \ldots \beta_n$ with $s(\beta_{i+1}) = e(\beta_i)$.)

I.5.5 QUIVER WITH RELATIONS

(a) Let e and f be vertices of $Q$. A *relation* $\rho$ on $Q$ is an element $\rho = \sum c_\omega \omega \in KQ$ where the $\omega$ are paths between two fixed vertices. If $\{\rho_\nu\}_\nu$ is a set of relations on $Q$ then $(Q, \{\rho_\nu\}_\nu)$ is a *quiver with relations*, or *"bound quiver"*.

(b) If $\omega = (r|\alpha_1, \ldots, \alpha_n|s)$ is a path in $Q$ and $\underline{V} = (V_i, \varphi_\alpha)$ is a representation of $Q$, then "$\omega$ acts on V" via the linear transformation $\omega(\underline{V}) = \varphi_{\alpha_n} \cdots \varphi_{\alpha_1}$. More generally, if $\rho$ is a relation on $Q$, say $\rho = \sum c_i\omega_i$ where $c_i \in K$ and each $\omega_i$ is a path then $\rho(V) = \sum c_i\omega_i(V)$.

(c) Given a quiver with relations $(Q, \{\rho_\nu\})$ and a representation $\underline{V} = (V_i, \varphi_\alpha)$ of $Q$ then $\underline{V}$ is a *representation of* $(Q, \{\rho_\nu\})$ if for all $\nu$ we have $\rho_\nu(\underline{V}) = 0$.

(d) The category of representations of $(Q, \{\rho_\nu\})$ is equivalent to mod $KQ/I$ where I is the ideal of $KQ$ generated by $\{\rho_\nu\}$.

We write $(Q,I)$ instead of $(Q, \{\rho_\nu\})$.

I.5.6 **REMARKS** Suppose $\Lambda = KQ/I$ where I is an ideal contained in $(KQ^+)^2$.

(a) Each vertex e of $Q$ corresponds to a primitive idempotent of $\Lambda$, hence to the simple module $e\Lambda/eJ$, which we call $S(e)$ or $S_e$. The simple modules for distinct vertices are non-isomorphic.

(b) The arrows of $Q$ generate the radical of $\Lambda$. In particular, the number of arrows $e \to f$ is the same as the multiplicity of $S_f$ as a composition factor of $eJ/eJ^2$ which is equal to dim Hom$(eJ, S_f)$.

(c) The quiver $Q$ is connected if and only if the algebra $\Lambda$ is indecomposable. (See for example [P].) If this is the case then one also says that $\Lambda$ *is connected*.

I.5.7 The following structure theorem characterizes basic algebras:

**THEOREM** (Gabriel) *Any basic finite dimensional K-algebra is of the form* $KQ/I$ *for a unique quiver* $Q$ *and some ideal* $I$ *with* $(KQ^+)^n \subseteq I \subseteq (KQ^+)^2$, *for some* $n \geq 2$.

Given an arbitrary finite-dimensional algebra $\Lambda$; by I.5.6, the quiver $Q$ describing the basic algebra of $\Lambda$ may be found as follows: $Q_0$ is identified with the set $\{S_i\}$ of isomorphism types of simple $\Lambda$-modules, and the number of arrows $i \to j$ is equal to the dimension of $\mathrm{Ext}^1(S_i, S_j)$.

If $\Lambda$ is an arbitrary K-algebra we say that $Q$ *is the quiver of* $\Lambda$ if the basic algebra of $\Lambda$ is isomorphic to $KQ/I$, as in I.5.7.

## I.6 *The quiver of a symmetric algebra*

Suppose $\Lambda$ is a symmetric algebra; we wish to study the quiver $Q$ of $\Lambda$. By I.3.3 we may assume that $\Lambda$ is basic.

I.6.1 *If there is an arrow* $e \overset{\alpha}{\to} f$ *in* $Q$, *then there must be a path* $f \to \ldots \to e$ *in* $Q$.

Proof: Since the quiver contains an arrow $e \to f$, we have that $e\Lambda f \neq 0$, hence $f\Lambda e \neq 0$, by I.3.7. The space $f\Lambda e$ is spanned by the residues of the paths $f \to \ldots \to e$.

We note an immediate consequence:

I.6.2 *Assume that* $Q$ *is connected. Given any two vertices* $e$, $f$ *in* $Q$ *then there is a path* $e \to \ldots \to f$.

I.6.3 *The following numbers are the same:*

*(a) The number of arrows* $j \to k$.

*(b) The multiplicity of* $S_k$ *as a composition factor of* $P_jJ/P_jJ^2$.

*(c) The multiplicity of* $S_j$ *as a composition factor of* $soc_2P_k/soc\ P_k$.

Proof: For (a) and (b), this is just the definition of the quiver. Comparing (b) and (c) we have to show that $Hom_\Lambda(\Omega S_j, S_k)$ and $Hom_\Lambda(S_j, \Omega^{-1}S_k)$ have the same dimension. Since $S_j$ and $S_k$ are simple, we know that $\underline{Hom}_\Lambda(-, S_k) = Hom_\Lambda(-, S_k)$ and $\underline{Hom}_\Lambda(S_j, -) = Hom_\Lambda(S_j, -)$. Now, it is well-known that $\underline{Hom}_\Lambda(X, Y) \cong \underline{Hom}_\Lambda(\Omega X, \Omega Y)$ for arbitrary modules X, Y of a symmetric algebra.

I.6.4 CONVENTION If $\alpha$ is an arrow of $Q$ terminating at e, let $R_\alpha := \{x \in e\Lambda:\ \alpha x = 0\}$. We shall often identify $R_\alpha$ with $\Omega(\alpha\Lambda)$, by taking the projective cover $\pi(x) = \alpha x$, $\pi: e\Lambda \to \alpha\Lambda$. Moreover, we identify $\Omega^{-1}(\alpha\Lambda)$ with $e\Lambda/\alpha\Lambda$.

I.6.5 (a) Let e be some vertex of $Q$; and suppose $\{\alpha_i\}$ and $\{\beta_j\}$ are the arrows of $Q$ starting and ending at e respectively.
Then $\mathrm{rad}\ e\Lambda = \Sigma\ \alpha_i\Lambda$; and also $\mathrm{soc}\ e\Lambda = \cap R_{\beta_j}$, in particular $\Omega^{-1}(S_e)$ is isomorphic to the submodule $(\beta_1, \beta_2, \ldots, \beta_r)\Lambda$ of $\oplus\, e_j\Lambda$ where $e_j$ is the starting point of $\beta_j$.

(b) We identify $\Omega^2 S_e$ with the module $\{(x_1, x_2, \ldots, x_s) \in \oplus f_i\Lambda:\ \Sigma\ \alpha_i x_i = 0\}$; here $f_i$ is the end point of $\alpha_i$.

Dually, we take for $\Omega^{-2}S_e$ the module $(\oplus\, e_i\Lambda)/\ (\beta_1, \dots, \beta_r)\Lambda$.

## I.7 *Auslander-Reiten sequences*

We will summarize basic facts on Auslander-Reiten theory. Details may be found in $[AR_{1,2}]$, also in $[G_2]$ or [Be], $[L_3]$.

Assume that $\Lambda$ is a finite-dimensional algebra.

I.7.1 (a) Suppose X and Y are $\Lambda$-modules. A homomorphism $f\colon X \to Y$ is said to be *irreducible* provided it is neither a split monomorphism nor a split epimorphism, and given any factorization $f = g\,h$ then either g is a split epimorphism or h is a split monomorphism.

(b) If f is irreducible then f is either 1-1 or onto but not both. This is clear, consider the factorization

$$\begin{array}{ccc} X & \xrightarrow{f} & Y \\ & \searrow \quad \nearrow & \\ & \mathrm{Im}\, f & \end{array}$$

We shall use this freely.

I.7.2 EXAMPLES (1) *Let P be indecomposable projective. Then the irreducible maps with domain P are precisely the inclusion maps of non-zero direct summands of rad P.*

Proof: Let first rad $P = X \oplus X'$ and let $f\colon X \to P$ be an inclusion. Suppose we have a factorization

$$\begin{array}{ccc} X & \xrightarrow{f} & P \\ & h\searrow \quad \nearrow g & \\ & Z & \end{array}$$

Assume that g is not a split epimorphism; then g is not onto since P is projective. We know that rad P is the only maximal submodule of P, therefore $g(Z) \subsetneq \mathrm{rad}\, P$. Define $\psi\colon Z \to X$ to be the map $\psi = f^{-1}\pi g$ where $\pi\colon \mathrm{rad}\, P \to X$ is the canonical projection. We also identify the summand X of rad P with f(X). Then $\psi\, h(x) = f^{-1}\pi g h(x) = f^{-1}\pi(f(x), 0) = f^{-1}f(x) = x$ for any $x \in X$.

Conversely, assume that $X \xrightarrow{f} P$ is an arbitrary irreducible map. Then f is not onto; otherwise f would split. Thus f is 1-1 (see I.7.1(b)), and then im f $\subsetneq$ rad P, since this is the only maximal submodule of P. Then we have the factorization

$$\begin{array}{ccc} X & \xrightarrow{f} & P \\ & {}_{f}\searrow \quad \nearrow_{j} & \\ & \text{rad } P & \end{array}$$

where j is the inclusion map. Now, f is irreducible, so this factorization must be trivial. Clearly, j is not a split epi, and hence f must be a split mono.

(2) *If E is indecomposable injective then f: E $\rightarrow$ Y is irreducible if and only if Y is a non-zero direct summand of E/soc E with f the canonical projection.*

This is dual to (1).

<u>I.7.3</u> A non-split exact sequence $0 \rightarrow X \xrightarrow{f} Y \xrightarrow{g} Z \rightarrow 0$ with X, Z indecomposable, is an *Auslander-Reiten sequence* (or *almost split sequence*) provided

(AR) For any h: X $\rightarrow$ X' which is not a split monomorphism, there exists some h': Y $\rightarrow$ X' such that h' f = h.

This is equivalent to the dual property:

(AR*) For any map k: Z' $\rightarrow$ Z which is not a split epimorphism, there exists some k': Z' $\rightarrow$ Y such that k = g k'.

We write briefly "AR-sequence".

The concept of irreducible maps and AR-sequences may be defined more generally. An AR-sequence, if it exists, is essentially unique up to isomorphism; this is not hard to show. In general, AR-sequences need not exist. However they do for finite-dimensional algebras:

<u>I.7.4</u> THEOREM [AR] *Suppose $\Lambda$ is a finite-dimensional algebra.*

*(a) If Z is an indecomposable non-projective $\Lambda$-module then there exists an AR-sequence ending in Z.*

*(b) If X is an indecomposable non-injective $\Lambda$-module then there exists an AR-sequence starting in X.*

One defines X = $\tau$Z, and $\tau$Z is the *Auslander-translate* of X. Similarly Z = $\tau^{-1}$X. The operation $\tau$ ($\tau^{-1}$) is defined on the set of indecomposable non-projective [non-injective] modules. In the special case of symmetric algebras, we have the following well-known fact (see for example [$G_2$]):

I.7.5 *If the algebra* $\Lambda$ *is symmetric then* $\tau \cong \Omega^2$.

I.7.6 THEOREM *Given an AR-sequence* $0 \to X \to Y \to Z \to 0$.
*(a) The irreducible maps starting at* $X$ *are of the form* $X \to Y'$ *where* $Y'$ *is a non-zero direct summand of* $Y$, *say* $Y = Y' \oplus Y''$, *and* $f = \begin{bmatrix} f' \\ f'' \end{bmatrix}$ *for some* $f''$.
*(b) The irreducible maps ending at* $Z$ *are of the form* $g'\colon Y' \to Z$ *where* $Y' \neq 0$ *and* $Y = Y' \oplus Y'$; *and* $g = [g', g'']$ *for some* $g''$.

A proof may be found in [AR] or [$G_2$]. Now assume that $\Lambda$ is symmetric. Let P be an indecomposable projective $\Lambda$-module with P/rad P = S, then S $\cong$ soc P. The following fact is extremely important:

I.7.7 *There is always an AR-sequence of the form*
(*) $0 \to \text{rad } P \to \text{rad } P/S \oplus P \to P/S \to 0$.
*This is the only AR-sequence where P occurs.*
We call this a "standard sequence". It will be useful to know that whenever an AR-sequence has a projective middle term, P say, then the cokernel is P/soc P and the kernel is rad P.

## I.8 *The Auslander-Reiten quiver*

I.8.1 Suppose X and Y are $\Lambda$-modules, and let $X = \Sigma \oplus X_j$ and $Y = \Sigma \oplus Y_i$ with $X_j$ and $Y_i$ indecomposable. We fix such decompositions, and we write homomorphisms f: X $\to$ Y in form of matrices $f = [f_{ij}]$ where $f_{ij}\colon X_j \to Y_i$. Then the radical of Hom(X, Y) is by definition the space

$$\mathfrak{r}(X,Y) = \{ f = [f_{ij}]\colon \text{no } f_{ij} \text{ is an isomorphism} \}.$$

Let $\mathfrak{r}^2(X, Y) = \sum_Z \mathfrak{r}(X, Z)\,\mathfrak{r}(Z, Y)$. For X and Y indecomposable, a map $f : X \to Y$ is irreducible if and only if $f \in \mathfrak{r}(X, Y)$ but $f \notin \mathfrak{r}^2(X, Y)$. In this situation, $\mathrm{Irr}(X, Y) := \mathfrak{r}(X, Y)/\mathfrak{r}^2(X, Y)$ is the space of irreducible maps. We define

$$a_{X,Y} = \dim \mathrm{Irr}(X, Y).$$

I.8.2 (a) The *Auslander-Reiten quiver* of $\Lambda$, denoted by $\Gamma(\Lambda)$, is, by definition, the quiver whose vertices are the isomorphism classes of indecomposable $\Lambda$-modules. Moreover, the number of arrows $[X] \to [Y]$ is equal to $a_{X,Y}$.
(b) The *stable Auslander-Reiten quiver* $\Gamma_s(\Lambda)$ of an algebra $\Lambda$ is obtained from $\Gamma(\Lambda)$ by removing all $[\tau^{-k}P]$ and all $[\tau^{k}E]$ where P is projective and E is injective, $k \geq 0$, and the adjacent arrows. In the special case when $\Lambda$ is self-injective, one removes only the vertices [P] for P projective.

Thus, if Y is indecomposable and non-projective then $a_{X,Y}$ is the multiplicity of X as a direct summand of E where $0 \to \tau Y \to E \to Y \to 0$ is the AR-sequence terminating in Y. Similarly, if Z is indecomposable and not injective then $a_{Z,X}$ is the multiplicity of X as a direct summand of F where $0 \to Z \to F \to \tau^{-1}Z \to 0$ is the AR-sequence starting in Z.

If Y is indecomposable projective then by I.8.2, $a_{X,Y}$ is equal to the multiplicity of X as a direct summand of rad P; and if E is indecomposable injective then $a_{E,X}$ = the multiplicity of X as a direct summand of E/soc E. In particular

(1) $a_{X,Y} = a_{\tau Y, X}$

(2) The AR-quiver is locally finite.

In general, the quiver $\Gamma(\Lambda)$ is a disjoint union of connected components; usually the number of components is infinite, and each component is infinite. In fact, finiteness is very rare, by the following result due to M. Auslander [Au]:

I.8.3 THEOREM *Let $\Lambda$ be an indecomposable algebra. If $\Gamma(\Lambda)$ has a component with modules of bounded length, then these are all indecomposable modules, and $\Lambda$ is of* *finite representation type.*

There is a fundamental result of C. Riedtmann concerning the graph structure of components of $\Gamma_s(\Lambda)$:

<u>I.8.4</u> **THEOREM** *Let $\Theta$ be a connected component of $\Gamma_s(\Lambda)$. Then there is a directed tree $T$ and an admissible group of automorphisms $\Pi \subseteq \mathrm{Aut}\ \mathbb{Z}T$ such that $\Theta \cong \mathbb{Z}T/\Pi$. The tree class of $T$ is defined by $\Theta$ uniquely up to isomorphism, and $\Pi$ is unique up to conjugation in $\mathrm{Aut}\ \mathbb{Z}T$.*

For details, see [$Ri_1$] or [Be, 2.29.6].

In the context here, the following types of components will occur:

(1) Tubes: $T = A_\infty$ = ·——·——·——· ... ∞

Let $\tau$ = translation to the left and $\Pi = \langle \tau^k \rangle$. Then, by definition, $\mathbb{Z}T/\Pi$ is a k-tube. The number k is the "rank".

$\mathbb{Z}T$

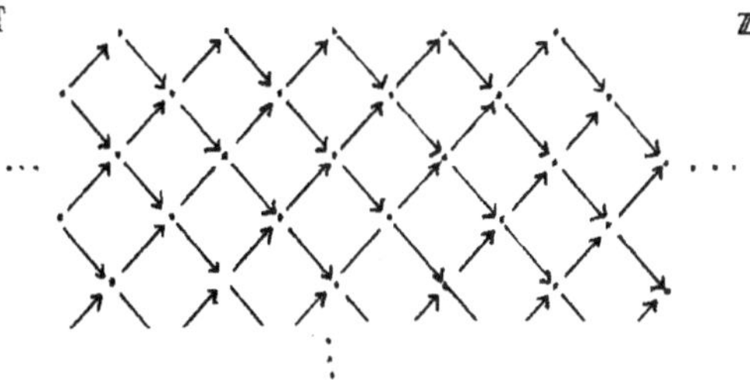

$\mathbb{Z}T/\Pi$

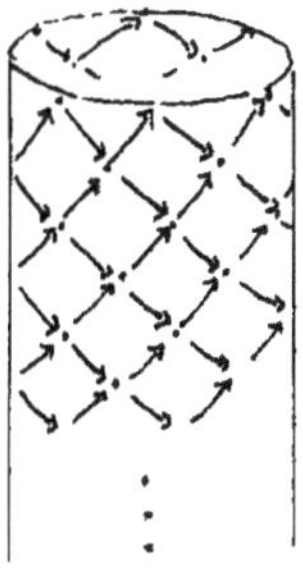

(2) $T = A_\infty^\infty$ ∞ ... ·——·——·——· ..... ∞

$\Pi = 1$, $\Theta = \mathbb{Z}T$. Every vertex of $\Theta$ has two predecessors and two successors:

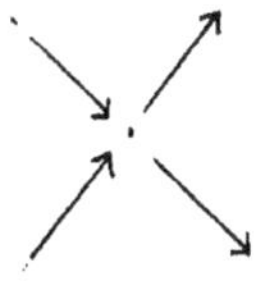

(3) $T = \tilde{A}_n$

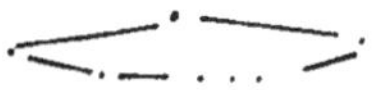

$\Theta = \mathbb{Z}T$, $\Pi = 1$. This component may be visualized by taking $\mathbb{Z}A_\infty^\infty$ and identifying, horizontally, the m-th row and the m+n-th row, for each m. The quivers $\mathbb{Z}\tilde{A}_n$ for $n \geq 2$ and $\mathbb{Z}A_\infty^\infty$ are locally isomorphic.

(4) $T = D_\infty =$ 

$\Theta = \mathbb{Z}T$ and $\Pi = 1$.

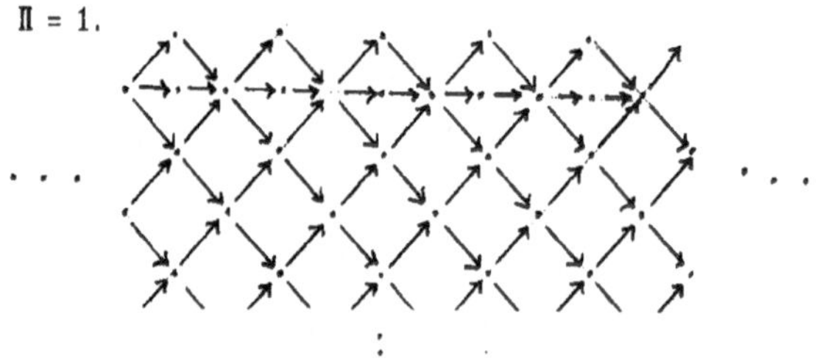

(5) $T = \tilde{D}_n$ 

$\Theta = \mathbb{Z}T$, $\Pi = 1$.

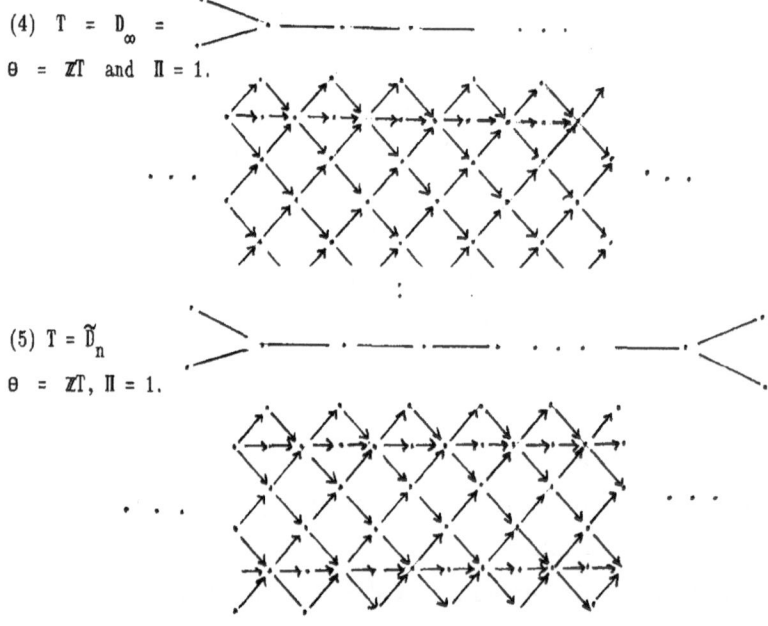

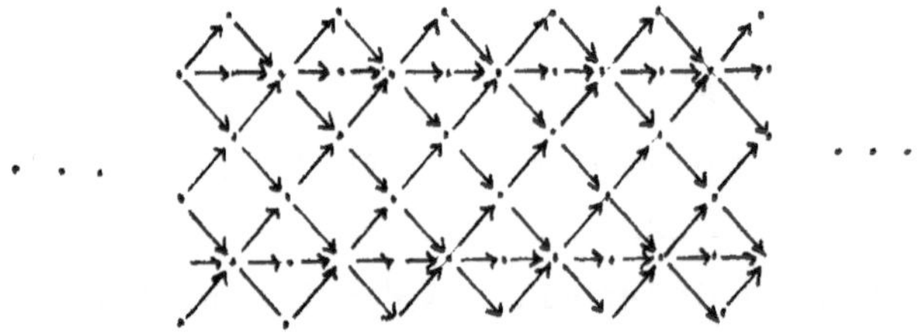

I.8.4.1 The Auslander-translate induces a graph isomorphism of a component of $\Gamma_s(\Lambda)$; in the usual representations of the components (as above), it is the shift to the left by one. In the case when $\Lambda$ is self-injective then $\Omega$ acts as a graph-automorphism; in particular, if $\Lambda$ is symmetric then $\Omega^2 \cong \tau$. We note that in this case $\Omega$-periodicity and $\tau$-periodicity is the same.

In I.8.4, the multiplicities are neglegted. We will now take them into account:

I.8.5 A *valued graph* is a graph together with a pair of positive integers $(a_{ij}, a_{ji})$ for each edge $i - j$. We omit the label when $a_{ij} = a_{ji} = 1$.

The *Cartan matrix* $C_T$ of a valued graph is the matrix whose rows and columns are indexed by the vertices of the graph, and with entries

$$c_{ij} = \begin{cases} 2 & \text{if } i = j \\ -a_{ij} & \text{if } i - j \text{ is an edge} \\ 0 & \text{otherwise.} \end{cases}$$

Given a component $\Theta$ of $\Gamma_s(\Lambda)$, with tree class T; then we define a valued graph on T if we set $a_{ij}$ = dim Irr($M_i$ $M_j$), see I.8.1.This is well-defined, by I.8.2(1).

EXAMPLE $\tilde{A}_{12}$ = $\cdot \overset{(2,2)}{\text{———}} \cdot$

The correspondent stable AR-component is of the form

This occurs for the Kronecker algebra (I.4.3), see Chapter II.7.3.

Let T be a valued tree. A *subadditive function* on T is, by definition, a function f: T $\rightarrow \mathbb{N}$ such that for all j $\in$ T

$$\sum_{i \in T} f(i)c_{ij} \geq 0$$

The funcion f is said to be *additive* if $\sum_{i \in T} f(i)c_{ij} = 0$, for all j.

The following characterization is important:

I.8.6 THEOREM [HPR] *Let T be a connected valued tree and f a subadditive function on T. Then*

*(a) T is either a Dynkin diagram, a Euclidean diagram or one of* $A_\infty^\infty$, $A_\infty$, $B_\infty$, $C_\infty$, $D_\infty$.

*(b) If d is not additive then T is a Dynkin diagram or* $A_\infty$.

*(c) If d is unbounded then T is* $A_\infty$.

In special circumstances one can use the length of modules to construct subadditive functions. We need the following application:

I.8.6.1 COROLLARY [HPR] *Let C be an infinite component of* $\Gamma_s(\Lambda)$. *If C contains a* $\tau$*-periodic module then the tree class of C is* $A_\infty$.

I.8.6.2 If T is an infinite Dynkin diagram $\neq A_\infty$ then every subadditive function on T is a multiple of a given (bounded) additive function.

In particular, if T = $A_\infty^\infty$ then it is constant, and if T = $D_\infty$ then it takes only two values, namely u and 2u, where u occurs "at the end" (see for example [Be 2.30.5]).

We wish to apply these results to stable AR-components.

I.8.7 Given a component of $\Gamma_s(\Lambda)$. A *subadditive function* d for $\Theta$ is, by definition, a function d: $\Theta_0 \to \mathbb{N}$ satisfying

$$d(M) + d(\tau M) \geq \sum_{Y \in M^-} d(Y)a'_{YM}$$

Such a function is *additive*, provided we always have equality, for all M in $\Theta_0$.

For example, $d(M) = \dim_K M$ is a subadditive function on any component $\Theta$; this follows from the existence of AR-sequences; and it is additive if and only if $\Theta$ is a component of $\Gamma(\Lambda)$.

However this will not immediately give rise to a subadditive function on the valued tree of $\Theta$.

In the special case of group algebras, it was proved by P. Webb [W] that there are always subadditive functions on a component $\Theta$ of $\Gamma_s(KG)$ from which one can construct subadditive functions on the associated valued tree. Then the tree must belong to the list in I.8.7.

Recently, in [Ok], T. Okuyama has given a new proof of this result. He made use of subadditive functions which he defined in terms of the Green ring. We will describe his idea now; and later we will use it again for various results on components. Since we work with a wider class of algebras, we will give a different formulation; see also [ES].

I.8.8 Let $\Lambda$ be a finite dimensional self-injective algebra. We fix an arbitrary $\Lambda$-module Y and define $d_Y$: $\Gamma_s(\Lambda) \to \mathbb{N}$ by

$$d_Y(M) := \dim \underline{\mathrm{Hom}}_\Lambda(Y, M).$$

Then we have the following:

LEMMA [ES] *Suppose that* $d = d_Y$ *is as above, and assume that*

(*) $0 \to M \xrightarrow{j} E \xrightarrow{p} M \to 0$ *is an AR-sequence. Then*

*(a) If M is not a summand of Y then* $d(M) + d(\tau M) \geq d(E)$.

*(b) If in addition* $\Omega M$ *is not a summand of Y then equality holds in (a).*

*In particular, if* $\Theta$ *is a component of* $\Gamma_s(\Lambda)$ *such that no indecomposable summand of Y belongs to* $\Theta$ *or* $\Omega\Theta$ *then* $d_Y$ *is a subadditive function on* $\Theta$.

Proof: (a) Apply the functor $\mathrm{Hom}_\Lambda(Y, \ )$ to (*). Since (*) is almost split and since M is not a direct summand of Y, we obtain an exact sequence

$$0 \to (Y, \tau M)_\Lambda \xrightarrow{j^*} (Y, E)_\Lambda \xrightarrow{p^*} (Y, M)_\Lambda \to 0.$$ Clearly, $j^*$ and $p^*$ preserve $\mathcal{P}(\ ,\ )$, the spaces of maps which factor through projective modules. Moreover, if $\eta \in \mathcal{P}(Y, M)$ then there is some $\rho \in \mathcal{P}(Y, E)$ with $p^*(\rho) = \eta$. Consequently we have a sequence $0 \to \mathcal{P}(Y, \tau M) \xrightarrow{j^*} \mathcal{P}(Y, E) \xrightarrow{p^*} \mathcal{P}(Y, M) \to 0$ which is exact at the ends, and $\mathrm{im}\, j^* \subseteq \ker p^*$. Therefore (a) holds.

(b) Suppose $\Omega M$ is not a summand of Y. We have to show the following:

(**) Let $\varphi: Y \to \tau M$ such that $j\varphi$ belongs to $\mathcal{P}(Y, E)$. Then $\varphi \in \mathcal{P}(Y, \tau M)$.

Take such $\varphi$, and assume for contradiction that $\varphi \notin \mathcal{P}(Y, \tau M)$. Then there is an indecomposable summand, $Y_1$ say, of Y such that $\varphi_1 = \varphi|Y_1$ does not belong to $\mathcal{P}(Y_1, \tau M)$. Let X be the push-out of $\varphi_1$ along an injective hull. That is, there is a commutative diagram with exact rows

$$\begin{array}{lcccccccc} (***) & 0 & \to & \tau M & \xrightarrow{u} & X & \xrightarrow{v} & \Omega^{-1}Y_1 & \to 0 \\ & & & \varphi_1 \uparrow & & \underline{\rho} \uparrow & & 1 \uparrow & \\ & 0 & \to & Y_1 & \xrightarrow{\iota} & I & \xrightarrow{\pi} & \Omega^{-1}Y_1 & \to 0 \end{array}$$ with I injective.

We will show that the top row is an almost split sequence. Then we are done since by the uniqueness of AR-sequences, $\Omega^{-1}Y_1 \cong M$, a contradiction to the hypothesis.

Explicitly, we may take $X = (\tau M \oplus I)/X_0$ where $X_0 = \{ (\varphi_1 y, \iota y): y \in Y_1 \}$, and u, v and $\underline{\rho}$ are canonical maps.

First we claim that (***) does not split: Otherwise, there is some $\psi \in \mathrm{Hom}_\Lambda(X, \tau M)$ such that $\psi u = \mathrm{id}_{\tau M}$ and we have that $\varphi_1 = (\psi u)\varphi = \psi(\underline{\rho} \circ \iota)$ and $\varphi_1$ factors through I. Since I is also projective, we have $\varphi_1 \in \mathcal{P}(Y_1, \tau M)$, a contradiction.

Now let $\eta: \tau M \to Z$, and assume that $\eta$ is not a split monomorphism. Since (*) is an AR-sequence, there exists some $\rho \in \mathrm{Hom}_\Lambda(E, Z)$ such that $\eta = \rho \circ j$. Now $j \circ \varphi_1$ belongs to $\mathcal{P}(Y_1, E)$ and $\Lambda$ is self-injective, therefore $j \circ \varphi_1$ must factor through the injective hull I of $Y_1$. Let $\lambda \in \mathrm{Hom}_\Lambda(I, E)$ with $\lambda \circ \iota = j \circ \varphi_1$; then we have a well-defined map $\underline{\rho} : X \to Z$ given by $\underline{\rho}[(m,a) + X_0] = \eta m - \rho\lambda a$, and moreover $\eta = \underline{\rho}\, u$. This shows that (***) is almost split.

<u>I.8.9</u> COROLLARY *Suppose that $Y \cong \tau Y$ and that no summand of $Y$ or $\Omega Y$ belongs to $\Theta$.*

*Then $d_Y$ induces a subadditive function on the tree of $\Theta$.*

Proof: This is true since $d_Y$ is constant on $\tau$-orbits:
We have $\underline{\mathrm{Hom}}_\Lambda(Y, M) \cong \underline{\mathrm{Hom}}_\Lambda(\tau Y, \tau M) \cong \underline{\mathrm{Hom}}_\Lambda(Y, \tau M)$.

I.8.10 The definition of $d_Y$ is the same as in [Ok]. Okuyama considers the bilinear form ( , ) on the Green ring a(KG) of the group algebra KG given by (M, N) = dim $\mathrm{Hom}_{KG}(M, N)$; and he defines s $\epsilon$ a(KG) to be $s = Y + \Omega^{-1}Y - I$ where $0 \to Y \to I \to \Omega^{-1}Y \to 0$ is a fixed injective resolution. Then he studies the function $M \to (s, M)$; we have that $(s, M) = d_Y(M)$.

Then Okuyama shows that for a component $\Theta$ of $\Gamma_s(KG)$, there is always a module Y such that $\tau Y \cong Y$ and $d_Y \neq 0$ on $\Theta$. Hence, by I.8.9, the component admits a non-zero subadditive function. This gives a new proof for Webb's Theorem, using I.8.6.

I.8.11 Let $\Lambda$ be a basic algebra, and suppose $\Lambda = \bigoplus_{i=1}^{n} e_i\Lambda$ with $e_i\Lambda$ indecomposable. Put $I = \{ i : e_i\Lambda$ is injective$\}$, and let $S = \bigoplus_{i\epsilon I} \mathrm{soc}\; e_i\Lambda$. Then $\Lambda$ and $\Lambda/S$ have almost the same indecomposable modules; in fact, $\mathrm{ind}\,\Lambda = \mathrm{ind}(\Lambda/S) \cup \{e_i\Lambda: i \in I\}$. Moreover, the AR-quiver of $\Lambda/S$ is obtained from $\Gamma(\Lambda)$ by removing the modules $e_i\Lambda$ with $i \in I$. [$G_2$]

In the special case when $\Lambda$ is injective then S = soc $\Lambda$, and $\Gamma(\Lambda/S)$ is the stable AR-quiver of $\Lambda$.

In particular, if $\Lambda_1$ and $\Lambda_2$ are injective (or even symmetric ) algebras such that $\Lambda_1/\mathrm{soc}\,\Lambda_1 \cong \Lambda_2/\mathrm{soc}\,\Lambda_2$ then $\Lambda_1$ and $\Lambda_2$ have the same stable AR-quiver.

I.8.12 Suppose $\Lambda$ is an algebra and $\Gamma = \Gamma(\Lambda)$. Then $\Gamma(\Lambda^{op})$ is obtained from $\Gamma$ by reversing all arrows. This is clear since $\mathrm{Hom}_K(\ ,K)$ is a duality between mod $\Lambda$ and mod $\Lambda^{op}$. In particular projective vertices of $\Gamma$ correspond to injective vertices of $\Gamma(\Lambda^{op})$. It follows that in the case of a self-injective algebra $\Lambda$, the standard AR-sequences (see I.7.7) have the same positions in $\Gamma$ and in $\Gamma(\Lambda^{op})$.

I.8.13 REMARKS, NOTATION (a) We write usually M instead of [M] for the vertex of

$\Gamma(\Lambda)$. Following [$Ri_1$], we denote by $M^-$ the set of predecessors of [M] in $\Gamma(\Lambda)$, and $M^+$ are its successors. If M is non-projective then $M^- = (\tau M)^+$. Moreover, given indecomposable modules M and Z, where M is non-projective and Z is not injective, if $M^- \cap Z^+$ contains more than two modules then $Z \cong \tau M$.

(b) Suppose M is indecomposable and non-projective, then $\beta(M)$ is the number of predecessors of M in $\Gamma_s(\Lambda)$. We say that M lies *at the end* of its component if $\beta(M) = 1$ or 3. (For the motivation, see I.8.4). Moreover, we say that a module is "of type d" if it lies at the end of a component $\cong \mathbb{Z}D_\infty$ or $\mathbb{Z}D_n$ (or $\mathbb{Z}D_n/\langle\tau^k\rangle$). If $\beta(M) = 2$ we sometimes call M "of type a".

(c) A component of $\Gamma(\Lambda)$ is regular if it is also a component of $\Gamma_s(\Lambda)$. We note that a component $\cong \mathbb{Z}\Delta$ where $\Delta$ is Euclidean, cannot be regular; see for example [W], 2.4.

## I.9 *Some wild algebras*

Let W be the algebra K< X, Y >. For various quivers T we wish to establish functors F: mod W $\to$ mod KT ( or mod KT/I if there are relations) which reflect isomorphisms and preserve indecomposables; and moreover, $F \cong - \otimes B$ where B is a bimodule as in I.4.4. We will also have that $F(M)e \neq 0$ for all primitive idempotents e of KT and $0 \neq M \in$ mod W; that is, F(M) is sincere.

We will deal with some hereditary algebras, which are well-known to be wild, and with some one-relation algebras. The functors are mostly taken from [$R_3$]. Actually, Ringel gives a classification of the one-relation algebras according to their representation type.

A module of W is a vector space V together with two linear transformations $\varphi$, $\psi$ representing X, Y respectively. We will define F(V) as a representation of the quiver T ( or of (T, I) ). In each case, it is clear how F is defined on morphisms, and we will not write this down. To verify that F preserves indecomposability and reflects isomorphism is usually elementary; we will give details only for some quivers.

I.9.0 (a) *The algebras KT are wild for the following quivers T:*

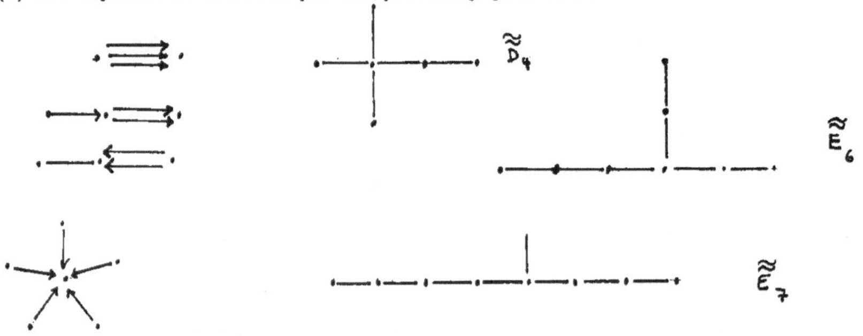


(b) *Any algebra of the form KT/I is wild when*

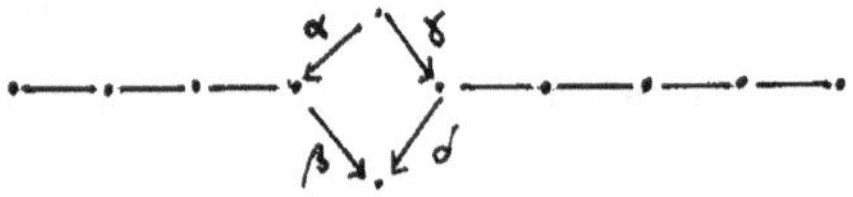


*and* $I = \langle \alpha\beta - \gamma\delta \rangle$

*and* $I = \langle \beta\gamma \rangle$.

We note that the orientation of arrows does not matter for the last three trees in (a), and also for the arrows $\neq$ α, β, γ, δ, see for example [$DF_1$] or [$R_4$]. The algebras in (b) are denoted by $C\tilde{\tilde{E}}_7$ and $C\tilde{\tilde{D}}_4$ in [$R_3$].

I.9.1 *Let* $T = e \overset{\alpha,\ \beta,\ \gamma}{\rightrightarrows} f$. *Then T is wild:*

Define F: mod W $\to$ mod KT as follows: Given a W-module V = (V, $\varphi$, $\psi$), set

$$F(V) = V \overset{1,\ \varphi,\ \psi}{\rightrightarrows} V$$

The image of F is the subcategory of $\mathcal{R}(T)$ consisting of all representations M on which α is an isomorphism.

(i) *F takes indecomposables to indecomposables*: Suppose that $F(V) = M_1 \oplus M_2$; let $M_i = [V_i' \overset{\alpha_i}{\rightrightarrows} V_i'']$ for i = 1, 2. Then $V = V_1' \oplus V_2' = V_1'' \oplus V_2''$. Since $\alpha_1 \oplus \alpha_2 \cong \mathrm{id}$ we deduce that $\alpha_1$ and $\alpha_2$ must be isomorphisms, and $\dim V_1' = \dim V_1''$. After changing bases if necessary we may assume that $\alpha_1 = \alpha_2 = \mathrm{id}$, and we identify the corresponding spaces. Then $M_1 = F(V_1)$ and $M_2 = F(V_2)$ where $V_i = (V_i', \varphi_i, \psi_i)$

and $V = V_1 \oplus V_2$.

(ii) *F reflects isomorphisms:* Suppose $F(V_1) \cong F(V_2)$ in $\mathcal{I}(T)$. Then there is a commutative diagram

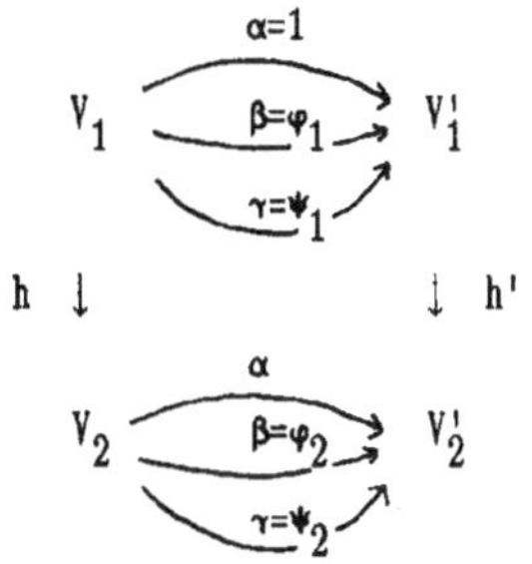


where h, h' are vector space isomorphisms. After choosing appropriate bases we may write h = id and identify $V_1$ and $V_2$. Moreover, as in (ii) we may assume $V_i = V_i'$ and $\alpha = 1$. With these conventions, it follows that h' = id since $h'\alpha = \alpha h'$. Moreover, $\beta h = h'\beta$ implies that $\varphi_1$ is identified with $\varphi_2$; and similarly $\psi_1 = \psi_2$.

(iii) *A bimodule representing F:* Let B be the free W-modue of rank 2 generated by $b_e$ and $b_f$. We define a KT-left action on B as follows: KT has a vector space basis $\mathcal{B} = \{e, f, \alpha, \beta, \gamma\}$. Set $(wb_e)e = wb_e$ and $(wb_f)f = wb_f'$

$$(wb_e)\alpha = wb_f, \quad (wb_e)\beta = Xwb_f \text{ and } (wb_e)\gamma = Ywb_f$$

and otherwise $(wb_e)\sigma = (wb_f)\sigma = 0$, for $\sigma \in \mathcal{B}$.

*Then* $F \cong - \otimes B$: Let $V \in \text{mod } W$. Then $V \otimes B = V \otimes b_e \oplus V \otimes b_f$ as vector spaces: This is clear since, for example, $v \otimes Xb_e = vX \otimes b_e \in V \otimes b_e$. In particular, $\dim V = \dim V \otimes b_e = \dim V \otimes b_f$. We have that $[v \otimes b_e]\alpha = v \otimes b_f$, moreover $[v \otimes b_e]\beta = v \otimes Xb_f = vX \otimes b_f = \varphi(v) \otimes b_f$, and similarly $[v \otimes b_e]\gamma = \psi(v) \otimes b_f$. This shows that $V \otimes B \cong F(V)$, with $[F(V)]_e = V \otimes b_e$ and $[F(V)]_f = V \otimes b_f$.

We note that the algebra KT in I.9 1 may be defined as a matrix algebra as follows: Take a 3-dimensional vector space V over K, and let $\Lambda = \begin{bmatrix} K & 0 \\ V & K \end{bmatrix}$. Then $\Lambda \cong KT$, where e corresponds to $\begin{bmatrix} 0 & 0 \\ 0 & 1 \end{bmatrix}$, f to $\begin{bmatrix} 1 & 0 \\ 0 & 0 \end{bmatrix}$ and as arrows one may take matrices of the form $\begin{bmatrix} 0 & 0 \\ v_i & 0 \end{bmatrix}$ where $\{v_1, v_2, v_3\}$ is a basis of V.

$\underline{\text{I.9.2}}$ *The quiver* $Y = 0 \xrightarrow{\alpha} 1 \underset{\gamma}{\overset{\beta}{\rightrightarrows}} 2$ *is wild:*

Define a functor F: mod W $\to$ mod KY as follows: Given a W-module $(V, \varphi, \psi)$, set

$$F(V) \; = \; V \xrightarrow{\begin{bmatrix}0\\1\end{bmatrix}} V \oplus V \;\underset{\begin{bmatrix}0 & 1\\ \varphi & \psi\end{bmatrix}}{\overset{\text{id}}{\rightrightarrows}}\; V \oplus V.$$

(i) *F preserves indecomposability:* Let $F(V) = U \oplus W$ where $U = \left[ U_0 \to U_1 \rightrightarrows U_2 \right]$ and $W = \left[ W_0 \to W_1 \rightrightarrows W_2 \right]$. Here $V = U_0 \oplus W_0$. Since $\beta = 1$ we know that $U_1 = U_2$ and $W_1 = W_2$; and since $\alpha = \begin{bmatrix}0\\1\end{bmatrix}$ we have $U_0 \subseteq U_1$ and $W_0 \subseteq W_1$, actually $U_0 \oplus W_0$ is the subspace $\{ \begin{bmatrix}0\\v\end{bmatrix} \}$ of $[V \oplus V]_1$. Since $\gamma$ maps $U_1$ into $U_1$, we deduce that $U_1$ contains the space $\Gamma_\psi(U_0) := \{ \begin{bmatrix}u\\ \psi u\end{bmatrix} : u \in U_0 \}$, that is, the graph of $\psi$ on $U_0$. Moreover, $U_1$ contains $\beta(U_0) = \{ \begin{bmatrix}0\\u\end{bmatrix} : u \in U_0 \}$. The spaces $\Gamma_\psi(U_0)$ and $\beta(U_0)$ have zero intersection; and by dimensions, their sum is $U_1$. A similar decomposition holds for $W_1$. We claim that $\psi$ maps $U_0$ into $U_0$:
We have that $[u \;\; \psi u]^T$ lies in $U_1$ for $u \in U_0$. Therefore $\gamma[u \;\; \psi u]^T$ also lies in $U_1$; and this is of the form $[\psi u \;\; *\,]^T$. Using the direct sum above it has a unique expression $[u' \;\; \psi u']^T + [0 \;\; u'']^T$ for $u'$ and $u'' \in U_0$. Hence $\psi u = u' \in U_0$, as required.

This implies now that also $U_1 = U_0 \oplus U_0$; and it follows that $\varphi$ maps $U_0$ into $U_0$ ( take $\begin{bmatrix}0\\u\end{bmatrix}$ ). Similarly $W_1 = W_0 \oplus W_0$, and $\varphi$, $\psi$ map $W_0$ into $W_0$. Consequently $V = U_0 \oplus W_0$, a direct sum as W-modules.

(ii) *The functor F reflects isomorphisms:* Let $F(V) \cong F(U)$; then there are bijective linear transformations $f_0$, $f_1$ and $f_2$ such that the following diagram commutes:

$$\begin{array}{ccccc}
V & \xrightarrow{\begin{bmatrix}0\\1\end{bmatrix}} & V \oplus V & \underset{\begin{bmatrix}0 & 1\\ \varphi & \psi\end{bmatrix}}{\overset{\text{id}}{\rightrightarrows}} & V \oplus V \\
f_0 \downarrow & & \downarrow f_1 & & \downarrow f_2 \\
U & \xrightarrow{\begin{bmatrix}0\\1\end{bmatrix}} & U \oplus U & \underset{\begin{bmatrix}0 & 1\\ \tilde\varphi & \tilde\psi\end{bmatrix}}{\overset{\text{id}}{\rightrightarrows}} & U \oplus U
\end{array}$$

We take $f_0$ as an identification of V and U; then $f_1$ identifies $0 \oplus V$ with $0 \oplus U$. In particular we may write $f_1$ as a matrix of the form $\begin{bmatrix} A & 0 \\ B & I \end{bmatrix}$; and $f_2$ has the same matrix since $f_2\ \beta = \beta\ f_1$ and $\beta = \mathrm{id}$. Moreover, this matrix must commute with the action of $\gamma$; consequently $B = 0$ and $A = I$ , and then $\psi = \tilde{\psi}$ and $\varphi = \tilde{\varphi}$. This proves that $V \cong U$.

(iii) *A bimodule B representing F, as in I.4.4*:

Take for B the free left W-module of rank 5 with basis $b_0$, $b_1$, $c_1$, $b_2$, $c_2$. Define a right KY-action as follows:

$$b_i e_j = \begin{cases} 0 & \text{if } i \neq j \\ b_i & \text{if } i = j \end{cases}.$$

Define $b_0\alpha = c_1$, $\quad b_1\beta = b_2 \quad b_1\gamma = b_2$

$$c_1\beta = c_2, \qquad c_1\gamma = Xb_2 + Yc_2$$

*Then* $F \cong - \otimes B$: The underlying space of $V \otimes B$ is $(V \otimes b_0) \oplus (V \otimes b_1) \oplus (V \otimes c_1) \oplus (V \otimes b_2) \oplus (V \otimes c_2)$. We have $(v \otimes b_0)\alpha = v \otimes c_1$, that is, $,\alpha = \begin{bmatrix} 0 \\ 1 \end{bmatrix} : V \otimes b_0 \rightarrow V \otimes b_1 \oplus V \otimes c_1$. It is clear that $\beta = 1 : V \otimes b_1 \oplus V \otimes c_1 \rightarrow V \otimes b_2 \oplus V \otimes c_2$. Moreover $(v \otimes b_1)\gamma = v \otimes b_2$ and $(v \otimes c_1)\gamma = v \otimes Xb_2 + v \otimes Yc_2 = vX \otimes b_2 + vY \otimes c_2 = \varphi(v) \otimes b_2 + \psi(v) \otimes c_2$, and the matrix of $\gamma$ is $\begin{bmatrix} 0 & 1 \\ \varphi & \psi \end{bmatrix}$, as required.

A consequence of I.9.1 is that the quiver $\cdot \leftarrow \cdot \leftleftarrows \cdot$ is wild as well; this is equal to $Y^{op}$.

One might like some motivation for the definition of the functor F in I.9.2. First, it is enough to consider representations where $\varphi_\alpha$ is 1-1: In general, if $\underline{V} \in \mathcal{R}(Y)$ then $\underline{V}$ is a direct sum

$$\left[\ker \varphi_\alpha \rightarrow 0 \rightrightarrows 0\right] \oplus \left[\overline{V}_0 \xrightarrow{\overline{\varphi}_\alpha} V_1 \underset{\varphi_\gamma}{\overset{\varphi_\beta}{\rightrightarrows}} V_2\right]$$ where $\overline{\varphi}_\alpha$ is 1-1, and the first summand is just a direct sum of copies of the simple Y-module corresponding to vertex 0.

Now assume that $V_0 \subset V_1$ and that $\alpha$ is the inclusion; and write $V_1 = V' \oplus V_0$. Consider the representations of Y in which $\beta$ acts as an isomorphism; for these we may identify $V_1$ and $V_2$ in such a way that $\beta = 1$. Then we think of $\gamma$ as a matrix; taking a "rational canonical form" and $V' = V_0$, we get $\gamma = \begin{bmatrix} 0 & 1 \\ \varphi & \psi \end{bmatrix}$.

Then $\underline{V}$ is given by the space $V_1$ together with a pair of matrices and may be considered as a $K\langle X, Y\rangle$-module. Moreover, $\underline{V}$ is indecomposable if and only if the $K\langle X, Y\rangle$-module $(V_1, \varphi, \psi)$ is indecomposable.

<u>I.9.3</u> *The quiver $\tilde{\mathbb{D}}_4$ is wild:* We take $\tilde{\mathbb{D}}_4$ with the orientation

$$\begin{array}{ccccccc} & & 2 & & & & \\ & & \delta_2 \downarrow & & & & \\ 3 & \xrightarrow{\delta_3} & c & \xleftarrow{\delta_1} & 1 & \xleftarrow{\delta_0} & 0 \\ & & \delta_4 \uparrow & & & & \\ & & 4 & & & & \end{array}$$

Let Y be the wild quiver in I.9.2; we will establish an appropriate functor F from Y-mod onto a subcategory of $\tilde{\mathbb{D}}_4$-mod. Then taking the composition of this functor with that in I.9.2 will show that $\tilde{\mathbb{D}}_4$ is wild.

Given a representation $\underline{V} = [\, U \xrightarrow{\kappa} V \overset{\varphi}{\underset{\psi}{\rightrightarrows}} W \,]$ of the quiver Y, define $F(\underline{V})$ be the representation

$$\begin{array}{ccccccc} & & G(\varphi) & & & & \\ & & \downarrow & & & & \\ 0 \oplus W & \rightarrow & V \oplus W & \xleftarrow{\begin{bmatrix}1&0\\0&0\end{bmatrix}} & V \oplus 0 & \xleftarrow{\begin{bmatrix}\kappa&0\\0&0\end{bmatrix}} & U \oplus 0 \\ & & \uparrow & & & & \\ & & G(\psi) & & & & \end{array}$$

Here $G(\varphi)$ denotes the graph of $\varphi$.

(i) *F preserves indecomposability:* Let $F(\underline{V}) = \underline{X} \oplus \underline{Y}$. Then $W = X_3 \oplus Y_3$, $V = X_1 \oplus Y_1$, from the action of $\alpha_1$ and $\alpha_3$ on $F(\underline{V})$; moreover $U = X_0 \oplus Y_0$ and $\kappa$ is a sum of the maps for $\alpha_0$. We have now $G(\varphi) = X_2 \oplus Y_2$ and from the definition of $\alpha_2$ on $F(\underline{V})$ we have $(v, \varphi v) = (x_2, x_2') + (y_2, y_2')$, for any $(x_2, x_2') \in X_2$ and $(y_2, y_2') \in Y_2$; here $v = x_2 + y_2$. Note that $x_2'$, $y_2'$ are uniquely determined by $x_2$, $y_2$ respectively. Taking $y_2 = y_2' = 0$ shows $x_2' = \varphi(x_2)$; and similarly $y_2' = \varphi(y_2)$. This proves that $X_2$ =

$G(\varphi|_{X_1})$ and $Y_2 = G(\varphi|_{Y_1})$. The same arguments apply to $\psi$; and it follows that $\underline{X} = F(\underline{X}')$ and $\underline{Y} = F(\underline{Y}')$ where $\underline{V} = \underline{X}' \oplus \underline{Y}'$.

(ii) *F reflects isomorphisms:* Suppose there is an isomorphism $\underline{f} : F(\underline{V}) \to F(\underline{W})$. Then $f_1$ identifies $V_1$ with $W_1$, and $f_3$ identifies $V_3$ with $W_3$; and then $\kappa_V$ is the same as $\kappa_W$, via $f_0$. Moreover there is a commutative diagram with isomorphism $f_2$ and $f_4$

$$\begin{array}{ccc} \{ (v, \varphi_V v) \} & \xrightarrow{f_4} & \{ (v, \varphi_W v) \} \\ \downarrow j & & \downarrow j \\ V_1 \oplus V_2 & \xrightarrow{id} & V_1 \oplus V_2 \\ \uparrow j & & \uparrow j \\ \{ (v, \psi_V v) \} & \xrightarrow{f_2} & \{ (v, \psi_W v) \} \end{array}$$

where j is the inclusion map. Hence $f_4(v, \varphi_V v) = (v, \varphi_V v)$; and since the only element of the form $(v, *)$ in $G(\varphi_W)$ is $(v, \varphi_W v)$, we deduce that $\varphi_V v = \varphi_W v$. Similarly $\psi_V = \psi_W$, and consequently $\underline{V} \cong \underline{W}$.

(iii) *A bimodule representing F:* Let $Y = f_0 \xrightarrow{\alpha} f_1 \overset{\beta}{\underset{\gamma}{\rightrightarrows}} f_2$; and denote the primitive idempotents of $\widetilde{\mathfrak{D}}_4$ by $e_0, e_1, \ldots, e_c$. Take for B the KY-left module with free basis $\{ b_0, b_1, b_2, b_3, b_4, b_c \}$; we define a right $\widetilde{\mathfrak{D}}_4$-action on B as follows: Let $w \in KY$; we take the multiplication with idempotents to be zero except that

$$\begin{array}{lll} (wb_0)e_0 = wf_0b_0, & & (wb_c)e_c = w(f_1 + f_2)b_c, \\ (wb_1)e_1 = wf_1b_1, & & (wb_2)e_2 = wf_1b_2 + w\gamma b_2, \\ (wb_3)e_3 = wf_2b_3, & & (wb_4)e_4 = wf_1b_4 + w\beta b_4. \end{array}$$

For arrows we define

$$\begin{array}{lll} (wb_0)\delta_0 = w\alpha b_1, & & \\ (wb_1)\delta_1 = wf_1b_c, & & (wb_2)\delta_2 = (wb_c)e_c, \\ (wb_3)\delta_3 = wf_2b_c, & & (wb_4)\delta_4 = (wb_c)e_c, \end{array}$$

and $(wb_i)\delta_j = 0$ otherwise.

This defines a $\widetilde{\mathfrak{D}}_4$- module; the only property which needs checking is that $e_2$ and $e_4$ act indeed as idempotents.

*Then* $F \cong - \otimes B$: Suppose that $\underline{V}$ is a KY-module. Then, similarly as in I.9.1 and I.9.2, we have a decomposition of vector spaces in the form $\underline{V} \otimes B = \Sigma \oplus (\underline{V} \otimes b_i)$. From the action of the idempotents, we have

$\underline{V} \otimes b_0 = Vf_0 \otimes b_0 = U \otimes b_0$ and $\underline{V} \otimes b_1 = V \otimes b_1$; and

$\underline{V} \otimes b_c = \underline{V}(f_1 + f_2) \otimes b_c = V \otimes b_c \oplus W \otimes b_c$; and

$\underline{V} \otimes b_2 = \underline{V}(f_1 + \gamma) \otimes b_2 = \{ (v,\psi v): v \in V \} \otimes b_2 = G(\psi) \otimes b_2$;

and similarly $\underline{V} \otimes b_4 = G(\varphi) \otimes b_4$. Moreover, $\underline{V} \otimes b_3 = W \otimes b_3$. Considering the arrows, we have, for example, $(\underline{V} \otimes b_2)\delta_2 = G(\psi) \otimes b_2\delta_2$; and $(v, \psi v) \otimes (f_1 + f_2)b_c = (v, \psi v)(f_1 + f_2) \otimes b_c = (v, \psi v) \otimes b_c$; therefore $\delta_2$ is an inclusion map. Similarly one checks that for the other arrows the action is the same as on $F(\underline{V})$.

I.9.4 *The quiver* $\tilde{\tilde{E}}_6$ *is wild:* We consider $\tilde{\tilde{E}}_6$ with the orientation

$$\begin{array}{ccccccccccc}
 & & & & 0'' & & & & & & \\
 & & & & \downarrow \delta_1'' & & & & & & \\
 & & & & 1'' & & & & & & \\
 & & & & \downarrow \delta_2'' & & & & & & \\
0' & \xrightarrow{\delta_0'} & 1' & \xrightarrow{\delta_1'} & c & \xleftarrow{\delta_2} & 2 & \xleftarrow{\delta_1} & 1 & \xleftarrow{\delta_0} & 0
\end{array}$$

Let $Y$ be the wild quiver in I.9.2; we will define an appropriate functor $F$ from $\mathcal{R}(Y)$ onto a subcategory of $\mathcal{R}(\tilde{\tilde{E}}_6)$. Then the composition of this functor with the one in I.9.2 shows that $\tilde{\tilde{E}}_6$ is wild. Following $[R_1]$, we use the notation $XYZ$ for the direct sum of vectorspaces $X \oplus Y \oplus Z$.

Given $\underline{V} = \left[ U \longrightarrow V \rightrightarrows W \right]$, define $F(\underline{V})$ to be the $\tilde{\tilde{E}}_6$-representation

$$\begin{array}{ccccccccccc}
 & & & & 0G(\varphi) & & & & & & \\
 & & & & \downarrow j & & & & & & \\
 & & & & 0WV & & & & & & \\
 & & & & \downarrow j & & & & & & \\
G(\psi)0 & \rightarrow & VW0 & \rightarrow & VWV & \leftarrow & V0V & \leftarrow & \Delta(V0V) & \xleftarrow{\Delta\kappa} & \Delta(U0U)
\end{array}$$

Here $\Delta(X0X) = \{ (x,0,x): x \in X \}$. As maps, we take $\alpha_0 = \Delta\kappa$, and canonical inclusions otherwise.

(i) *F preserves indecomposability:* Let $\underline{M} = F(\underline{V})$ where $\underline{V} \in Y$-mod, and suppose that $\underline{M} = \underline{X} \oplus \underline{Y}$. Then we may recover the spaces $V$ and $W$ since $M_{2'} \cap M_2 = V00$ and $M_{2''} \cap M_2 = 00V$ and moreover $M_{2'} \cap M_{2''} = 0W0$. We must verify that the decompositions of $V00$ and $00V$ are the same. Write $M_{2'} \cap M_2 = (X_c \oplus Y_c)00$ and also $M_{2''} \cap M_2 = 00(\bar{X}_c \oplus \bar{Y}_c)$. Then by dimensions, $X_2 = X_c0\bar{X}_c$ and $Y_2 = Y_c0\bar{Y}_c$. The space

$M_1$ is also a direct sum, namely $M_1 = X_1 \oplus Y_1$ where $X_1 = \Delta(X0X)$ and $Y_1 = \Delta(Y0Y)$. Since $X_1 \subseteq X_2$ we deduce that $X \subseteq X_c$ and also $X \subseteq \tilde{X}_c$, similarly for Y. By dimensions, equality must hold. That is, $X_c$ and $\tilde{X}_c$ are identified and also $Y_c$ and $\tilde{Y}_c$. Studying the spaces at vertex 0 shows that $X_0 = \Delta(U_X 0 U_X)$ and $Y_0 = \Delta(U_Y 0 U_Y)$ with $U = U_X \oplus U_Y$, and $\kappa = \kappa_X \oplus \kappa_Y$. Moreover, $M_{2'} \cap M_{2''} = 0(W_X \oplus W_Y)0$; and $M_{1''} = G(\varphi_X) \oplus G(\varphi_Y)$ where $\varphi_X$ is the restriction of $\varphi$ to X, and so on. We see that $\underline{X} = F(\underline{A})$ and $\underline{Y} = F(\underline{B})$ where $\underline{V} = \underline{A} \oplus \underline{B}$.

(ii) *F reflects isomorphisms:* Suppose there is an isomorphism $\underline{f}$: $F(\underline{V}) \to F(\underline{Z})$ where $\underline{Z} = (X \to Y \rightrightarrows Z)$. Then $F(\underline{V})_{2'} \cap F(\underline{V})_2 = V00$ which is identified with $Y00$; similarly $00V$ corresponds to $00Y$; and by the arguments above we identify $V00$ with $00V$ and $Y00$ with $00Y$; similarly U corresponds to X, and then $\kappa_V = \kappa_Z$. Moreover $F(\underline{V})_{2'} \cap F(\underline{V})_{2''} = 0W0$ which we consider the same as $0Z0$; and then the corresponding maps in V and Z are the same; that is, $\underline{V} \cong \underline{Z}$.

Similarly as in I.9.3 one may construct a bimodule B such that $F \cong - \otimes B$.

<u>I.9.5</u> $\tilde{E}_7$ *is wild:* We consider $\tilde{E}_7$ with the orientation

$$\begin{array}{ccccccccccccccc} & & & & & & 1'' & & & & & & & & \\ & & & & & & \downarrow & & & & & & & & \\ 1' & \to & 2' & \to & 3' & \to & c & \leftarrow & 3 & \leftarrow & 2 & \leftarrow & 1 & \leftarrow & 0. \end{array}$$

Let Y be the wild quiver in I.9.2; we will define an appropriate functor F from Y-mod into a subcategory of $\tilde{E}_7$-mod. Given $\underline{V} = \left[ U \to V \rightrightarrows W \right] \in$ Y-mod, define F(V) as follows (with the notation as explained in I.9.4):

$$\begin{array}{ccccccccccccccc} & & & & & & 0G(\varphi) & & & & & & & & \\ & & & & & & \downarrow & & & & & & & & \\ G(\psi)0 & \to & VW0 & \to & VWV & \to & 0WV & \leftarrow & VWV & \leftarrow & V0V & \leftarrow & \Delta(V0V) & \leftarrow & \Delta(U0U) \end{array}$$

We omit details, they are similar to I.9.4.

<u>I.9.6</u> *The "five-subspace" quiver Q =* 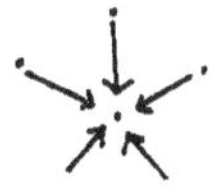 *is wild:*

Define a functor F as follows: If $(V, \varphi, \psi)$ is a K<X, Y>-module, let F(V) be the representation of *Q*

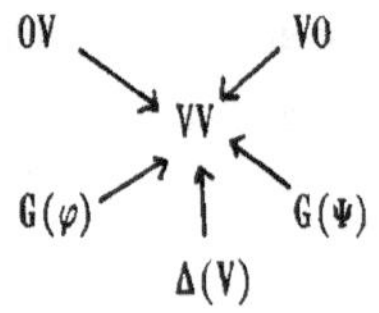


<u>I.9.7</u> *The algebras*

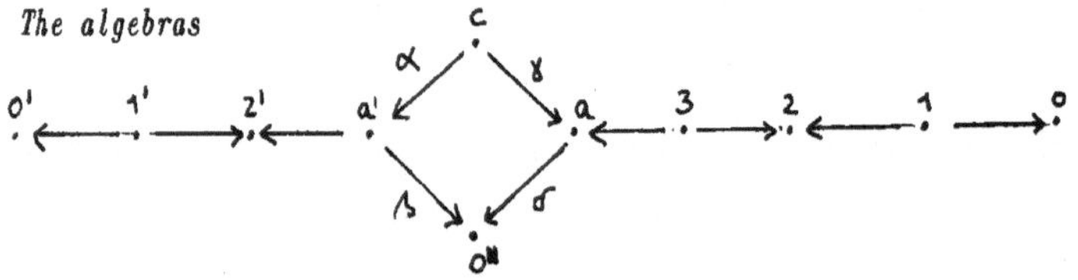


*with relation* $\alpha\beta = \gamma\delta$ *are wild:*

We define an appropriate functor F: $\mathcal{I}(\tilde{\tilde{E}}_7) \to \mathcal{I}(Q)$ and then take the composition with the functor in I.9.5.

We take $\tilde{\tilde{E}}_7$ with the orientation

$$\begin{array}{ccccccccccccccc} 0' & \xleftarrow{\delta_0'} & 1' & \xrightarrow{\delta_1'} & 2' & \xleftarrow{\delta_2'} & 4 & \xleftarrow{\delta_3} & 3 & \xrightarrow{\delta_2} & 2 & \xleftarrow{\delta_1} & 1 & \xrightarrow{\delta_0} & 0 \\ & & & & & & \downarrow{\scriptstyle \delta_0''} & & & & & & & & \\ & & & & & & 0'' & & & & & & & & \end{array}$$

Let $\underline{V} \in \mathcal{I}(\tilde{\tilde{E}}_7)$. Then $F(\underline{V})$ is defined to be the module

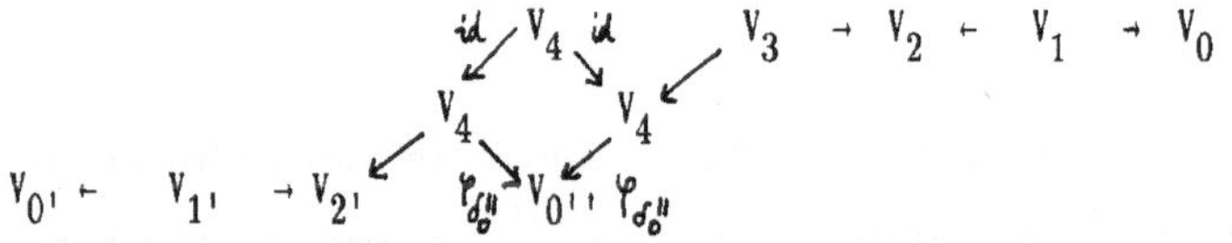


where all other maps are the ones given by $\underline{V}$. We will not go into details.

<u>I.9.8</u> *The algebra*

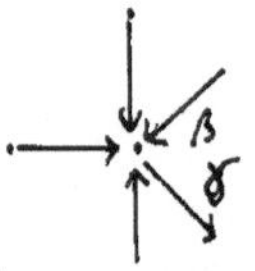


*with* $\beta\gamma = 0$ *is wild:*

This is reduced to $\tilde{\tilde{D}}_4$, see see $[R_3]$ p. 193.

## I.10 *Coverings*

Our practical problem is to recognize whether an algebra is wild. For this purpose it is convenient to use coverings; we will give a short introduction but only for simple special cases.

Suppose R is a basic algebra of the form $R = K\mathcal{Q}/I$ where $\mathcal{Q}$ is a finite quiver and I an ideal of $K\mathcal{Q}$ with $I \subseteq (K\mathcal{Q}^+)^2$. We will only consider the following two situations:

(A) I is generated by a set of 0-relations.

(B) $I = \langle \beta\gamma - \delta\eta \,;\, \rho_i \rangle$ where $\{\rho_i\}$ is a set of 0-relations; and $\beta, \gamma, \delta, \eta$ are arrows.

We will illustrate the concepts of coverings using the following two examples:

<u>I.10.0</u>

$(\Lambda_1)$ 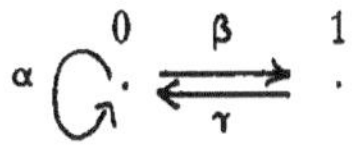

$I_1 = \langle \{ \alpha^3, \gamma\alpha, \alpha\beta \} \rangle.$

$(\Lambda_2)$ -"- $I_2 = \langle \{ \alpha^2 - \beta\gamma, \gamma\beta \} \rangle.$

<u>I.10.1</u> Let $\mathcal{Z}$ and $\mathcal{Q}$ be quivers. Recall that a map $\pi: \mathcal{Z} \to \mathcal{Q}$ is a *covering* if $\pi$ is a quiver morphism which satisfies the following:

(*) Let $a \in \mathcal{Q}_0$ and $x \in \mathcal{Z}_0$ with $\pi(x) = a$. Then $\pi$ induces a bijection $s^{-1}(x) \to s^{-1}(a)$ and also a bijection $e^{-1}(x) \to e^{-1}(a)$.

A Galois covering $\pi: \mathcal{Z} \to \mathcal{Q}$ with group G is a covering together with a group G of automorphisms of $\mathcal{Z}$ such that each fibre of $\pi$ is a transitive G-space, and G acts freely [that is, $gx \neq x$ for every vertex x in $\mathcal{Z}$ and each $g \neq 1$ in G].

Concerning existence of coverings, if no further conditions have to be satisfied then we have:

<u>I.10.2</u> Given some quiver $Q$, fix a vertex $x_0$. Recall that the (topological) universal cover $\tilde{Q}$ of $Q$ with base point $x_0$ may be thought of as the quiver whose vertices are all walks starting at $x_0$. The arrows of $\tilde{Q}$ are defined as follows:

Let $w = (z|\alpha_{m-1}, \ldots, \alpha_1|x_0)$ be a walk, here $\alpha_i$ or $\alpha_i^{-1}$ is an arrow.

(a) If $y \xrightarrow{\alpha_m} z$ is an arrow in $Q$ [but $\alpha_m \neq \alpha_{m-1}^{-1}$] we define an arrow of $\tilde{Q}$ by

$$\tilde{\alpha}_m : w \to (y|\alpha_m, \alpha_{m-1}, \ldots, \alpha_1|x_0)$$

(b) If $z \xrightarrow{\alpha_m} y$ is an arrow in $Q$ [but $\alpha_m \neq \alpha_{m-1}^{-1}$] we define an arrow of $\tilde{Q}$ by

$$\tilde{\alpha}_m : (y|\alpha_m^{-1}, \alpha_{m-1}, \ldots, \alpha_1|x_0) \to w.$$

Then $\tilde{Q}$ is an infinite tree.

Define $\pi: \tilde{Q} \to Q$ by $\pi(w) = e(w)$ if $w$ is a walk ending in the vertex $e(w)$ of $Q$, and $\pi(\tilde{\alpha}) = \alpha$ for any arrow $\alpha$ of $Q$. Then $\pi$ is a Galois covering, where the group G is the fundamental group of $Q$.

<u>I.10.3</u>(a) Let $\mathcal{I}$ and $Q$ be quivers and $\pi: \mathcal{I} \to Q$ be a covering. Then $\pi$ extends to a (unique) surjective algebra homomorphism $F: K\mathcal{I} \to KQ$ of the path algebras. This is clear since a path algebra is the free algebra whose the primitive idempotents are the vertices, and where the generators of the radical are the arrows of the quiver.

(b) *The covering of a quiver with relations:* Let $\pi: \mathcal{I} \to Q$ be a covering and L, I be ideals of $K\mathcal{I}$ and $KQ$ respectively such that F takes L onto I. Then the map F defined above induces a surjective algebra homomorphism $K\mathcal{I}/L \to KQ/I$. We say that $\pi: (\mathcal{I}, L) \to (Q, I)$ is a covering of bound quivers.

EXAMPLE (1) $A_1 = KQ/I_1$ as in I.10.0. Let $\tilde{Q}$ be the (topological) universal cover:

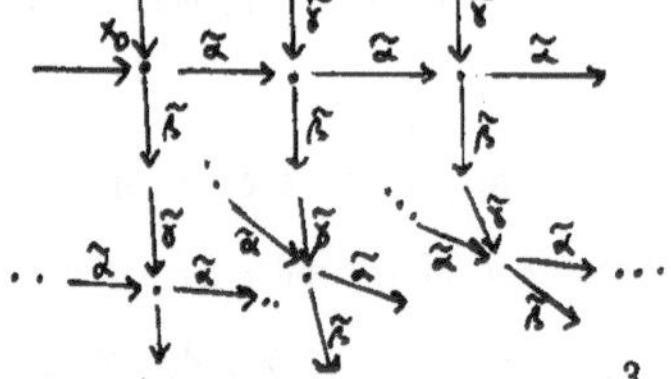

Take $\tilde{I}_1$ = the ideal generated by all possible paths $\tilde{\alpha}^3$, $\tilde{\gamma}\tilde{\alpha}$, $\tilde{\alpha}\tilde{\beta}$ .

(2) $A_2 = KQ/I_2$ as in I.10.0. Take the covering $\mathcal{I}$:

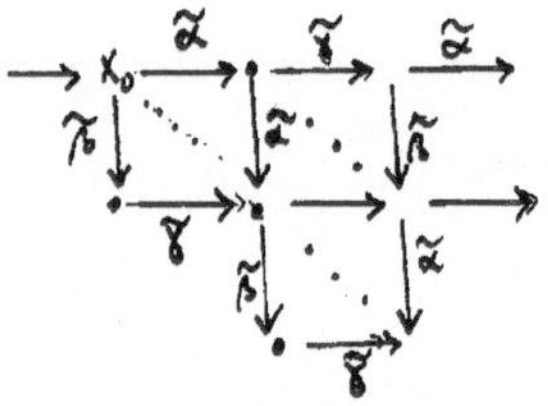

We take $\tilde{I}_2 = \langle \tilde{\alpha}^2 - \tilde{\gamma}\tilde{\beta},\ \tilde{\beta}\tilde{\gamma} \rangle$.

I.10.4 It has been proved in [MP] that every finite-dimensional algebra $\Lambda = KQ/I$ has a universal cover (as an algebra), which is a Galois covering, which however depends on the presentation $(Q, I)$. For the two types of algebras we consider here, we will give the explicit description, following [MP].

(A) Let $\pi: \tilde{Q} \to Q$ be the (topological) universal cover of $Q$, as in I.10.2. Suppose that $\{ \rho_\nu \}$ is a set of zero-relations in $KQ$ which generates the ideal I. We lift this set in the obvious way to a set $\{ \tilde{\rho}_\nu \}$ of zero-relations of $K\tilde{Q}$. Let $\tilde{I}$ be the ideal generated by $\{ \tilde{\rho}_\nu \}$; then we have a covering $\pi: (\tilde{Q}, \tilde{I}) \to (Q, I)$ of bound quivers. This is a Galois covering where the group is the fundamental group $\Pi_1(Q, x_0)$.

The other type is a general version of our second example.

(B) Consider an algebra $\Lambda = KQ/I$ where we fix a set of generators $\mathcal{I} = \{ \beta\gamma - \delta\eta;\ \rho_\nu \}$ of I where the $\rho_\nu$ are zero-relations. Fix a vertex $x_0$ of $Q$ and let $\tilde{Q}$ be the topological universal cover of $Q$ with base point $x_0$, as in I.10.2. Let $p: \tilde{Q} \to Q$ be the natural map given by the action of the fundamental group $\Pi_1(Q, x_0)$. Let $N = N(Q, \mathcal{I}, x_0)$ be the normal subgroup of $\Pi_1(Q, x_0)$ generated by all elements of the form $\sigma^{-1}(\beta\gamma)^{-1}(\delta\eta)\sigma$ where $\sigma$ is a path from $x_0$ to the starting point x of $\beta\gamma$, $\delta\eta$. Then $\mathcal{Z}$ is defined to be the orbit quiver $\tilde{Q}/N$ and the map $\pi: \mathcal{Z} \to Q$ is given by the action of $\Pi_1(Q, x_0)/N$. The relation $\beta\gamma - \delta\eta$ can be lifted to $\mathcal{Z}$, and L is the ideal generated by the liftings of $\beta\gamma - \delta\eta$ and the $\rho_\nu$. This is a Galois covering where the group is $\Pi_1(Q, x_0)/N$.

I.10.5 THE PUSH-DOWN In the following we will consider only modules whose total dimension is finite. Let $\pi: (\mathcal{Z}, L) \to (Q, I)$ be a covering of bound quivers. The corresponding algebra maps constructed above induce a functor

$F_\lambda: (\mathcal{Z}, L)\text{-mod} \to (Q, I)\text{-mod}$, or equivalently $\text{mod}(K\mathcal{Z}/L) \to \text{mod}(KQ/I)$ :

Let $M \in (\mathcal{Z}, L)$-mod, then $V = F_\lambda(M)$ is defined as follows: For any vertex $a \in Q_0$, set

$$V_a := \bigoplus_{\pi x = a} M_x$$

Suppose $a \xrightarrow{\alpha} b$ is an arrow in $\mathcal{Q}$. For any $x \in \mathcal{I}_0$ with $\pi x = a$ there is a unique arrow $\tilde{\alpha}$ of $\mathcal{I}$ starting at $x$ with $\pi(\tilde{\alpha}) = \alpha$ (and necessarily $\pi$ takes the end point of $\tilde{\alpha}$ to $b$). Then $V_\alpha : V_a \to V_b$ is the linear transformation which takes $m \in M_x$ to $M_{\tilde{\alpha}}(m)$.

EXAMPLE $\quad \mathcal{I} = \left[ \overset{0}{\cdot} \overset{\tilde{\alpha}}{\underset{\tilde{\beta}}{\rightrightarrows}} \overset{1}{\cdot} \right] \xrightarrow{\pi} \left[ \alpha \circlearrowleft \cdot \circlearrowright \beta \right] = \mathcal{Q}$ where

$\pi(0) = a = \pi(1)$ and $\pi(\tilde{\alpha}) = \alpha$, $\pi(\tilde{\beta}) = \beta$.

Let $M = K \underset{\tilde{\beta}=c}{\overset{\tilde{\alpha}=1}{\rightrightarrows}} K$. Then $F_\lambda(M)$ has underlying space $K \oplus K$; and $\alpha$, $\beta$ are the linear transformations whose matrices are $\begin{bmatrix} 0 & 0 \\ 1 & 0 \end{bmatrix}$ and $\begin{bmatrix} 0 & 0 \\ c & 0 \end{bmatrix}$ respectively.
(This shows that the push-down functor is not full in general.)

PROPERTIES OF THE PUSH-DOWN Let $\pi : (\mathcal{I}, L) \to (\mathcal{Q}, I)$ be a covering and $F_\lambda$ the corresponding push-down. We allow the possibility that $I$ or $L$ is zero.

(a) *The " pull-up "* If $V$ is an $(\mathcal{Q},I)$-module then $F.V = V\,F^{op}$ is a module for $(\mathcal{I}, L)$, denoted by "pull-up". Then $F_\lambda$ is left-adjoint to $F$ (see [$G_3$]).

Assume now that $\pi$ is a Galois covering, with group $G$.

(b) *Translates of modules:* Let $M \in (\mathcal{I},L)$-mod and $g \in G$. Then the translate ${}^gM$ is the representation of $(\mathcal{I},L)$ with

$$({}^gM)_x = M_{g^{-1}x} \qquad (x \in \mathcal{I}_0) \quad \text{and}$$
$$({}^gM)_\alpha = M_{g^{-1}\alpha} \qquad (\alpha \in \mathcal{I}_1).$$

(c) LEMMA [$G_3$; 3.2] *For each $M \in (\mathcal{I},L)$-mod and each $g \in G$, we have $F_\lambda {}^gM \cong F_\lambda M$ and $\bigoplus_{h\in G} {}^hM \cong F.F_\lambda M$ canonically.*

(d) LEMMA [$G_3$; 3.5] *Suppose that $M \in (\mathcal{I},L)$-mod is indecomposable and that ${}^gM \not\cong M$ if $1 \neq g \in G$. Then $F_\lambda M$ is indecomposable. Moreover, each $V \in (\mathcal{I}, L)$-mod such that $F_\lambda V \cong F_\lambda M$ is isomorphic to ${}^gM$ for some $g \in G$.*

Proof: Assume that $F_\lambda M = N \oplus N'$ where $N \neq 0$. Then we have that $\Sigma\, {}^gM \cong F.F_\lambda M =$

$F.N \oplus F.N'$. Since the ${}^gM$ are pairwise non-isomorphic, we must have that $F.N \cong \oplus {}^hM$ and $F.N' \cong \Sigma\ {}^fM$ for some subset H of G. On the other hand, we have $F.N = {}^gF.N \cong \oplus {}^{gh}M$ for each $g \in G$. This implies $H = gH$ for each $g \in G$, hence $H = G$ and $N' = 0$. Assume that $F_\lambda W \cong F_\lambda M$. Then w is isomorphic to a direct summand $\Sigma\ {}^gM$ of $F.F_\lambda W \cong F.F_\lambda M \cong \oplus {}^gM$. Since $F_\lambda W \cong \oplus F_\lambda({}^gM)$ is indecomposable, we must have that $H = \{g\}$ and $W \cong {}^gM$ for some $g \in G$.

<u>I.10.6</u> We shall now study the restrictions of push-down functors to subcategories. A representation M of a quiver T is said to be *sincere* if $M(x) \neq 0$ for each vertex x of T.

PROPOSITION *Let $\pi: (\tilde{Q}, L) \to (Q, I)$ be a Galois covering of bound quivers, $T \subset \tilde{Q}$ a finite subquiver and $L' = KT \cap L$, and suppose that $\pi$ maps $L'$ onto I. Let $(T, L')_s$-mod be the full subcategory of $(T, L')$-mod generated by the sincere objects. Then*

*(i) The restriction $F_\lambda : (T,L')_s\text{-mod} \to (Q, I)\text{-mod}$ preserves indecomposability and reflects isomorphism.*

*(ii) There is a finitely generated $KT/L'$- $KQ/I$- bimodule B which is free as a left $KT/L'$-module such that on $(T, L')_s$-mod, $F_\lambda \cong - \otimes_{KT/L'} B$.*

Proof: (1) <u>$F_\lambda$ preserves indecomposability:</u> Suppose M is an indecomposable $(T, L')_s$-module. Then we consider M as a $\tilde{Q}$-module (with $M_x = 0$ for x not in T ). By I.10.5 it suffices to show that ${}^gM \neq M$ for $1 \neq g \in G$. If $1 \neq g$ then ${}^gT \neq T$ and hence the supports of ${}^gM$ and M are distinct; hence ${}^gM \neq M$.

(2) <u>$F_\lambda$ reflects isomorphisms:</u> Let $F_\lambda(W) \cong F_\lambda(M)$ where W and M are sincere objects in $(T, L')$-mod. Then, considering W, M as $\tilde{Q}$-modules as above, we have by I.10.5(d) that $W \cong {}^gM$ for some $g \in G$. Hence T = the support of M = the support of W = the support of ${}^gM$, and ${}^gT = T$; consequently $g = 1$ and $W \cong M$.

(3) <u>The bimodule B</u>: Define B to be the free $KT/L'$-module $B = \Sigma \oplus (KT/L')b_x$ with basis $\{ b_x: x \in T_0 \}$. We define a right $(Q,I)$-action on B as follows (see also the examples in I.9): Let $e \in Q_0$, $w \in KT/L'$ and $x \in T_0$. We denote by $f_x$ the idempotent of $KT/L'$ corresponding to x; and we set

$$(wb_x)e = \begin{cases} 0 & \text{if } \pi x \neq e \\ wf_x b_x & \text{otherwise} \end{cases}$$

Suppose $e \xrightarrow{\alpha} f$ is an arrow in $Q$. If $x \in T_0$ with $\pi x = e$ and $\tilde{\alpha}$ is an arrow in T starting at x then define $(wb_x)\alpha = (w\tilde{\alpha})b_y$ where y is the endpoint of $\tilde{\alpha}$; and set $(wb_x)\alpha = 0$ otherwise.

We claim that this is a $(Q, I)$-action: Suppose $\rho \in I$; it suffices to consider $\rho \in e[KQ]f$ for $e, f \in Q_0$. By the hypothesis, there are vertices $x, y \in T_0$ with $\pi(x) = e$ and $\pi(y) = f$, and $\tilde{\rho} \in L'$ such that $\pi(\tilde{\rho}) = \rho$. From the definition of B, we have for $w \in KT/L'$ that $(wb_x)\rho = (\omega\tilde{\rho})b_y = 0$.

Now let $M \in$ T-mod; we will show that $F_\lambda(M) \cong M \otimes_{KT} B$, canonically.

The module $M \otimes_{KT} B$ has underlying space $\bigoplus_{x \in T_0} (M \otimes b_x)$. This is clear, since for any arrow $\tilde{\alpha}$ of T we have that $M \otimes \tilde{\alpha}b_x = M\tilde{\alpha} \otimes b_x \subseteq M \otimes b_x$. Let $e \in Q_0$; then $(M \otimes b_x)e = 0$ for $\pi x \neq e$. On the other hand, if $\pi x = e$ then $(M \otimes b_x)e = M \otimes f_x b_x = Mf_x \otimes b_x = M_x \otimes b_x$. So we may identify $(M \otimes B)e$ with $\bigoplus_{\pi x = e} M_x = F_\lambda M$, as required.

Now consider the action of an arrow $e \xrightarrow{\alpha} f$ in $Q$. Let $x \xrightarrow{\alpha} y$ be an arrow in T with $\pi x = e$, $\pi\tilde{\alpha} = \alpha$ and hence $\pi y = f$. Then

$$(M \otimes b_x)\alpha = M \otimes \tilde{\alpha}b_y = M\tilde{\alpha} \otimes b_y = Mf_x\tilde{\alpha} \otimes b_y = M_x\tilde{\alpha} \otimes b_y = M_{\tilde{\alpha}}(M_x) \otimes b_y$$

and this is just the action of $\alpha$ on the module $[F_\lambda M]_x$.

COROLLARY *With the notation as above, if $(KT/L')_s$ is wild then so is $KQ/I$.*

Proof: Suppose there is a functor $- \otimes_{K\langle X,Y\rangle} M : \text{mod } K\langle X, Y\rangle \to \text{mod}(KT/L')$ which reflects isomorphisms and preserves indecomposability whose image consists only of sincere modules; then take the composition of this functor and $F_\lambda$.

<u>I.10.7</u> APPLICATION Given $R = KQ/I$ with I satisfying (A) or (B) at the beginning of I.10. The problem is to find out whether R is wild. We consider the universal cover $\pi: (\tilde{Q}, L) \to (Q, I)$ and look for either:

(a) finite trees $T \subset \tilde{Q}$ such that $KT \cap L = 0$, in case $I = 0$. If there is such tree where $KT_s$ is wild then so is R, by the Corollary.

(b) Finite subquivers $T \subset \tilde{Q}$ such that $L' = L \cap KT = \langle \tilde{\beta}\tilde{\gamma} - \tilde{\delta}\tilde{\eta} \rangle$, and $\pi(L') = I$. If

$[KT/L']_S$ is wild then so is R.

Appropriate candidates for T and for (KT/L') are given in I.9.0. Actually, the explicit constructions provide functors whose image consists of sincere objects.

(By a result of Drozd [D], an algebra $\Lambda$ is wild if and only if the category $(\text{mod}\ \Lambda)_S$ is wild; so we need not worry about non-sincere objects.)

<u>I.10.8</u> *Any algebra is wild whose quiver contains one of the following quivers or their duals:*

*(i)* 

*(ii)* 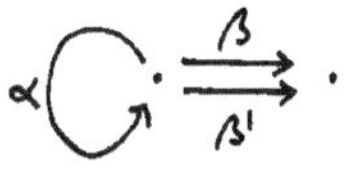

*(iii)* 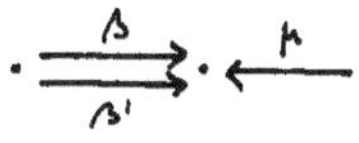

*(iv)* 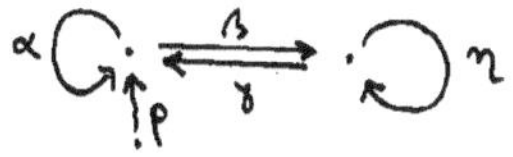

*(v)* 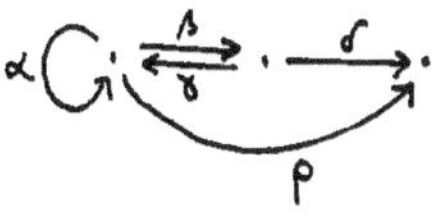

*(vi)* 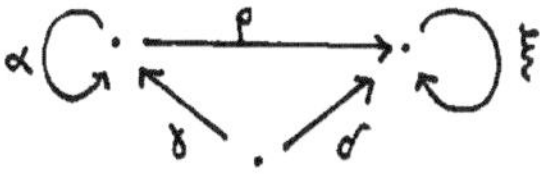

Proof: The algebra $\Lambda/J^2$ is generated by zero-relations. So it has a universal cover as described in I.10.4. In each case we find a tree $\cong \tilde{\tilde{E}}_7$ in the universal cover of $\Lambda/J^2$, namely

(i) 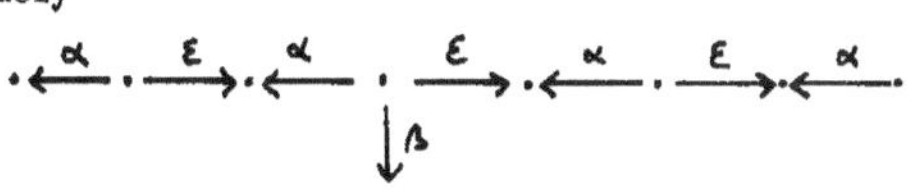

(ii) 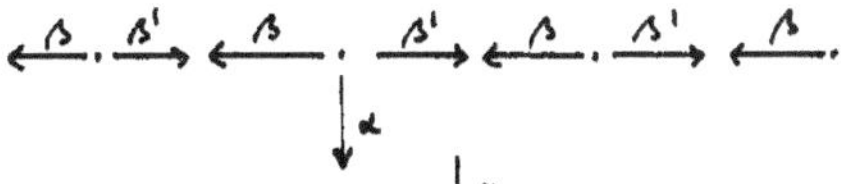

(iii) 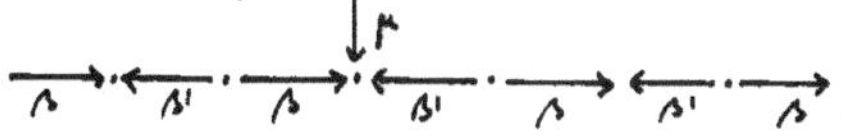

(iv) 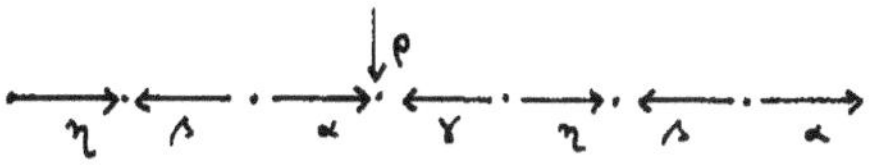

(v) 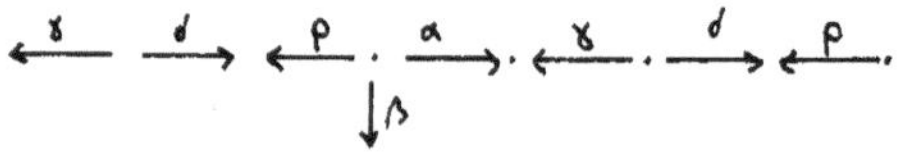

(vi) 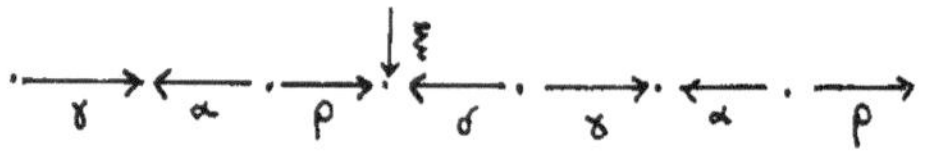

Hence $\Lambda/J^2$ is wild, by I.10.7, and so is $\Lambda$, by I.4.7.

I.10.9 LEMMA *Let $\Lambda$ be an algebra which is not wild and let $Q$ be the quiver of $\Lambda$. Suppose that for each vertex $f$ of $Q$, at least two arrows start and two arrows end at $f$. Given any arrow $\alpha$ of $Q$ then there is at most one arrow $\beta$ such that $\alpha\beta$ is not involved in any relation.*

Proof: We may assume that $\Lambda$ is basic. Suppose $\alpha\beta$ and $\alpha\gamma$ are both not involved in any relation. Let $I := J^3 + \text{span}\{\ \delta\rho\colon \delta, \rho$ are arrows and $\delta \neq \alpha\ \}$. This is an ideal of $\Lambda$, and the factor algebra $\Lambda/I$ may be generated by zero-relations. Hence $\Lambda/I$ has a universal cover, as described in I.10.4.

By the hypothesis, $\Lambda$ has enough arrows, and hence the universal cover contains a tree of the form $\tilde{\tilde{E}}_7$, namely

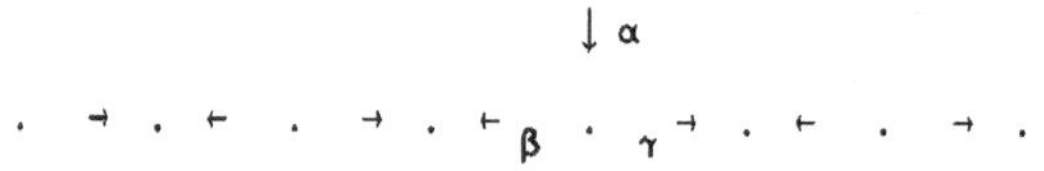

Consequently the algebra $\Lambda/I$ is wild, by I.10.7, and so is $\Lambda$, by I.4.7.

I.10.10 *The following local algebras $\Lambda$ are wild:*

*(a)* $\Lambda = K\langle X, Y, Z\rangle/J^2$.

*(b)* $\Lambda = K[X, Y]/J^3$.

*(c)* $\Lambda = K\langle X, Y\rangle/I$ *where $I$ is generated by* $XY$, $X^2-Y^2$ *and* $J^3$.

*(d)* $\Lambda = K\langle X, Y\rangle/I$ *where $I$ is the ideal generated by* $Y^2$, $XY - cYX$ *and* $J^3$ $(c \neq 0)$.

Proof: (a) The algebra is generated by zero-relations and has a universal cover $Q$ as described in I.10.4. The quiver of $\Lambda$ consists of one vertex and three loops; hence $Q$ contains $E_7$, namely 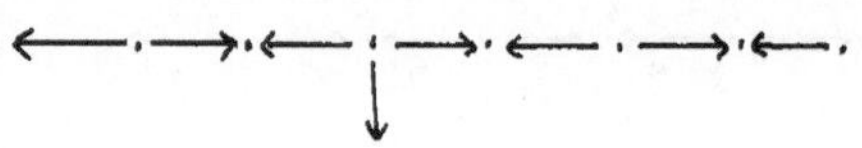

Therefore Λ is wild.

(b) The algebra satisfies condition (B) in I.10; hence it has a universal cover $(Q, I)$. This contains the quiver

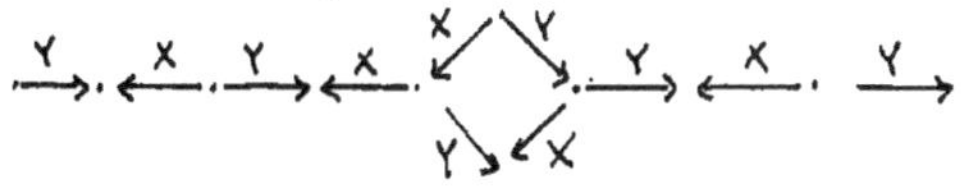

where the square is commutative. Hence Λ is wild, by I.9.7 ( and I.10.7).

Part (c) is similar to (b); here we use the quiver

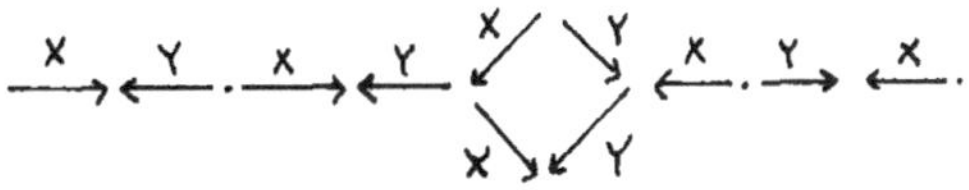

In (d) take

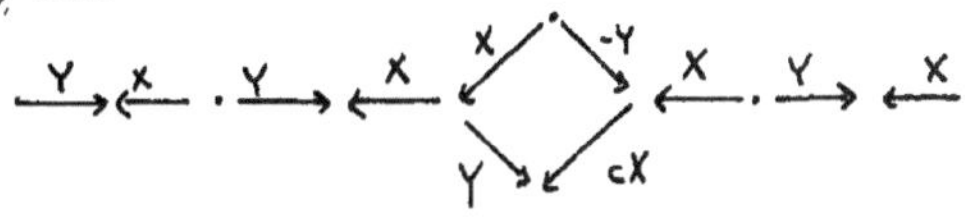

Now we will use coverings to show that a number of algebras are not of finite type.

<u>I.10.11</u> *Suppose* Λ *is an algebra with quiver* $Q$ *such that for each vertex* e *of* $Q$ *at least two arrows start and at least two arrows end at* e. *Then* Λ *is not of finite type.*

Proof: We may assume that Λ is basic. Consider the algebra $\Lambda/J^3$; it has the same quiver as Λ and may be defined by zero-relations only. It has a universal cover which contains infinite lines of the form

$$\ldots . \leftarrow \; . \; \rightarrow . \leftarrow \; . \; \rightarrow . \leftarrow \; . \; \rightarrow . \ldots .$$

Hence $\Lambda/J^3$ has indecomposable modules of arbitrary dimension and so does Λ.

<u>I.10.12</u> *Let* Λ *be any algebra with quiver* of the *form*

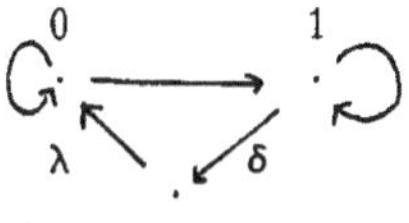


*If* δλ *is not involved in a relation then* Λ *is of infinite type.*

Proof: Let $e = e_0 + e_1$; then the quiver of the algebra eΛe is of the form

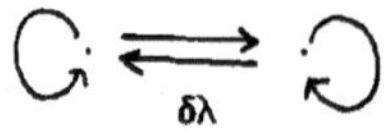

. By I.10.11, $e\Lambda e$ is of infinite type, and so is $\Lambda$, by I.4.7.

# II *Special biserial algebras and the local semidihedral algebra*

In this chapter we introduce special biserial algebras and describe the classification of their indecomposable representations; and we also determine all irreducible maps. The results go back to [GP], [$R_2$], [BS]; for a detailed account, see [BR]. Then we describe the classification of the indecomposable modules for the local semidihedral algebras; this is recent work by Crowley-Boevey [$CB_2$]. We use his results to determine the graph structure of the stable Auslander-Reiten quiver for the local semidihedral algebras (in II.10).

II.1.1 DEFINITION The algebra $\Lambda$ is *special biserial* provided the basic algebra $KQ/I$ satisfies the following conditions:

(1) Any vertex of $Q$ is starting point of at most two arrows.

(1*) Any vertex of $Q$ is end point of at most two arrows.

(2) Given an arrow $\beta$, there is at most one arrow $\gamma$ with $s(\beta) = e(\gamma)$ and $\beta\gamma \notin I$.

(2*) Given an arrow $\gamma$, there is at most one arrow $\beta$ with $s(\beta) = e(\gamma)$ and $\beta\gamma \notin I$.

This definition was introduced in [SW]; many algebras of finite type are special biserial [Ku], [GR], but also the algebras occuring in [GP]. The Kronecker algebra in I.4.4 is special biserial, and also the algebra $(A_1)$ in I.10. We note that special biserial algebras are biserial.

II.1.2 DEFINITION The algebra $\Lambda$ is a *string algebra* if its basic algebra is of the form $KQ/I$ where I is generated by zero relations and in addition satisfies the conditions in II.1.1.

In particular, a string algebra is special biserial. A string algebra is usually not self-injective.

II.1.3 *Suppose* $\Lambda$ *is special biserial. For studying indecomposable non-projective modules and irreducible maps, one may assume that* $\Lambda$ *is a string algebra.*

To see this, without loss of generality, $\Lambda$ is basic. Let $L = \{e \in Q_0$: $\Lambda e$ is injective and not uniserial$\}$, a finite set; and let $S_0 = \bigoplus_{e \in L} \operatorname{soc} \Lambda e$. Then the algebra $\bar{\Lambda} = \Lambda/S_0$ is a string algebra; this is easy to see. Concerning indecomposable modules one shows that $\operatorname{ind} \Lambda = \operatorname{ind} \bar{\Lambda} \cup \{ \Lambda e: e \in L \}$; and moreover, the AR-quiver of $\bar{\Lambda}$ is equal to $\Gamma(\Lambda) - \{ \Lambda e: e \in L \}$; in particular $\Gamma(\bar{\Lambda})$ contains $\Gamma_s(\Lambda)$, see I.8.11.

If one knows $\Gamma(\bar{\Lambda})$ then one can reconstruct $\Gamma(\Lambda)$. Since $\Lambda e/\operatorname{soc} \Lambda e$ is a projective $\bar{\Lambda}$-module whose radical is a direct sum of two non-zero modules, it corresponds to a projective vertex of $\Gamma(\bar{\Lambda})$ with two predecessors. These vertices are visible, and one attaches $[\Lambda e]$ there.

Our aim is to study the non-projective indecomposable modules of a special biserial algebra $\Lambda$. By II.1.3 we may assume that $\Lambda$ is a string algebra; and moreover, we take also $\Lambda$ to be basic, say $\Lambda = KQ/I$, satisfying II.1.1.

First we introduce the parameter sets.

II.2 (a) **WORDS, INVERSES, ROTATIONS** First we assume tat $Q$ is an arbitrary quiver. As alphabet, we take the arrows of $Q$ and their formal inverses. Given an arrow $\beta$ of $Q$, denote by $\beta^{-1}$ a formal inverse for $\beta$, and we set $s(\beta^{-1}) = e(\beta)$ and $e(\beta^{-1}) = s(\beta)$, and we write $(\beta^{-1})^{-1} = \beta$.

By a *word* w we mean a sequence $w_1w_2\ldots w_n$ where the $w_i$ are of the form $\beta$ or $\beta^{-1}$ ($\beta \in Q_1$), and where $s(w_i) = e(w_{i+1})$ for $1 \leq i \leq n$. We define

$$(w_1w_2\ldots w_n)^{-1} = w_n^{-1}\ldots w_2^{-1}w_1^{-1}$$
$$s(w_1w_2\ldots w_n) = s(w_n)$$
$$e(w_1w_2\ldots w_n) = e(w_1).$$

For each vertex u of $Q$, there is an empty word $1_u$ of length 0 with $e(1_u) = s(1_u)$, and $(1_u)^{-1} = 1_u$. A *rotation* of w is a word of the form $w_{i+1}\ldots w_nw_1\ldots w_i$. The *product* of two words is defined by placing them next to each other, provided that the resulting sequence is a word. If the product of w with itself is defined, one has also arbitrary powers $w^j$ of w.

On the set of words, we consider two equivalence relations. Firstly, $\sim$ denotes the relation which identifies w with $w^{-1}$; and secondly, we define $\sim_r$ to be the equivalence relation which identifies each word with its rotations and their inverses (if defined).

From now, assume $\Lambda = KQ/I$ where I satisfies II.1.1 and 2.

(b) STRINGS Let $\mathcal{S}t$ be a set of representatives of words w under the relation $\sim$ where either $w = w_1w_2\ldots w_n$ and $w_i \neq w_{i+1}^{-1}$ for $1 \leq i \leq n$ and no subpath of w or its inverse belongs to I, or $w = 1_u$. We call the elements of $\mathcal{S}t$ strings.

(c) BANDS Consider the set of words $w = w_1w_2\ldots w_n$ of positive length such that $w_i \neq w_{i+1}^{-1}$ and $w_n \neq w_1^{-1}$ whose powers are defined and which are not themselves powers, and such that $w^m$ has no subword lying in I. Let $\mathcal{B}a$ be the set of representatives of such words under $\sim_r$. We call the elements of $\mathcal{B}a$ bands.

<u>II.2.1</u> COMPOSITION OF STRINGS (a) Let C and D be strings of length $\geq 1$. Then the composition of C and D is defined by placing them next to each other, provided that the resulting sequence is a string. The composition of $1_u$ and D is defined if e(D) = u, then let $1_uD = D$. Similarly if s(C) = u then $C1_u = C$.

(b) EXAMPLE Let $\Lambda$ be the Kronecker algebra in I.4.3. If $C = XY^{-1}$ and $D = XY^{-1}X$ then $CD = XY^{-1}XY^{-1}X$. On the other hand, if $E = YX^{-1}$; then the composition of C and E is not defined.

(c) We note the following property. Let C be a string of length $\geq 1$; then there is at most one arrow $\beta$ with $C\beta$ a string and at most one arrow $\gamma$ with $C\gamma^{-1}$ a string. (In [BR] the formal set-up with signs guarantees that this also holds for strings of length zero.)

## II.3 *Indecomposable modules*

Suppose $\Lambda$ is a string algebra. For each string and band w we shall define an algebra $C_w$, and a functor $G_w$: $C_w$-mod $\to$ $\Lambda$-mod. If V is a $C_w$-module we describe $G_w(V)$ as a representation of a quiver $Q_w$. On morphisms, the functors are defined in the obvious way.

(a) STRINGS If $w = w_1w_2\ldots w_n$ ( or $w = 1_u$) is a string of length $n \geq 0$, let $C_w = K$, and let $Q_w$ be the quiver with underlying graph $A_{n+1}$

where the edge labelled with $w_i$ is an arrow pointing to the left if $w_i$ is an arrow, and to the right otherwise. Then $G_w(V)$ for $V \in K$-mod is defined to be the representation of $Q_w$ where at each vertex the space is V, and for each arrow, the map is the identity map.

EXAMPLE Let $\Lambda$ be the Kronecker algebra and $w = XY^{-1}X$. Then $Q_w$ is the quiver

$\cdot \xleftarrow{X} \cdot \xrightarrow{Y} \cdot \xleftarrow{X} \cdot$ . For V = K, the module $G_w(K)$ has total dimension 4. Considering $G_w(V)$ as a $\Lambda$-left module, X and Y are represented by the matrices

$$\begin{bmatrix} 0&1&0&0 \\ 0&0&0&0 \\ 0&0&0&1 \\ 0&0&0&0 \end{bmatrix} \text{ and } \begin{bmatrix} 0&0&0&0 \\ 0&0&0&0 \\ 0&1&0&0 \\ 0&0&0&0 \end{bmatrix}.$$

(b) BANDS Suppose now $w = w_1w_2\ldots w_n$ is a band. We may assume that $w_1$ is an arrow, by rotating and possibly inverting. We let $C_w = K[x, x^{-1}]$; and for $Q_w$ we take the circular quiver

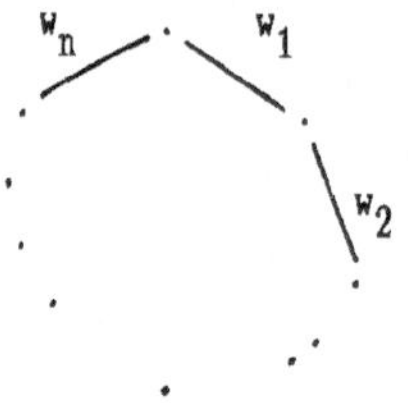


where the edge labelled with $w_i$ is an arrow pointing anticlockwise if and only if $w_i$ is an arrow. Let $V \in K[x, x^{-1}]$-mod. Then $G_w(V)$ is defined to be the representation of $Q_w$ with vector space V at each vertex, where the linear transformation representing the arrow $w_1$ is x, and at every other arrow, the map is the identity.

EXAMPLE Let $\Lambda$ be the Kronecker algebra. Then there is only one band, namely $w = YX^{-1}$; and $Q_w$ is of the form

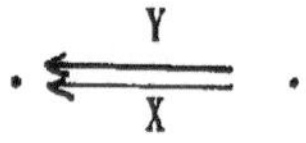


The total dimension of $G_w(V)$ is 2dim V. Considering $G_w(V)$ as a $\Lambda$-module, X and Y are given by matrices

$$X \to \begin{bmatrix} 0 & I \\ 0 & 0 \end{bmatrix} \text{ and } Y \to \begin{bmatrix} 0 & x \\ 0 & 0 \end{bmatrix}.$$

We can now state the main result.

THEOREM *Suppose $\Lambda$ is a string algebra. Then the modules $G_w(V)$ for $w \in St \cup Ba$ and $V \in$ ind $C_w$ form a complete set of representatives for the indecomposable $\Lambda$-modules.*

The above theorem is essentially due to Gelfand and Ponomarev. In [GP] they have considered a special case, however the proof generalizes without serious difficulty; see [$R_2$] and also [$DF_3$]. We shall briefly describe how the classification is obtained. Before doing this, we note the following consequence (see II.1.3):

II.3.1 *Any special biserial algebra is tame or of finite type.*

To establish the bimodules required by the definition I.4.2, it suffices to consider bands since the number of indecomposable strings of a given dimension is finite. This is straightforward, similar to I.4.3. For a characterization of finite-type special biserial algebras, see II.8.

II.3.2 *The proof of the Theorem.* Let $I = St \cup Ba$, for each $w \in I$ we have an algebra $C_w$ and a functor $G_w : C_w\text{-mod} \to \Lambda\text{-mod}$. Then one constructs functors $F_w$: $\Lambda\text{-mod} \to C_w\text{-mod}$, for each $w \in I$, and the method of [GP] (as presented in [$R_2$]) establishes the following result:

PROPOSITION *The funcors $F_w$, $S_w$ ($w \in I$) satisfy the following:*
*(i) $F_w G_w \cong id_{C_w\text{-mod}}$ and $F_v S_w = 0$ for $w \neq v$.*
*(ii) The set $\{ F_w : w \in I \}$ is locally finite and reflects isomorphisms.*
*(iii) For every $M$ in $A$-mod and every $w \in I$, there is a map $\gamma_{w,M}: S_w F_w(M) \to M$ such that $F_w(\gamma_{w,M})$ is an isomorphism.*

The Main Theorem follows from this proposition (by [$R_2$] Ch. 3, Lemma). For the construction of the functors we refer to [$R_2$] or [BR].

II.3.3 The proposition above has the following consequences:

*(iv) For any* $\Lambda$*-module* $M$*, the map* $(\tau_{w,M})_w : \oplus G_wF_w(M) \to M$ *is an isomorphism.*

*(v) Let* $M$ *be indecomposable and* $w \in I$*. Then either* $F_w(M) = 0$ *or* $M \cong G_wF_w(M)$*:*

Suppose $F_w(M) \neq 0$. The map $\tau_{w,M}$ in (iii) is an isomorphism between $F_wG_wF_w(M)$ and $F_w(M) \neq 0$, and since $F_w$ reflects isomorphisms, we deduce that $\tau_{w,M}$ is an isomorphism.

We note that (ii) and (v) imply:

*(vi) If* $X$*,* $Y$ *are in* $C_w$*-mod and* $f: G_w(X) \to G_w(Y)$ *is a map with* $F_w(f)$ *an isomorphism, then* $f$ *is an isomorphism.*

## II.4 *Irreducible maps for band modules*

We shall now prove that the functors $G_w$ in II.3 preserve irreducibility of maps. The following is due to [BR]:

II.4.1 LEMMA *Let* $\mathcal{A}$*,* $\mathcal{B}$ *be Krull-Schmidt categories, and* $G : \mathcal{A} \to \mathcal{B}$*,* $F: \mathcal{B} \to \mathcal{A}$ *be additive functors with the following properties:*

*(a)* $FG \cong id_{\mathcal{A}}$.

*(b) If* $M$ *is indecomposable in* $\mathcal{B}$ *then either* $F(M) = 0$ *or else* $M$ *is isomorphic to* $GF(M)$.

*(c) If* $X$*,* $Y$ *are in* $\mathcal{A}$*, and* $f: G(X) \to G(Y)$ *is a map in* $\mathcal{B}$ *with* $F(f)$ *an isomorphism in* $\mathcal{A}$*, then* $f$ *is an isomorphism.*

*Then the image of an irreducible map in* $\mathcal{A}$ *under* $G$ *is irreducible in* $\mathcal{B}$.

Proof: (1) Let $M$ be indecomposable in $\mathcal{B}$. Then either $F(M) = 0$ or else $F(M)$ is indecomposable: Let $F(M) \neq 0$. According to (b), we have $M \cong GF(M)$. Assume $F(M) = X_1 \oplus X_2$ with $X_1$, $X_2$ both non-zero. Then $GF(M) = G(X_1) \oplus G(X_2)$, and both $G(X_1)$, $G(X_2)$ are non-zero, according to (a). This contradicts the indecomposability of $M$.

Now let $\tau: X \to Z$ be irreducible in $\mathcal{A}$. Since $\tau$ is not split, it follows from (a) that $G(\tau)$ cannot be split mono or epi. Let $f: G(X) \to M$ and $g: M \to G(Z)$ be maps in $\mathcal{B}$ with $G(\tau) = fg$. Then $FG(\tau) = F(f)\ F(g)$. Now, $FG(\tau)$ is irreducible, by (a), thus $F(f)$ is split mono or $F(g)$ is split epi.

We consider the case of $F(f)$ being split mono. Decompose $M = \bigoplus_{j\in J} M_j$ with $M_j$ indecomposable, write $f = [f_j]_{j\in J}$ with $f_j : M \to M_j$; then $F(f) = [F(f_j)]_{j\in J}$. Since $F(f)$ is split mono and all $F(M_j)$ are indecomposable or zero, there is a subset $J'$ of $J$ such that the map $[F(f_j)]_{j\in J'}$: $FG(X) \to \bigoplus_{j\in J'} F(M_j)$ is an isomorphism, and, moreover, all $F(M_j)$ are non-zero. According to (b), we can assume $M_j = G(Y_j)$ for some $Y_j$ in $\mathcal{A}$, for all $j \in J'$. Let $Y = \bigoplus_{j\in J'} Y_j$; and apply (c) to the map $f' = [f_j]_{j\in J'}$ : $G(X) \to G(Y)$. Since $F(f)$ is an isomorphism in $\mathcal{A}$, it follows that $f'$ is an isomorphism in $\mathcal{B}$. This shows that $f$ is split mono.

A similar argument in the case that $F(g)$ is split epi shows that then $g$ is split epi.

## II.4.2 *Finite-dimensional $K[x,x^{-1}]$-modules and their AR-sequences.*

Let $C = K[x,x^{-1}]$. A finite-dimensional $C$-module is the same as a finite-dimensional vector space $V$ together with an invertible linear transformation $X$ of $V$. We assume that $K$ is algebraically closed; then $V$ is indecomposable if and only if $X$ is similar to an indecomposable Jordan matrix $J_n(\lambda)$ with eigenvalue $\lambda \in K^*$, where $n = \dim V$. We write $V = V_n(\lambda)$ for this module.

LEMMA *The module $V_n(\lambda)$ has an AR-sequence of the form*

$$(*) \quad 0 \to V_n(\lambda) \xrightarrow{f} V_{n-1}(\lambda) \oplus V_{n+1}(\lambda) \xrightarrow{g} V_n(\lambda) \to 0, \qquad n = 1, 2, 3, \ldots .$$

*In particular, the component of $V_n(\lambda)$ is a tube of rank 1, containing all modules $V_k(\lambda)$.*

Proof: For each $n$, we fix a generator $w_n$ of $V_n(\lambda)$ such that $\{ w_n, (x-\lambda)w_n, \ldots, (x-\lambda)^{n-1}w_n\}$ is a $K$-basis of $V_n(\lambda)$ according to which the matrix of $X$ has Jordan form. Define a map $g$: $V_{n-1}(\lambda) \oplus V_{n+1}(\lambda) \to V_n(\lambda)$ by setting $g(w_{n-1}, 0) = (x-\lambda)w_n$ and $g(0, w_{n+1}) = w_n$. Then $g$ is surjective, and $\ker g = \langle (w_{n-1}, (x-\lambda)w_{n+1}) \rangle \cong V_n(\lambda)$. Hence we obtain a short exact sequence, with $f$ the inclusion map. The sequence is clearly non-split. Suppose $\eta$: $M \to V_n$ is not a split epimorphism. As usual, it suffices to consider the case when $M$ is indecomposable and $\eta \neq 0$. Then the action of $x$ on $M$ has also $\lambda$ as an eigenvalue, hence $M \cong V_k(\lambda)$ for some $k \geq 1$. Since $M$ is cyclic, $\eta$ is determined by $\eta(w_k)$; let $\eta(w_k) = \Sigma\, c_i w_n(x-\lambda)^i$ with $c_i \in K$.

Suppose first that $\eta$ is not onto, that is, $c_0 = 0$. Define a C-module homomorphism $\psi: M \to V_{n-1} \oplus V_{n+1}$ by $\psi(w_k) = (\Sigma\, c_i w_{n-1}(x-\lambda)^{i-1}, 0)$. This is well-defined, and $g\,\psi = \eta$. If $\eta$ is ono, then $\eta$ is not 1-1 and therefore $k \geq n+1$. Now define $\psi(w_k) = (0,\ \Sigma\, c_i w_{n+1}(x-\lambda)^i)$; and again this gives rise to a well-defined C-module homomorphism satisfying $g\,\psi = \eta$.

II.4.3 *Let* $\Lambda$ *be a string algebra and* $G_w(V)$ *a band module, where* $V = V_n(\lambda)$. *Then there is an AR-sequence*

$$(**)\quad 0 \to G_w(V) \to G_w[V_{n-1}(\lambda) \oplus V_{n+1}(\lambda)] \to G_w(V) \to 0.$$

*In particular,* $G_w(V)$ *lies in a 1-tube; and moreover, the component of* $G_w(V)$ *consists of the modules* $G_w[V_k(\lambda)]$ *for* $k \geq 1$.

Proof: Applying $G_w$ to the AR-sequence of $V_n(\lambda)$ in II.4.2 gives an exact sequence (**). Using II.4 with $\mathcal{A} = C_w$-mod and $\mathcal{B} = \Lambda$-mod, we see that the maps occring are irreducible; in particular $\tau V \cong V$, by I.8.13; and (**) is an AR-sequence, by the uniqueness. Note that we obtain the whole component since we have determined the AR-sequences for all modules $G_w[V_n(\lambda)]$.

The method in II.4 breaks down for strings since the parametrizing category K-mod does not have irreducible maps.

## II.5 *Irreducible maps for string modules*

Our aim is to determine irreducible maps for string modules, where the algebra is a string algebra. We write M(C) instead of $G_C(K)$, for C a string.

<u>II.5.1</u> *Some canonical maps* Let C be a string. We say that "C starts on a peak" provided there is no arrow $\beta$ with $C\beta$ a string, and that C "starts in a deep" provided that there is no arrow $\gamma$ with $C\gamma^{-1}$ a string.

Dually, we say that the string C ends on a peak provided there is no arrow $\beta$ with $\beta^{-1}C$ a string, and that C ends in a deep provided there is no arrow $\gamma$ with $\gamma C$ a string. Note that C ends on a peak if and only if $C^{-1}$ starts on a peak, and C ends in a deep if and only if $C^{-1}$ starts in a deep.

Let C, D be strings and $\beta$ an arrow such that $C\beta D$ is a string. Then there is a canonical embedding of M(C) into M($C\beta D$) as follows: Let C be of length n and D of length m, then $C\beta D$ is of length n + m + 1. By construction, M($C\beta D$) is given by n + m + 2 base vectors $z_0, z_1, \ldots, z_{n+m+1}$ on which the algebra operates according to the shape of $C\beta D$. The subspace with basis $z_0, z_1, \ldots, z_{n+1}$ is then a submodule of the form M(C). We call the inclusion map the "*canonical embedding*". Moreover, the corresponding quotient M($C\beta D$)/M(C) is just M(D), and we call the induced map M($C\beta D$) M(D) the "*canonical projection*".

Similarly, if C and D are strings and $\beta$ is an arrow such that $D\beta^{-1}C$ is a string then there is a canonical embedding M(C) $\to$ M($D\beta^{-1}C$), and the corresponding quotient M($D\beta^{-1}C$)/M(C) is just M(D); the induced map M($D\beta^{-1}C$) M(D) is called the canonical projection.

Let C be a string, $C = c_1c_2\ldots c_n$, $n \geq 0$. We say that C is directed if all $c_j$ are arrows, and C is "inverse" if all $c_j^{-1}$ are arrows.

<u>II.5.2</u> ***Hooks and Co-hooks***

(1) Assume that C is a string, not starting on a peak, say $C\beta$ is a string for some arrow $\beta$. Then there is a unique directed string D such that $C\beta D^{-1}$ is a string starting in a deep. We denote $C\beta D^{-1}$ by $C_h$.

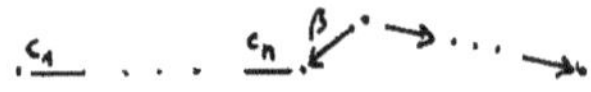

(2) Assume that C is a string, not ending on a peak; say $\beta^{-1}C$ is a string for some arrow $\beta$. Then there is a unique directed string D such that $D\beta^{-1}C$ is a string ending in a deep. We denote $D\beta^{-1}C$ by ${}_hC$.

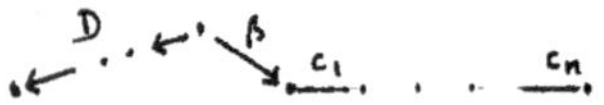


(3) Assume that C does not start in a deep, say $C\gamma^{-1}$ is a string for some arrow $\gamma$. Then there is a unique directed string D such that $C\gamma^{-1}D$ is a string starting on a peak. We denote $C\gamma^{-1}D$ by $C_c$.

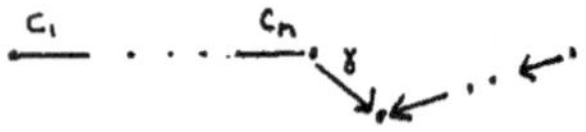


(4) Assume that C does not end in a deep, say $\gamma C$ is a string for some arrow $\gamma$. Then there is a unique directed string D such that $D^{-1}\gamma C$ is a string ending on a peak, and we denote $D^{-1}\gamma C$ by ${}_cC$.

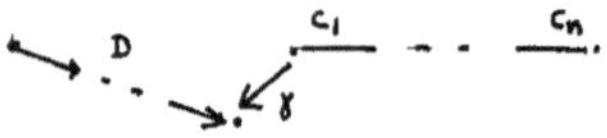


Note that $(C_h)^{-1} = {}_hC^{-1}$ and $(C_c)^{-1} = {}_cC^{-1}$.

<u>II.5.3</u> PROPOSITION *The canonical maps* $M(C) \to M(C_h)$, $M(C) \to M({}_hC)$ *and* $M(C_c) \to M(C)$, $M({}_cC) \to M(C)$ *are irreducible.*

For a proof, see [BR].

## II.6 *AR-sequences for string algebras*

Suppose that $\Lambda$ is a string algebra; our aim is to show that we have in II.5.3 found all irreducible maps ending at string modules. First, we will study irreducible maps ending at projectives or starting at injectives; then we will describe all AR-sequences of string modules.

<u>II.6.1</u> *Irreducible maps ending at projectives* Let u be a vertex; then P(u) is a string module; Take the ($\leq$ two) directed strings of maximal length starting at u.

Denote these by $C_1$ and $C_2$, say $C_1 = \beta_r \ldots \beta_1$ and $C_2 = \gamma_s \ldots \gamma_1$: then $P(u) = M(C_1 C_2^{-1})$.

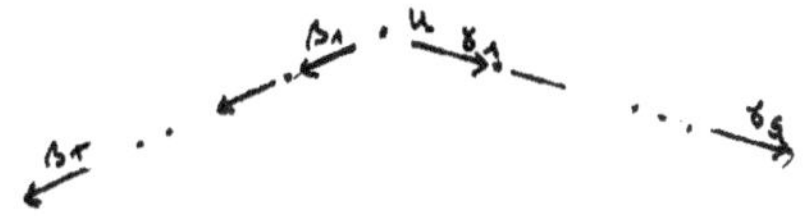

Moreover, since $C_1$ and $C_2$ are maximal directed strings, it follows that $C_1 C_2^{-1}$ both starts and ends in a deep. If $C_1$ and $C_2$ are both of length zero then $P(u)$ is simple, and no irreducible maps end at $P(u)$. Assume now that $C_1$ has length $r \geq 1$. Then $M(\beta_r \ldots \beta_2)$ is a direct summand of rad $P(u)$. (If $r = 1$ this module is simple.) Moreover, the inclusion map $M(\beta_r \ldots \beta_2) \to M(C_1 C_2^{-1})$ is the canonical embedding $M(C) \to M(C_h)$. Similarly, if $C_2$ has length $s \geq 1$ then the string module $M(\gamma_2^{-1} \ldots \gamma_s^{-1})$ is a direct summand of rad $P(u)$, and the corresponding inclusion map is the canonical embedding $M(\gamma_2^{-1} \ldots \gamma_s^{-1}) \to M({}_h(\gamma_2^{-1} \ldots \gamma_s^{-1}))$.
These are all irreducible maps ending at $P(u)$, by I.7.2.

Similarly, the indecomposable injective module $E(u)$ corresponding to the vertex $u$ is $M(D_1^{-1} D_2)$ where $D_1$ and $D_2$ are direct of maximal length ending at vertex $u$. Then $D_1^{-1} D_2$ both starts and ends on a peak. The dual consideration shows that $E(u)/\text{soc}\, E(u)$ is the direct sum of at most two uniserial modules, and that the corresponding canonical epimorphisms are the irreducible maps starting at $E(u)$.

## <u>II.6.2</u> *Canonical exact sequences*

(1) For any arrow $\beta$ in the quiver, we have an AR-sequence
$0 \to \tau(\Lambda u/\Lambda\beta) \to X \to \Lambda u/\Lambda\beta \to 0$, where the middle term is indecomposable. We will prove this in IV.4.2 (and IV.4.3). The middle term $X$ is the module $M(B)$ where $B$ is the word $C^{-1}\beta D^{-1}$ with $C$, $D$ as follows:

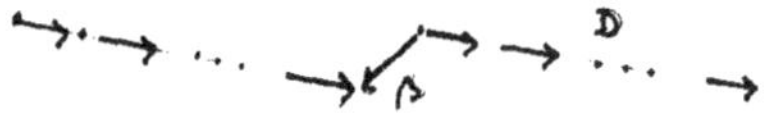

Consider now a string $C$ with neither $M(C)$ injective, nor isomorphic to any $\Lambda u/\Lambda\beta$.

(2) First, assume that $C$ neither starts nor ends on a peak. Thus both ${}_hC$ and $C_h$ and also ${}_hC_h$ are defined. We call the exact sequence

$$0 \to M(C) \xrightarrow{[u\ u]} M({}_hC) \oplus M(C_h) \xrightarrow{\begin{bmatrix} u \\ -u \end{bmatrix}} M({}_hC_h) \to 0$$

a canonical exact sequence.

(3) Next, assume that C does not start on a peak but ends on a peak; then $C_h$ is not defined. We claim that $C = (\gamma_r^{-1}\ldots\gamma_1^{-1})\gamma_0 D$ with $r \geq 0$ and D a string:
Since C ends on a peak, it is of the form $(\gamma_r^{-1}\ldots\gamma_1^{-1})C_1$ and no $\gamma^{-1}C$ is a string. If $C_1$ would have length zero then M(C) would be of the form $\Lambda u/\Lambda\beta$ for some $\beta$ which we have excluded.

Therefore $C = {}_cD$. With C also D does not start on a peak, and $D_h$ is defined, and there is the following exact sequence

$$0 \to M(C) \xrightarrow{[p\ u]} M(D) \oplus M(C_h) \xrightarrow{\begin{bmatrix} u \\ -p \end{bmatrix}} M(D_h) \to 0,$$

called a canonical exact sequence.

(4) If C starts on a peak but does not end on a peak we can smilarly write $C = D_c$ for some D and have the following exact sequence

$$0 \to M(C) \xrightarrow{[u\ p]} M({}_hC) \oplus M(D) \xrightarrow{\begin{bmatrix} p \\ -u \end{bmatrix}} M({}_hD) \to 0.$$

(5) Suppose C both starts and ends on a peak. We assume that M(C) is not injective. It follows that $C = {}_cD_c$, and there is the following exact sequence

$$0 \to M(C) \xrightarrow{[p\ p]} M(D_c) \oplus M({}_cD) \xrightarrow{\begin{bmatrix} p \\ -p \end{bmatrix}} M(D) \to 0,$$ which we call canonical.

<u>II.6.3</u> PROPOSITION *The canonical exact sequences are the AR-sequences containing string modules.*

Proof: We have already seen that (1) is an AR-sequence. Consider now one of the others, it has two indecomposable middle terms. By I.8.13 it suffices to show that the maps defining the sequence are irreducible. This is clear by I.7.6 (and the Krull-Schmidt theorem) if the summands in the middle are non-isomorphic. This is the case in (3) and (4), by the lengths. So we are left with the following:

(i) $M({}_hC) \cong M(C_h)$ when C neither starts nor ends on a peak.

(ii) $M(D_c) \cong M({}_cD)$ when D neither starts nor ends in a deep.

We deal with the first case, the second is dual; and we are done by the following:

<u>II.6.3</u> LEMMA *Let $\Lambda = KQ/I$ be a string algebra, and assume I satisfies II.1.2. Suppose (*) $\beta_0 \neq \beta_1$, $e(\beta_0) = e(\beta_1)$ and $s(\beta_0) = s(\beta_1)$, and that $\gamma\beta_i \in I$ for all arrows $\gamma$.*

*Let $C = (\beta_0\beta_1^{-1})^s$ for some $s \geq 0$. Then*

*(a) ${}_hC = (\beta_0\beta_1^{-1})^{s+1} = C_h$, moreover*

*(b) the space $Irr(M(C), M(C_h))$ is 2-dimensional, and*

*(c) A basis of $Irr(M(C), M(C_h))$ is given by the residue classes of $M(C) \to M({}_hC)$ and $M(C) \to M(C_h)$.*

*Conversely, assume C neither starts of ends on a peak, and assume $M({}_hC) \cong M(C_h)$. Then there are arrows $\beta_0$, $\beta_1$ with (*) such that $C = (\beta_0\beta_1^{-1})^s$ for some $s \geq 0$.*

This is proved in [BR], chapter 3.

II.6.4 *The stable AR-quiver of an indecomposable self-injective special biserial algebra of infinite type.* If $\Lambda$ is special biserial and self-injective then $\Gamma_s(\Lambda) = \Gamma(\Lambda/\mathrm{soc}\ \Lambda)$, and $\Lambda/\mathrm{soc}\ \Lambda$ is a string algebra, see II.1.3. We have seen in II.4 that band modules form tubes of rank 1. The only string modules whose AR-sequence has an indecomposable middle term are those in II.6.2(1). The number of these modules is equal to the number of arrows of the quiver of $\Lambda$, hence is finite; and therefore these modules must be $\tau$-periodic, and their AR-sequences lie at ends of tubes, by I.8.6.1. The number of such tubes is finite; we call suc a tube "exceptional" if it has rank > 1.

All other string modules , that is, all non-periodic modules, lie in components in which every module has two predecessors.

## II.7 *AR-components for some special biserial algebras*

In this chapter we will determine the graph structure of the stable AR-quiver for a number of symmetric special biserial algebras. These include as special cases when char K = 2 the group algebras of the Klein 4-group, of $A_4$ and $S_4$ and the dihedral 2-groups. The examples are well-known.

II.7.0 By the result of the first part of this chapter, we have the following:
*1-tubes* All bands lie in 1-tubes, and there is one tube for each word and each $0 \neq \lambda \in K$ (II.4).
*Exceptional tubes* The only tubes consisting of string modules are those having at the ends the maximal directed strings (II.6.4).
*Non-periodic components* All other strings form stable components in which two arrows start and two arrows end at each vertex.

It remains to determine the number and rank of the exceptional tubes and the tree class of non-periodic components.

For the proofs, it is convenient to apply the following more general results from [ES]:

II.7.1 THEOREM *Let A be a special biserial self-injective algebra.*
*The following are equivalent:*
*(i)* $\Gamma_s(A)$ *has a component of the form* $\mathbb{Z}\tilde{A}_{p,q}$.
*(ii)* $\Gamma_s(A)$ *is infinite but has no component of the form* $\mathbb{Z}A_\infty^\infty$.
*(iii) There are positive integers m, p, q such that* $\Gamma_s(A)$ *is a disjoint union of m components of the for* $\mathbb{Z}\tilde{A}_{p,q}$, *m components of the form* $\mathbb{Z}A_\infty/\langle \tau^p \rangle$, *m components of the form* $\mathbb{Z}A_\infty/\langle \tau^q \rangle$ *and infinitely many components of the form* $\mathbb{Z}A_\infty/\langle \tau \rangle$.
*(iv) A is representation-infinite domestic.*
*(v) A is representation-infinite of polynomial growth.*

For the definitions of the terms occuring see [ES] or [$R_4$].

II.7.2 **THEOREM** *Let $A$ be a special biserial self-injective algebra. The following conditions are equivalent:*

*(i) $\Gamma_s(A)$ has a component of the form $\mathbb{Z}A_\infty^\infty$.*

*(ii) $\Gamma_s(A)$ has infinitely many (regular) components of the form $\mathbb{Z}A_\infty^\infty$.*

*(iii) $\Gamma_s(A)$ is a disjoint union of a finite number of components of the form $\mathbb{Z}A_\infty/\langle \tau^n \rangle$ with $n > 1$, infinitely many components of the form $\mathbb{Z}A_\infty/\langle \tau \rangle$ and infinitely many components of the form $\mathbb{Z}A_\infty^\infty$.*

*(iv) $A$ is not of polynomial growth.*

II.7.3 **LEMMA** *Let $\Lambda$ be the Kronecker algebra. Then $\Gamma_s(\Lambda)$ has the following components:*

*(a) Infinitely many 1-tubes.*

*(b) One component of tree type $\tilde{A}_{12}$ (see I.8.5).*

For char $K = 2$, $\Lambda$ is isomorphic to the group algebra of the Klein 4-group, see I.4.3.

Proof: Recall that $\Lambda = K[X, Y]/(X^2, Y^2)$.

(1) Exceptional tubes: We have to consider only two maximal directed strings, namely $M(X) = \Lambda Y$ and $M(Y) = \Lambda X$. They have both $\Omega$-period 1; for example $\Omega(\Lambda X) \cong \{ z \in \Lambda: zX = 0 \} = \Lambda X$. Hence they belong to 1-tubes. This proves (a), by II.7.0. In particular, the strings lying in tubes are those with words

$$(XY^{-1})^k X, \quad (YX^{-1})^k Y \quad (k \geq 0).$$

(2) Non-periodic components: A complete set of representatives for the non-periodic strings is given by the following list:

$$1 \text{ and } (XY^{-1})^k, \ (X^{-1}Y)^k \text{ for } k \geq 1.$$

These are just the modules $\{ \Omega^n K : n \in \mathbb{Z} \}$, hence they form two $\tau$-orbits. Therefore there is only one component, and the associated tree can only have two vertices. The standard AR-sequence (I.7.7) is $0 \to \Omega K \to K \oplus K \oplus \Lambda \to \Omega^{-1}K \to 0$. Applying $\Omega^n$ gives AR-sequences $0 \to \Omega^{n+1}K \to \Omega^n K \oplus \Omega^n K \to \Omega^{n-1}K \to 0$, for $0 \neq n \in \mathbb{Z}$. Hence the AR-component is of the form

$$\begin{array}{c} \qquad\qquad\qquad\qquad\qquad\quad \Lambda \\ \ldots \ \Omega^n K \ \rightrightarrows \ \Omega^{n-1}K \ \ldots \ \ \Omega K \ \rightrightarrows \ K \ \rightrightarrows \ \Omega^{-1}K \ \ldots \rightrightarrows \ \Omega^{-n}K \ \rightrightarrows \ \ldots \end{array}$$

(with arrows $\Omega K \to \Lambda \to \Omega^{-1}K$)

which has stable type $\tilde{A}_{12}$.

II.7.4 LEMMA *Let $\Lambda$ be the algebra with quiver*

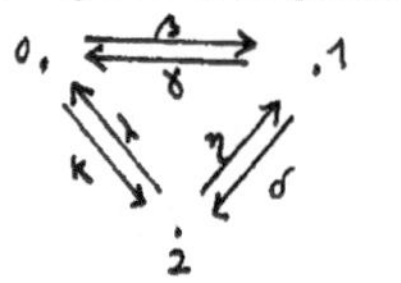

*and relations* $\beta\delta = 0 = \delta\lambda = \lambda\beta, \quad \gamma\kappa = 0 = \kappa\eta = \eta\gamma$

$$\beta\gamma = \kappa\lambda, \quad \gamma\beta = \delta\eta, \quad \eta\delta = \lambda\kappa.$$

*Then $\Gamma_s(\Lambda)$ has the following components:*

*(a) Infinitely many 1-tubes.*

*(b) Two 3-tubes.*

*(c) One non-periodic component of tree class $\tilde{A}_5$ (see I.8.4).*

If char K = 2 then $\Lambda$ is isomorphic to the group algebra of the alternating group $A_4$, see V.2.4.1.

Proof: (1) Exceptional tubes: There are six maximal directed strings, namely all $\omega$ where $\omega$ is an arrow. It follows immediately from the zero-relations that these modules break up into two $\Omega$-orbits, each of length 3. Each of these is therefore also a $\tau$-orbit; and we obtain (b).

(2) Non-periodic components: One argument is as follows: Consider the component $\Delta$ of the string $\beta\kappa^{-1}$. It is easy to calculate irreducible maps in $\Delta$, using II.5.2. We see that the components contains a subquiver

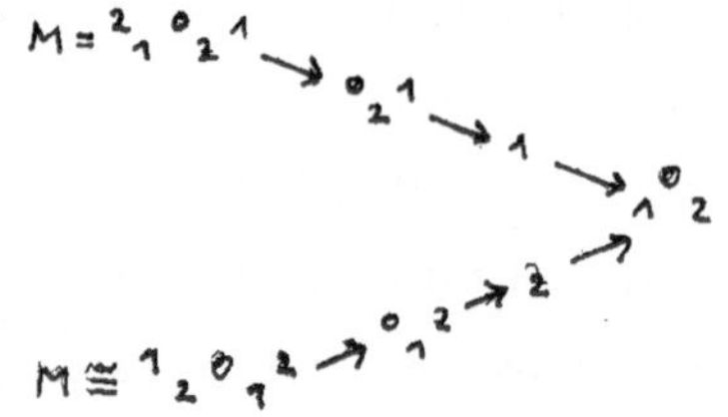

The identification shows that $\Delta \cong \mathbb{Z}\tilde{A}_{3,3}$, that is, $\Delta$ has tree class $\tilde{A}_5$. By II.7.1, this is the only non-periodic component.

Alternatively, it is easy to describe a complete list of strings. The non-periodic ones are the subwords w of even length of some word of the form $(\beta^{-1}\eta\lambda^{-1}\gamma\delta^{-1}\kappa)^k$ for some $k \geq 1$. It is easy to see that for any such word w, M(w) belongs to the $\Omega$-orbit of a simple module. So there are only six $\tau$-orbits. The standard AR-sequences are $0 \to \Omega S_i \to S_j \oplus S_k \oplus P_i \to \Omega^{-1}S_i \to 0$ for $\{i, j, k\} = \{0, 1, 2\}$. We see that there is one non-periodic component, namely

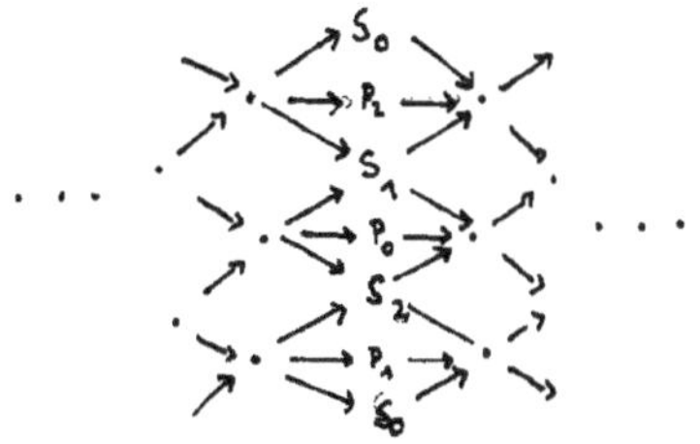


where the top- and bottom rows are identified. This has tree class $\tilde{A}_5$.

II.7.5 LEMMA *Let $\Lambda$ be the algebra with quiver*

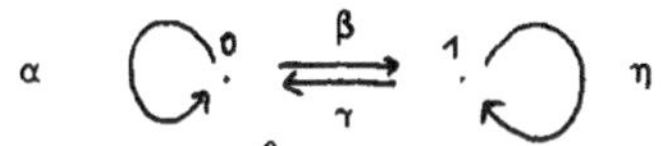


*and relations* $\alpha^2 = 0 = \beta\eta = \eta\gamma$, $\alpha\beta\gamma = \beta\gamma\alpha$, $\eta^2 = \gamma\alpha\beta$.

*Then $\Gamma_s(\Lambda)$ has the following components:*

*(a) Infinitely many 1-tubes.*

*(b) One 3-tube.*

*(c) Infinitely many components of tree class $A_\infty^\infty$ .*

If char $K = 2$ then $\Lambda$ is the basic algebra for the group algebra of the symmetric group $S_4$, see V.2.5.1.

Proof: (1) Exceptional tubes: There are four maximal directed strings, namely $\beta\gamma$, $\alpha\beta$, $\gamma\alpha$, $\eta$. One sees from the relations that $M(\beta\gamma)$ has $\Omega$-period 1 and lies in a 1-tube. The other three maximal directed string form one $\Omega$-orbit of length 3 which is also a $\tau$-orbit. This gives (b).

(2) Non-periodic components: Suppose (c) is not true. Then by II.7.2 and II.7.1, $\Gamma_s(\Lambda)$ has only one non-periodic component which is $\cong \mathbb{Z}\tilde{A}_{3,1}$. That is, given any non-periodic module $M$, the AR-component around $M$ is of the form

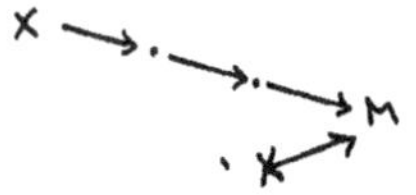


One finds that this is not the case. For example, take $M = P_0/S_0$, then the component

of M is of the form

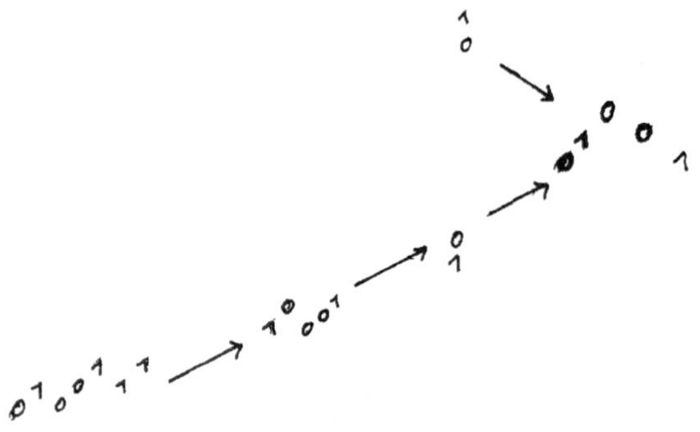


II.7.6 LEMMA *Let Λ be the local special biserial algebra* $\Lambda = K\langle X, Y\rangle/(X^2, Y^2, (XY)^k - (YX)^k)$. *Then* $\Gamma_s(\Lambda)$ *has the following components:*

*(a) Infinitely many 1-tubes.*

*(b) Infinitely many components of tree class* $A_\infty^\infty$ .

If char K = 2 and k = $2^n$ then Λ ≅ KD where D is the dihedral 2-group of order $2^{n+2}$, see III.13.

Proof: It has been proved in [$R_2$] that all bands have Ω-period ≤ 2, hence they lie in 1-tubes. Moreover, [BS] have determined the non-periodic components and proved (a). Alternatively, we apply the results of II.6 (see II.6.4) and observe that:

(1) Exceptional tubes: There are two maximal directed strings, namely $(\alpha\beta)^{m-1}\alpha$ and $(\beta\alpha)^{m-1}\beta$. The corresponding modules have both Ω-period 1; consequently all tubes of Λ have rank 1.

(2) Non-periodic components: The statement (b) follows from [ES], by the argument in II.7.5.

## II.8 *Some algebras of finite type*

We will now give a characterization of special biserial algebras of finite type and then investigate a number of particular algebras.

II.8.1 LEMMA *Suppose Λ is a special biserial algebra.*

*Then the following are equivalent:*

*(i) Λ is of finite type.*

*(ii) Λ has finitely many strings.*

*(iii) Λ has no bands.*

Proof: We may assume that Λ is indecomposable, and also that Λ is a string algebra, by II.1.3.

In II.7 we have seen that $\Gamma_s(\Lambda)$ has a stable component consisting of $\tau$-periodic strings. If (ii) holds then this component is finite, and Λ is of finite type, by Auslander's theorem I.8.3.

Now assume that Λ does not have bands; and suppose that there are strings of arbitrary length. Since the length of directed strings is bounded, it follows that there must be some string in which some subword $\alpha\beta^{-1}$ occurs at least twice, say the string is $\ldots\alpha\beta^{-1}w\alpha\beta^{-1}\ldots$ where w is a word. Then all powers of $\beta^{-1}w\alpha$ are strings, and we deduce that Λ has a band, a contradiction.

The implication (i) ⇒ (ii) is trivial; and (i) implies (iii) since for any band w there is a tube containing the indecomposable modules $G_w(V)$, by II.4.

II.8.2 *An algebra is of finite type if its quiver is of the form*

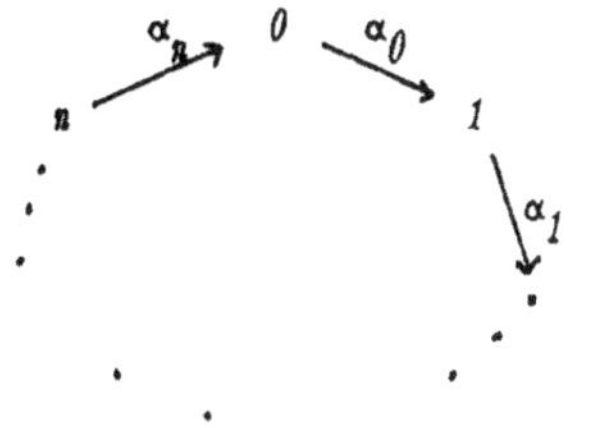


Proof: Supose Λ is such an algebra. Then Λ is trivially a string algebra; moreover there are only directed strings. The number of these is finite.

II.8.3 *Let Λ be the special biserial algebra with quiver* 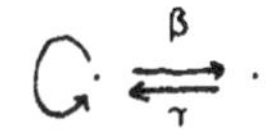


*defined by the relations*

$$\alpha\beta = 0 = \gamma\alpha, \quad \beta\gamma = 0 = \gamma\beta, \quad \alpha^n = 0 \text{ where } n \geq 2.$$

*Then Λ is of finite type.*

Proof: A complete list of strings is given by $\alpha^r$, $\alpha^r\gamma^{-1}$, $\beta^{-1}\alpha^r$, $\beta^{-1}\alpha^r\gamma^{-1}$, $0 \le r < n$.

II.8.4 *Let $\Lambda$ be the special biserial algebra with quiver* 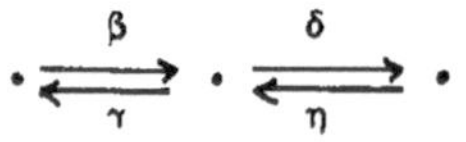

*defined by the relations*

$$\beta\delta = 0 = \eta\gamma \text{ and } (\gamma\beta)^k = (\delta\eta)^l.$$

*Then $\Lambda$ is of finite type if and only if $\{k, l\} = \{1, 1\}$ or $\{1, 2\}$.*

Proof: For the first part, it is enough to consider the case when k = 1. Assume first that l = 2. Then all strings are subwords of $\gamma(\delta\eta)^{-1}\beta$ or $\eta\delta\eta$, and $\Lambda$ is of finite type. If k = l = 1 then $\Lambda$ is a quotient algebra of the one with l = 2 and hence also of finite type.

Now suppose that k, l $\ge$ 2. Then the radical of the local algebra $e_0\Lambda e_0$ has independent generators $\gamma\beta$ and $\delta\eta$; hence $e_0\Lambda e_0$ is of infinite type and so is $\Lambda$ by I.4.7.

II.8.5 *Let $\Lambda$ be the special biserial algebra with quiver* 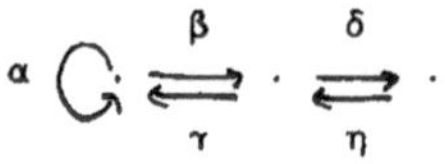

*defined by the relations*

$$\gamma\alpha = \alpha\beta = \beta\delta = \eta\gamma = 0, \quad (\gamma\beta)^k = (\delta\eta)^l, \ (\beta\gamma)^k = \alpha^s \quad \text{with } s \ge 2 \text{ and } k, l \ge 1.$$

*Then $\Lambda$ is of finite type if and only if $k = l = 1$.*

Proof: Suppose first that k = l = 1. Then every string is a subword of $\delta\gamma^{-1}\alpha^r\beta^{-1}\eta$ or of $\eta^{-1}\beta\alpha^{-r}\gamma\delta^{-1}$ for some r < s, and $\Lambda$ is of finite type, by II.8.1.

Now assume that k $\ge$ 2; then $J(e_1\Lambda e_1)$ has independent generators $\alpha$, $\beta\gamma$; and therefore $e_1\Lambda e_1$ is of infinite type. It follows from I.4.7 that $\Lambda$ is of infinite type. Now, if l $\ge$ 2 then there are strings $(\alpha\beta^{-1}\eta\delta\gamma^{-1})^m$ for all m $\ge$ 1, and $\Lambda$ is of infinite type.

II.8.6 *Let $\Lambda$ be the special biserial algebra with quiver* 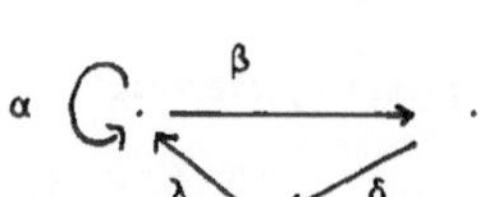

*defined by the relations*

$$\alpha\beta = 0 = \lambda\alpha, \quad (\beta\delta\lambda)^k = \alpha^s, \quad (\delta\lambda\beta)^k\delta = 0 \quad \textit{where } k \geq 1 \textit{ and } s \geq 2.$$

*Then* $\Lambda$ *is of finite type if and only if* $k = 1$.

Proof: Suppose $k = 1$, then every string is a subword of $\delta\beta(\alpha^{-t})\lambda\delta$ for some $t < s$. By II.8.1, $\Lambda$ is of finite type.

Now assume that $k \geq 2$. Then the local algebra $e_0\Lambda e_0$ has two independent generators, namely $\alpha$ and $\beta\delta\lambda$. Therefore $e_0\Lambda e_0$ is of infinite type and so is $\Lambda$, by I.4.7.

## II.9 *Indecomposable modules of the semidihedral algebra*

In this chapter we study representations of the local symmetric algebra

$\Lambda = \Lambda_m = K < a, b >/(b^2, (ab)^{m+1} = (ba)^{m+1}, a^2 = (ba)^m b , a^3 )$ for $m \geq 1$.

If char $K = 2$ and $m+1 = 2^n \geq 4$ then $\Lambda/\mathrm{soc}\,\Lambda \cong KD/\mathrm{soc}\,KD$ where $D$ is the semidihedral group of order $2^{n+2}$ [BD], see also Chapter III. Therefore we call $\Lambda$ the *"semidihedral algebra"*. B. Crowley-Boevey was able to give a very good parametrization of the indecomposable representations of $\Lambda$. We follow here his account $[CB_3]$.

In the following we fix an algebra $\Lambda = \Lambda_m$. There are two types of indecomposable representations of $\Lambda$, the "string modules" and the "band modules".

II.9.1 (a) WORDS The indecomposable modules are parametrized by words with underlying alphabet

$\{ a_i , b_j : i \in \mathbb{Z}$ and $-(m+1) \leq i \leq m+1 , j = \pm 1\}$.

We consider only words $w = w_1 w_2 \ldots w_n$ where the $w_i$ alternate between a's and b's. We use the notation $c_i$ to mean $a_i$ or $b_i$. Otherwise, the conventions are the ones described in II.2.

(b) There is a partial ordering on the set of words defined by

$w < w'$ provided (1) $w = w'c_i x$ with $i > 0$ and some word $x$, or

(2) $w' = wc_i x$ with $i < 0$ and some word $x$, or

(3) $w = xc_i y$, $w' = xc_j z$ where $i > j$ and $x$, $y$ and $z$ are words.

(c) Define now $\mathcal{W}$ to be the set of such words such that $w$ and $w^{-1}$ have no subwords of the form

(W1) $b_1 a_m b_1$

(W2) $a_{m+1} b_1$ or (W3) $b_1 a_{m+1}$

(W4) $a_i b_1 a_j$ for $i, j > 0$.

(d) STRINGS We use the words $1 \neq w \in \mathcal{W}$ which do not start or end with $b_i$. Let $\mathcal{S}t$ be a set of representatives of such words, with respect to the equivalence relation $\sim$ which identifies $w$ with its formal inverse. We call the elements of $\mathcal{S}t$ strings and say that a string is <u>symmetric</u> if $w = w^{-1}$, and otherwise $w$ is <u>asymmetric</u>.

(e) BANDS Consider the subset $\mathcal{W}'$ of $\mathcal{W}$ consisting of the non-trivial words $w$ of even length which are not powers. Let $\sim_r$ be the equivalence relation on $\mathcal{W}'$ which identifies $w$ with all rotations of $w$ and $w^{-1}$; and define $\mathcal{B}a$ to be a set of representatives. We call the elements of $\mathcal{B}a$ bands and say that a band $w \in \mathcal{B}a$ is <u>symmetric</u> if $w$ is equal to some rotation of $w^{-1}$, and otherwise $w$ is <u>asymmetric</u>.

In the following, for each string and band $w$ we shall define an algebra $C_w$, and a functor $G_w$: mod $C_w \to$ mod $\Lambda$. If $V$ is a $C_w$-module, we describe $G_w(V)$ by giving a quiver $\Psi$ and a representation of $\Psi$. This may be translated into a $\Lambda$-module as usual, see I.5. Any arrow where the map is not specified is represented by the identity.

<u>II.9.2</u> We fix two non-zero, unequal elements $\lambda$. $\mu \in K$.

We use a shorthand to describe $\Psi$, in which the following substitutions must be made:

(1) replace the labels $a_1$ and $b_1$ by $a$ and $b$ respectively;

(2) replace an arrow $\ldots V \leftarrow V \ldots$ labelled with $a_i$ $(i>1)$ by

$$\ldots V \xleftarrow{a} V_1 \xleftarrow{b} V_2 \xleftarrow{a} \ldots \xleftarrow{b} V_{2i-2} \xleftarrow{a} V \ldots$$

where each $V_i = V$;

(3) replace an arrow $\ldots V \leftarrow V \ldots$ labelled with $a_0$ by

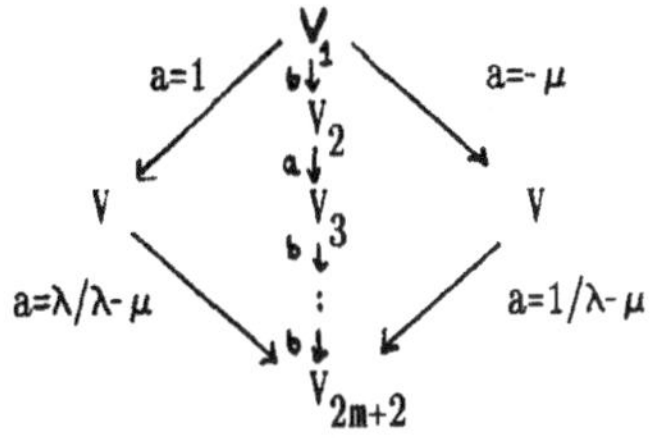


where each $V_i = V$; and

(4) if e is an idempotent endomorphism of V, replace

by

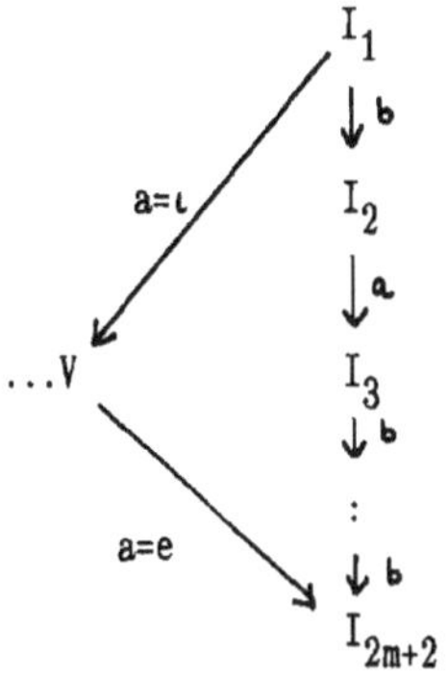


where each $I_i$ = im e and $\iota$ is the inclusion of $I_1$ in V.

II.9.3 *Indecomposable modules*

(a) ASYMMETRIC STRINGS If $w = w_1 w_2 \ldots w_n$ is an asymmetric string, let $C_w = k$ and define $G_w$ by the quiver

$$\begin{matrix} & \gamma_1 & & \gamma_2 & & \gamma_3 & & & \gamma_{n-1} & & \gamma_n & \\ V & - & V & - & V & - & \ldots & - & V & - & V \end{matrix}$$

in which if $w_i = c_j$ then $\gamma_i$ is labelled with $c_{|j|}$ and it points to the left if and only if $j > 0$ or $j = 0$ and $w_{i-1}^{-1} \ldots w_1^{-1} > w_{i+1} \ldots w_n$.

(b) SYMMETRIC STRINGS If w is a symmetric string then w is of the form $za_0z^{-1}$ for some word $z = z_1 \ldots z_n$. We set $C_w = K\langle e \colon e^2 = e \rangle$, and define $G_w$ by

$$V \overset{\gamma_1}{-} V \overset{\gamma_2}{-} V \overset{\gamma_3}{-} \cdots - V \overset{\gamma_n}{-} V \circlearrowleft e$$

where if $z_i = c_j$ then $\gamma_i$ is labelled with $c_{|j|}$ and it points to the left if and only if $j > 0$ or $j = 0$ and $z_{i-1}^{-1} \ldots z_1^{-1} > z_{i+1} \ldots z_n a_0 z^{-1}$.

(c) **ASYMMETRIC BANDS** If $w = w_1 \ldots w_n$ is an asymmetric band, by rotating and possibly inverting $w$ we may assume that $w_1 = b$. We set $C_w = K[x, x^{-1}]$, and define $G_w$ by the quiver

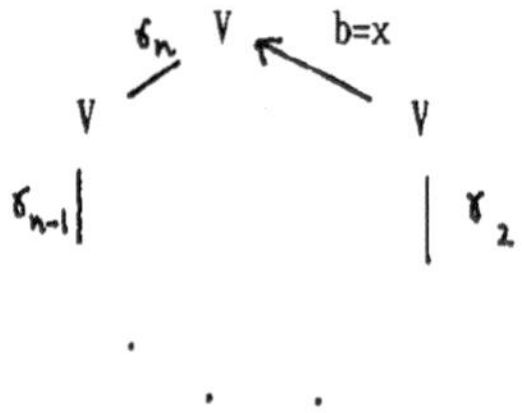


in which if $w_i = c_j$ then $\gamma_i$ is labelled with $c_{|j|}$ and it points anticlockwise if and only if $j > 0$ or $j = 0$ and $w_{i-1}^{-1} \ldots w_1^{-1} w_n^{-1} \ldots w_{i+1}^{-1} > w_{i+1} \ldots w_n w_1 \ldots w_{i-1}$.

(d) **SYMMETRIC BANDS** If $w$ is a symmetric band, by rotating $w$ we may assme that $w = z a_0 z^{-1} a_0$ for some word $z_1 \ldots z_n$. Let $C_w = K\langle e,f: e^2 = e,\ f^2 = f\rangle$, and define $G_w$ by

$$f \circlearrowright V \overset{\gamma_1}{-} V \overset{\gamma_2}{-} V \overset{\gamma_3}{-} \cdots \overset{\gamma_{n-1}}{-} V \overset{\gamma_n}{-} V \circlearrowleft e$$

where if $z_i = c_j$ then $\gamma_i$ is labelled with $c_{|j|}$ and it points to the left if and only if $j > 0$ or $j = 0$ and $z_{i-1}^{-1} \ldots z_1^{-1} a_0 z a_0 z_n^{-1} \ldots z_{i+1}^{-1} > z_{i+1} \ldots z_n a_0 z^{-1} a_0 z_1 \ldots z_{i-1}$.

**THEOREM** [$CB_3$] *Suppose $\Lambda$ is a local semidihedral algebra. Then the $\Lambda$-modules $G_w(V)$, with $w$ running through all strings and bands and $V$ running through a complete set of non-isomorphic indecomposable $C_w$-modules, provide a complete set of non-isomorphic indecomposable non-projective $\Lambda$-modules.*

The proof is based on the principle of functorial filtration, as in II.3.2.

**II.9.4 EXAMPLES**

(a) *An asymmetric string:* Let $w = a_m$ and $V = K$. Then $G_w(V)$ is the module

$$K \overset{a}{\leftarrow} K \overset{b}{\leftarrow} K \overset{a}{\leftarrow} \cdots \overset{a}{\leftarrow} K \overset{b}{\leftarrow} K \overset{a}{\leftarrow} K.$$

As a $\Lambda$-module, it is uniserial of dimension $2(m+1)$. Moreover, it is isomorphic to the left ideal $\Lambda b$ of $\Lambda$. Since $b^2 = 0$ and $2|\Lambda b| = |\Lambda|$ we see that this module has $\Omega$-period 1.

(b) *Symmetric strings* Consider the word $w = a_0$. The algebra $C_w$ has two indecomposable modules; let $E_0$ be the one on which e acts as zero and $E_1$ the other one. Then $G_w(E_0) = K$, and one finds easily that $G_w(E_1)$ is isomorphic to the left ideal $\Lambda a$.

(c) *Asymmetric band*: Let $w = b_1 a_{-1}$; and take for V the 1-dimensional $K[x,x^{-1}]$-module K on which x acts as multiplication by $\lambda \in K^*$. Then $M := G_w(V)$ is the 2-dimensional module

$$K \underset{\leftarrow\, a=1}{\overset{\leftarrow\, b=\lambda}{}} K.$$

We can see directly that $M \cong \Omega^2 M$: To obtain an explicit projective cover for M, take the element $\omega \in \Lambda$ given by

$$\omega = (ba)^m b + \lambda(ab)^m a.$$

Then $M \cong \Lambda\omega$; and right multiplication with $\omega$ gives a projective cover $\pi : \Lambda \to M$; and we identify ker $\pi$ with $\Omega M$. In fact, let $\zeta = b - \lambda a$. One checks that $\zeta\omega = 0$ and alsO $\omega\zeta = 0$. In particular, $\Lambda\zeta \subseteq \ker \pi$, and by dimension we deduce $\Lambda\zeta = \ker \pi = \Omega M$.

Similarly one defines a projective cover of $\Lambda\zeta$ whose kernel is equal to $\Lambda\omega$.

<u>II.9.5</u> In order to determine the stable AR-quiver of $\Lambda$ we will need to know some AR-sequences close to the standard sequence (I.7.7) in detail.

Let $H(\Lambda) = \text{rad } \Lambda/\text{soc } \Lambda$.

LEMMA *Let $U = \Lambda a/\Lambda a^2$ and $V := \Lambda b/\text{soc } \Lambda$. Then U, V are indecomposable and $\Omega^2 U \cong V$. Moreover, there is an AR-sequence*

$$0 \to V \to H(\Lambda) \to U \to 0.$$

Proof: (i) We identify $\Omega U$ with the set $\{x \in \Lambda: xa \in \Lambda a^2\}$; then evidently $\Lambda a \subseteq \Omega U$. Moreover, $|\Lambda a| = 2(m+1)+1 = |\Omega U|$, and equality must hold.

(ii) Similarly, we identify $\Omega(\Lambda a)$ with $\{x \in \Lambda: xa = 0\}$. Define a map $\psi: \Lambda b \to \Omega(\Lambda a)$ by $\psi(xb) = xba$. Let $xba = 0$. Expressing xb in terms of a basis one sees that then $xb = r(ab)^{m+1}$ for some $r \in K$. Consequently $\ker \psi = \text{soc } \Lambda$; and comparing lengths shows that $\psi$ is surjective.

(iii) We wish to determine the AR-sequence of U. It is isomorphic to the pull-back of some map in $\mathrm{Hom}_\Lambda(U, \Omega U) - \mathcal{P}(U, \Omega U)$. It is easy to check directly that $\underline{\mathrm{Hom}}_\Lambda(U, \Omega U)$ is 1-dimensional, and for any map $\theta \notin \mathcal{P}(U, \Omega U)$, the pull-back will be an AR-sequence, see [$G_2$]. For example, take $\theta$ to be the map induced by $(x \to xa)$. Then one verifies easily that the middle in the pull-back is isomorphic to $H(\Lambda)$.

LEMMA *The module $H(\Lambda)$ is indecomposable.*

Proof: One checks that the endomorphism ring of $H(\Lambda)$ is local.

## II.10 *Auslander-Reiten components for the semidihedral algebra*

Suppose $\Lambda = \Lambda_m$ is the local semidihedral algebra, as in II.9. In this chapter we will prove the following results:

II.10.1 PROPOSITION *The stable AR-quiver $\Gamma_s(\Lambda)$ has the following components:*
*(a) Non-periodic components isomorphic to $\mathbb{Z}D_\infty$ and $\mathbb{Z}A_\infty^\infty$*
*(b) Tubes of rank 1 and 2.*
*Moreover, in $\Gamma(\Lambda)$, the projective module $\Lambda$ is attached to the end of a $\mathbb{Z}D_\infty$-component, and rad $\Lambda$ has one predecessor in $\Gamma_s(\Lambda)$.*

II.10.2 PROPOSITION *The stable AR-quiver $\Gamma_s(\Lambda)$ has infinitely many components both of type $\mathbb{Z}A_\infty^\infty$ and of type $\mathbb{Z}D_\infty$.*

It has been proved in [$CB_3$] that the band modules all lie in tubes of rank 1 or 2; and this proves a part of II.10.1(b). This was a by-product of the classification: Since the proof of Theorem II.9.3 uses functorial filtration, that is, II.3.2, we have that the functors $G_w$ and $S_w$ satisfy II.4.1, and therefore the arguments of II.4 apply. It remains to study string modules; the idea is to investigate bounded (sub-)additive functions.

<u>II.10.3</u> (a) Let Y be the cyclic Λ-module Λ := Λb; then Λ is indecomposable. Consider the function $d_Y = d\colon \Gamma(\Lambda) \to \mathbb{N}$ defined by $d(M) = \dim \underline{\mathrm{Hom}}_\Lambda(Y, M)$ (see I.8.8). We have

(1) If Θ is a component with Y not in Θ then d is subadditive on Θ (I.8.8).

(2) $Y \cong \Omega Y \cong \tau Y$ (see II.9.4(a)). Hence if $Y \notin \Theta$ then d induces a subadditive function on the tree associated to Θ (I.8.9).

Our aim is now to show that d is uniformly bounded , and that d is additive on any non-periodic component.

<u>II.10.4</u> (a) "Frobenius reciprocity" : Let $\mathcal{A}$ be any algebra and $\mathcal{B} \subset \mathcal{A}$ be a subalgebra with $1_{\mathcal{B}} = 1_{\mathcal{A}}$. For $X \in \mathcal{B}$-mod and $M \in \mathcal{A}$-mod we have

$$\mathrm{Hom}_{\mathcal{B}}(X, M_{\mathcal{B}}) \cong \mathrm{Hom}_{\mathcal{A}}(\mathcal{A} \otimes_{\mathcal{B}} X, M).$$

Moreover, if $\mathcal{A}$ is a projective $\mathcal{B}$-module then this induces an isomorphism

$$\underline{\mathrm{Hom}}_{\mathcal{B}}(X, M_{\mathcal{B}}) \cong \underline{\mathrm{Hom}}_{\mathcal{A}}(\mathcal{A} \otimes_{\mathcal{B}} X, M).$$

This is well-known.

(b) We apply this when $\mathcal{A} = \Lambda$ and $\mathcal{B}$ is the subalgebra B generated by b. That is, B is the local algebra of dimension 2. We write K for the simple B-module, and we take Y and d as in II.10.3. Then $Y \cong \Lambda \otimes_B K$; consequently , by (a), we have that

$$d(M) = \dim \underline{\mathrm{Hom}}_B(K, M)$$

This is equal to the number a of trivial summands in a decomposition

$M_B = aK \oplus (\oplus B)$.

<u>II.10.5</u> LEMMA *Suppose that M is an indecomposable string module. Then, as a B-module, $M = aK \oplus (\oplus B)$ where a = 2 if M is asymmetric, and 1 otherwise.*
*If M is a band module then $M_B$ is free.*

Proof: Let w be the word associated to M, say $w = c_1c_2\dots c_n$.
Assume first that M is a string. For the purpose of this proof, we consider first for <u>every</u> word in $\mathcal{S}t$ the module $G_w(K) = M(w)$ defined in II.9.3. (We note that, if w is symmetric, then M(w) is isomorphic to the direct sum of the two indecomposable symmetric strings with word w.)

We show that $M(w)_B \cong 2K \oplus (\oplus B)$, and we proceed by induction on n. Since w does not start or end with $b_i$, it follows that n must be odd. Let n = 1. Then M(w) = $(K \overset{w}{-} K)$ where $w = a_j$; we may assume that $j \geq 0$ (since $w \sim w^{-1}$). Assume first that $j \neq 0$. Then M(w) is as in the example II.9.4(a), and we deduce that $M_B$ = $2K \oplus (\bigoplus_{j-1} B)$.

Note that the summands K may be taken as the spaces "at the end vertex of the underlying quiver". This is the reason why the inductive step works.

Now assume that $w = a_0$. Then M(w) is as described in II.9.2(3), with V = K. We see directly that $M(w)_B \cong 2K \oplus (\bigoplus_{m+1} B)$. Note that again the trivial summands may be taken as the spaces "at the end".

Now suppose $n \geq 3$. Then M(w) is of the form $K \overset{w'}{\longrightarrow} K \overset{b_i}{-} K \overset{w''}{\longrightarrow} K$. where w' and w'' are also strings. By the induction hypothesis, we know that the restrictions of M to the appropriate subquivers corresponding to w' and w''both are $\cong$ 2K $\oplus$ free; and moreover, the summands $\cong$ K can be taken at the end of this quiver. Then in M the "right" summand K of w' and the "left" summand of w'' are one copy of B, and hence $M_B \cong 2K \oplus (\oplus B)$ as B-modules, as required.

So far, this proves the statement for asymmetric strings completely. Moreover, it suffices to prove that for one of the indecomposable modules for each symmetric string the restriction to B is $\cong K \oplus B$.

We proceed again by induction on n. Let n = 1, then $w = a_0$. One of the modules with word $a_0$ is K, and the statement is trivially true for K. Now assume that $n \geq 3$. We take the module with word w on which e acts as 0. Then M may be written in the form $K \overset{w'}{\longrightarrow} K \overset{b}{\longrightarrow} K$ where $w = w'b_ia_0(w')^{-1}$. By the first part of the proof, we know that the restriction of M to the support of w' is as a B-module isomorphic to $2K \oplus \Sigma B$; and moreover the summands K may be taken at the end of the quiver. It follows therefore that $M_B \cong K \oplus (\oplus B)$, as required.
The proof for bands is similar; we omit details.

By I.8.8, we know that d is additive on components not containing the module Ab.

<u>II.10.6</u> *Proof of II.10.1*

The band modules of $\Lambda$ all lie in tubes of rank 1 or 2, see [$CB_3$].

(1) <u>The component $\Theta$ containing the module $\Lambda b$:</u> This is a 1-tube, see II.9.4(a). (Note that $\Lambda$ does not lie in $\Theta$ since $\Lambda/\mathrm{soc} \neq \mathrm{rad}\,\Lambda$.) We will show that $\Theta$ consists precisely of the (asymmetric) string modules $L_k = M(w_k)$ where $w_k$ is the word $(a_{m+1}b_{-1})^k a_{m+1}$.

We have that $L_0 \cong \Lambda b$. One sees easily by constructing a projective cover that $\Omega(L_k) \cong L_k$, and hence $L_k$ lies some tube of rank 1. Moreover, $d(L_k) = 2$, see II.10.5. If $L_k$ would not lie in $\Theta$ then d would be a bounded additive non-zero function on the component of $L_k$. This is not possible, by I.8.6; and hence $L_k$ must lie in $\Theta$.

By IV.4.2 and IV.4.3, the module $\Lambda b$ lies at the end of $\Theta$; moreover the lengths of the modules in $\Theta$ are strictly increasing, and $|L_k| = (k+1)|L_0|$, for $k \geq 0$. It follows inductively that $L_k$ is the module in the (k+1)-th row of $\Theta$.

(2) <u>Let $\Upsilon$ be a component of $\Gamma(\Lambda)$ which contains a string module but $\Lambda b$ does not lie in $\Upsilon$</u>. Then d is a non-zero subadditive function on $\Upsilon$, by II.10.3 and 5.

(2a) <u>Assume first that $\Upsilon$ does not contain $\Lambda$</u>. Then d is additive; and moreover, the tree class T of $\Upsilon$ is an infinite Dynkin diagram, by [W]. Since there is a bounded non-zero additive function on $\Upsilon$, T is not $A_\infty$. We assume that K is algebraically closed, this excludes $B_\infty$ and $C_\infty$; and T can only be $D_\infty$ or $A_\infty^\infty$ as required.

(2b) <u>Let $\Delta$ be the component containing $\Lambda$</u>. This contains the standard AR-sequence $0 \to \mathrm{rad}\,\Lambda \to H(\Lambda) \oplus \Lambda \to \Lambda/\mathrm{soc}\,\Lambda \to 0$. By II.9.5, $H(\Lambda)$ is indecomposable, hence $\Delta$ has an end. By (1) and II.10.5, d defines a bounded subadditive function on $\Delta$ and on $\Omega(\Delta)$, and $d \neq 0$ on these components (for example, $d(\mathrm{rad}\,\Lambda) = 1$).

We claim that the tree class T of $\Delta$ is $D_\infty$. By II.9.5 there is another AR-sequence having $H(\Lambda)$ as a middle term. This shows that either $T = \tilde{D}_n$ for some n, or $T = D_\infty$. Moreover, $\tilde{D}_n$ is excluded, by 3.10 in [ES] provided we show that

(*) there are infinitely many $\tau$-orbits of non-periodic modules with dim $\tau^n M$ unbounded.

We will postpone this; since it will be a by-product of the proof of II.10.2, see below.

We record a fact which has been proved on the way in (1) above:

II.10.6a COROLLARY *The only periodic string modules are the asymmetric strings with word* $(a_{m+1}b_{-1})^k a_{m+1}$.

II.10.7 REMARK Suppose KG is the group algebra of a finite group G, and let $P_1$ be the projective KG-module whose simple quotient is K. In Theorem F of [W], the possibilities for the tree class of $\Delta$ are determined where $\Delta$ is the stable part of the component of $\Gamma_s(KG)$ which contains $P_1$. The statement is not correct, the list should also include $D_\infty$. This occurs, for example, if G is the semidihedral 2-group and char K = 2. In this case, KG/soc KG $\cong \Lambda_m/\text{soc}\ \Lambda_m$ for m appropriate, see III.13; and by II.10.1, the tree class of $\Delta$ is $D_\infty$ (see I.8.11). The error arises from the fact that in the proof of 5.6 in [W] it was overlooked that there is a maximal subgroup which is generalized quaternion; in this case, $\Omega^2(\mathcal{X}) \cong \Omega^{-2}(\mathcal{X})$.

Also, the statement of Theorem 1.1 in [Li] is not correct; the proof used Webb's result.

II.10.8 *Proof of II.10.2* Let $W = G_w(K)$ where w is the asymmetric band $w = b_1a_{-1}$, see II.9.4. Recall that dim W = 2; in fact $W = \left[ K \underset{a\,=\,1}{\overset{b\,=\,\lambda}{\leftleftarrows}} K \right]$ $(\lambda \neq 0)$, and $\tau W \cong W$. Therefore the function $d_W$ defined by $d_W(M) = \dim \underline{\text{Hom}}_\Lambda(W, M)$ (see II.10.3) is subadditive on any component which does not contain W; in particular on any non-periodic component. Moreover, $d_W$ induces a subadditive function on the tree of $\Theta$ (I.8.9). Now, this tree is neither Dynkin nor $A_\infty$, by II.10.1 (independent of the postponed proof of (*)). Hence by I.8.6, it follows that $d_W$ is additive. Moreover, $d_W$ is constant on a $\mathbb{Z}A_\infty^\infty$-component, and takes two values, u and 2u, on a $\mathbb{Z}D_\infty$-component, see I.8.6.2. Therefore, the statement will be proved by providing, for any integer $n \geq 1$

(I) a component $\cong \mathbb{Z}A_\infty^\infty$ on which the value of $d_W$ is $\geq n$; and

(II) a component $\cong \mathbb{Z}D_\infty$ on which the values of $d_W$ are $\geq n$.

(I) Let $X_k = G_w(K)$ be the asymmtric string with word $w = (a_1b_{-1})^k a$, $k \geq 1$; then dim $X_k = 2k+2$.

(1) $\underline{d_W(X_k) = 2k+1;\ \text{in particular}\ d_W \neq 0\ \text{on the component of}\ X_k:}$

It is visible that $J^2X_k = 0$, and that dim soc$(X_k) = k+1$. Actually, $X_k$ is the lift of a module for the Kronecker algebra.

(1.a) $\mathcal{P}(W, X_k) = 0$: Let $\varphi \in \mathcal{P}(W, X_k)$; then $\varphi$ factors through an injective hull $W \xrightarrow{j} \Lambda$; let $\psi \in \mathrm{Hom}_\Lambda(\Lambda, X_k)$ with $\psi j = \varphi$. Now, $W$ has Loewy length 2, so $j(W) \subseteq \mathrm{soc}_2(\Lambda)$. From the structure of $\Lambda_m$ we see that $\mathrm{soc}_2(\Lambda) \subseteq J^2$, for any $m \geq 1$. On the other hand, $J^2X_k = 0$ and therefore $\varphi(W) = \psi j(W) \subseteq \psi(\mathrm{soc}_2\Lambda) \subseteq \psi(J^2) = J^2(\mathrm{im}\psi) \subseteq J^2X_k = 0$ and $\varphi = 0$.

Hence $d_W(X_k) = \dim \mathrm{Hom}_\Lambda(W, X_k)$. This can be calculated easily, and one obtains (1).

Let $\Theta$ be the component containing $X_k$.

(2) The tree class of $\Theta$ is $A_\infty^\infty$:

Firstly, $\Theta$ is not a tube: We have determined the only strings which lie in tubes, see II.10.5; and $X_k$ is not one of them. Secondly, $\Theta$ does not belong to the component of rad $\Lambda$: We have that $d_W(\mathrm{rad}\ \Lambda) = d_W(K) = 1$, so the values of $d_W$ on the component of rad $\Lambda$ are 1 and 2 hence $\neq d_W(X_k)$.

Now, by II.10.1, the tree class of $\Theta$ is either $A_\infty^\infty$ or $D_\infty$. Assume for contradiction it is $D_\infty$. Then, by I.8.6.2, the values of $d_W$ on $\Theta$ are u and 2u, for some constant u, where u occurs at the end. On the other hand, $d_W(X_k) = 2k+1$ which is odd; this shows that $X_k$ must lie at the end. Remember now the additive function $d = d_Y$ defined in II.10.3. By II.10.5 we know that $d_Y(X_k) = 2$ since $X_k$ is an asymmetric string. Hence by I.8.6.2, we should have that $d_Y$ takes value 4 on $\Theta$. This is a contradiction to II.10.5.

This completes the proof of (I).

(II) Let $W$ be as before; we take now a sequence of symmetric strings. Let $Z_k = G_w(E_0)$ where $w = za_0z^{-1}$ and $z = (a_1b_{-1})^k$; here $E_0$ is the 1-dimensional module of the algebra $K[e: e^2 = e]$ on which e acts as zero.

Then $\dim Z_k = 2k+1$. It is visible that $J^2Z_k = 0$, and also $\dim \mathrm{soc}\ Z_k = k+1$.

(1) $d_W(Z_k) = 2k$: The argument in I (1.a) above applies here as well, and we obtain that $\mathcal{P}(W, Z_k) = 0$. Therefore $d_W(Z_k) = \dim \mathrm{Hom}_\Lambda(W, Z_k)$ which is easily calculated.

Let $\Theta$ be the component containing $Z_k$.

(2) The tree class of $\Theta$ is $D_\infty$: $Z_k$ is not one of the modules in II.10.6.1, hence $\Theta$ is not a tube. Suppose for contradiction that (2) is not true, then $\Theta$ is not the component of rad $\Lambda$; and then the tree cass can only be $A_\infty^\infty$. Then the additive function

$d_Y$ defined in II.10.3 is constant on $\Theta$. We have $d_Y(Z_k) = 1$ and then $d_Y \equiv 1$ on $\Theta$. This implies, by II.10.5 that all modules in $\Theta$ must be symmetric strings.

To obtain a contradiction we will prove that there is an asymmetric string L and an irreducible map $L \to Z_k$. Define L to be the asymmetric string $L = G_{w'}(K)$ where $w' = (a_{-m}b_1)w$ and $w = w_k$, the word for $Z_k$. Only one of the letters occuring is $a_0$, and it does not occur in the middle of w' since it is the middle of w. Hence w' is asymmetric. Now, L has a submodule $\cong \Lambda b$ and the quotient is isomorphic to $G_w(E_0) \oplus G_w(E_1)$ (here $E_0$ and $E_1$ are the 1-dimensional modules of $K[e: e^2 = e]$ on which e acts as 0 and 1 respectively). One shows now directly that the epimorphism $L \to Z_k$ ($\cong G_w(E_0)$) is irreducible, by the method similar to that in II.5.3.

<u>Proof of (*) in II.10.1</u>: The modules $X_k$ for k = 1, 2, 3, ... defined above lie in distinct non-periodic $\tau$-orbits.

<u>II.10.9</u> *Suppose w is a band.*

*(1) If w is asymmetric then any indecomposable module $G_w(V)$ lies in a 1-tube.*

*(2) Suppose w is symmetric, and $G_w(V)$ is indecomposable. Then $G_w(V)$ lies in a 2-tube if and only if the idempotents e, f act as 0 or 1 on V. Otherwise, $G_w(V)$ belongs to a 1-tube.*

Proof: Since the functors $G_w$ satisfy II.3.2, they preserve AR-sequences (II.4.1 and II.3).

(1) Suppose that w is asymmetric; then $G_w$ is defined on the category $K[x, x^{-1}]$-mod. The finite-dimensional indecomposable $K[x, x^{-1}]$-modules lie in 1-tubes (II.4.2).

(2) Assume that w is symmetric. Then $G_w$ is defined on the category A-mod where $A = K\langle e, f: e^2 = e, f^2 = f\rangle$. This is equivalent to the full subcategory of the four-subspace problem, in which the subspaces are paired, and within each pair they are complementary; see [$CB_2$]. The modules of this category may be found in [GP] or [$Br_3$], and they lie all in tubes of rank 1 or 2. For a precise answer, one may apply results from [$R_4$]. The 4-subspace category, that is, $\tilde{D}_4$, is of tubular type (2, 2, 2) (see p158 in [$R_4$]). That is, there are three 2-tubes and otherwise only 1-tubes (see p.121 in [$R_4$]).

Take $\widetilde{D}_4$ with the orientation

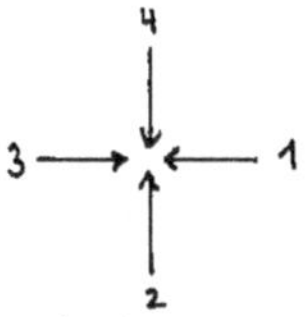


Consider the $D_4$-modules $(K; V_1, V_2, V_3, V_4)$ where $V_i = K$ for two values of i, $1 \leq i \leq 4$; and $V_i = 0$ otherwise. One checks that these modules have $\tau$-period 2. Moreover, they must lie at the end of a tube (they have only one submodule which is not periodic). Consequently these are all modules at the ends of the 2-tubes.

We can also see which of these belong to A-mod. Choose the notation such that $V_1$ = im e, $V_2$ = ker e; and $V_3$ = im f and $V_4$ = ker f. Then A-mod contains only the 2-tube having the modules (K; K, 0, K, 0 ) and (K; 0, K, 0, K) at the end. This gives the statement.

# III *Tame symmetric local algebras*

In this chapter we shall study arbitrary tame local symmetric algebras, following essentially the ideas in $[R_1]$. The first result is a classification, by generators and relations; and then we give a structural characterization. We will also determine the centers of these algebras.

Moreover, we will give presentations of the group algebras of the dihedral 2-groups, of the semidihedral groups and the quaternion groups over fields of characteristic 2.

Any tame local algebra must have a quiver of the form $Q$: X  Y (see I4.3(a) and I.10.10(a), and I.4.6).
Moreover, if $\Lambda$ is symmetric then soc $\Lambda$ is simple, and dim $\Lambda \geq 4$.

We have the following description in terms of generators and relations:

III.1 THEOREM *Assume that $\Lambda$ is a tame local symmetric algebra. Then $\Lambda \cong KQ/I$ where $I$ is one of the following ideals:*

(a) $I = ( X^m - Y^n, XY, YX )$ where $m \geq n \geq 2$ , $m + n > 4$.

(b) $I = ( X^2, Y^2, XY - YX )$.

(b') char K = 2, and $I = ( X^2, Y^2 - XY, XY - YX )$.

(c) $I = ( (XY)^k - (YX)^k, X^2, Y^2 )$ where $k \geq 2$.

(c') char K = 2 and $I = ( (XY)^k - (YX)^k, X^2 - (XY)^k, Y^2 - d(YX)^k, (XY)^kX , (XY)^kY )$ where $k \geq 2$ and $d \in K$.

(d) $I = ( (XY)^k - (YX)^k, X^2 - (YX)^{k-1}Y, Y^2, (XY)^kX )$, $k \geq 2$.

(d') char K = 2 and $I = ( (XY)^k - (YX)^k, X^2 - (YX)^{k-1}Y - c(XY)^k, Y^2 - d(XY)^k,$
$(XY)^kX )$ where $k \geq 2$ and $c, d \in K$, not both 0.

(e) $I = ( (XY)^k - (YX)^k, X^2 - (YX)^{k-1}Y, Y^2 - (XY)^{k-1}X, (XY)^kX )$, $k \geq 2$.

(e') char K = 2 and $I = ( (XY)^k - (YX)^k, X^2 - (YX)^{k-1}Y - c(XY)^k,$
$Y^2 - (XY)^{k-1}X - d(XY)^k, (XY)^kX, (YX)^kY )$ where $k \geq 2$ and $c, d \in K$, not both 0.

As a consequence, we obtain the following characterization:

III.1.1 COROLLARY *Suppose $\Lambda$ is tame, local and symmetric. Then exactly one of the following holds:*

*(i) $\Lambda/\mathrm{soc}\,\Lambda$ is special biserial.*

*(ii) $\Lambda/\mathrm{soc}\,\Lambda$ is isomorphic to $\Lambda_m/\mathrm{soc}\,\Lambda_m$ where $\Lambda_m$ is the semidihedral algebra.*

*(iii) The simple $\Lambda$-module is $\Omega$-periodic of period 4.*

(We expect that in (iii) actually all $\Lambda$-modules have $\Omega$-period dividing 4 and then $\tau$-period 1 or 2.) Clearly all algebras in III.1 are local and symmetric. Moreover:

III.1.2 *Suppose $\Lambda$ is any of the algebras listed in III.1. Then $\Lambda$ is tame.*

Proof: Clearly, $\Lambda$ is not of finite type (see I.4.3). It suffices to show that $\Lambda/\mathrm{soc}\,\Lambda$ is tame, since ind $\Lambda$ = ind $\Lambda/\mathrm{soc}\Lambda \cup \{$ indecomposable projectives $\}$. Assume first that $\Lambda$ satisfies III.1.1(i) then $\Lambda/\mathrm{soc}\,\Lambda$ is tame, by II.3.1. Moreover, the algebras $\Lambda_m$ are tame, by [BD] or [$CB_3$], see chapter II.9. Otherwise, if $\Lambda$ is an algebra in III.1(e) or (e') then, according to [BD], $\Lambda$ is also tame.

As an application, we will determine the centers of the algebras in the list. In particular, we obtain the following:

III.1.3 COROLLARY *Suppose $\Lambda$ is tame, local symmetric. Then either $\Lambda$ is commutative, orelse $\dim \Lambda = 4k$ and $\dim Z(\Lambda) = k + 3$ $(k \geq 1)$.*

When k = 2 we have therefore a special case of the result in [CK].

For the proof of III.1, we shall frequently use the following fact:

III.2 LEMMA *Let $\Lambda = KQ/I$ be local symmetric, such that $J^2$ is generated by $X^2$ and $Y^2$. Then $XY = 0$ if and only $YX = 0$.*

Proof: We have that $\Lambda$ is spanned by $1, X, X^2, \ldots, Y, Y^2, \ldots$ . Suppose that $XY = 0$.

We may write $YX = X^c w + Y^d z$ where w, z are units with $w \in K[X]$ and $z \in K[Y]$, and where c, d $\geq 2$. Then $0 = XYX = X^{c+1}w$ and hence $X^{c+1} = 0$. Since also $X^cY = 0$, it follows that $X^c \in$ soc Λ. Moreover $0 = YXY = wX^cY + Y^{d+1}z = Y^{d+1}z$, and we deduce $Y^{d+1} = 0$ . Since also $XY^d = 0$ , the element $Y^d$ must lie in soc Λ. This shows that $YX \in$ soc Λ. Let now ψ be a symmetrizing form of Λ, then $\psi[XY] = \psi[YX] = 0$, and XY spans an ideal contained in ker ψ. We deduce that $YX = 0$.

We will study first the algebras of smallest possible dimension.

<u>III.3</u> PROPOSITION *Assume that Λ is a 4-dimensional local symmetric algebra. Then Λ is isomorphic to one of the algebras in III.1 (b) or (b') (not both).*

Proof: Let X, Y be generators of rad Λ. Since Λ is symmetric of dimension 4, XY and YX lie in soc Λ, and since Λ is symmetric, XY = YX. In particular Λ is commutative.

<u>Case 1</u> char K ≠ 2.

(1) Assume first that XY = 0. Then also YX = 0; moreover we must have that $X^2 = aY^2$ for some $0 \neq a \in K$, since Λ is symmetric of dimension 4. Let $\mu \in K$ with $\mu^2 = a$, then $U = X + \mu Y$ and $V = X - \mu Y$ are generators whose square is 0, and we see that $\Lambda = K\langle U, V\rangle/(U^2, V^2, UV - VU)$; and consequently Λ is an algebra as in III.1(b).

(2) Now suppose XY ≠ 0. Then rad Λ has a basis { X, Y, XY}. We shall now investigate generators whose square is 0. We have $X^2 = aXY$ and $Y^2 = bXY$ where $a, b \in K$; without loss of generality b ≠ 0. Put $U = Y + tX$, then $U^2 = (b + 2t + t^2a)XY$.

If a = 0 we take t = -b/2, then X and U are generators which bring I to the form III.1(b). Now let a ≠ 0 ≠ b. We may assume that a = 1, and also that b ≠ 1; otherwise X and Y - X are generators whose product is zero, and we have the situation which we already dealt with in (1). Then the polynomial $t^2 + 2t + b$ has two distinct roots, and we have two independent generators of rad Λ whose square is 0, and hence Λ is as in III.1(b).

<u>Case 2</u> char K = 2.

Let $L = \{ w \in \Lambda: w^2 = 0 \}$, then L is an ideal an ideal of Λ which is non-zero. Also, L is not simple since the subring of Λ consisting of squares must have dimension ≤ 2. If L = rad Λ then Λ is as in III.1(b). Otherwise, we have that soc Λ $\subsetneq$ L $\subsetneq$ rad Λ. Then there are generators X, Y with $X \in L$ and $Y \notin L$. It follows that Λ = span { X, Y,

XY}, and $Y^2 = cXY$ for $0 \neq c \in K$. We may assume that $c = 1$, otherwise we replace X by cX. Then $\Lambda$ is as in III.1(b').

Comparing the dimensions of L we see that the algebras in (b) and (b') are non-isomorphic.

III.4 LEMMA *Let $\Lambda = KQ/I$ be symmetric. Then*
*(i) If $\Lambda$ is tame then $\dim J^2/J^3 \leq 2$.*
*(ii) If $\dim J^2/J^3 \leq 1$ then $\Lambda$ is an algebra as in III.1(b) or (b'), or as in III.1(a) with $n = 2$.*

Proof: (i) If $\dim J^2/J^3 \geq 3$ then $\Lambda$ has one of the algebras in I.10.10 as a quotient, and $\Lambda$ is wild.
(ii) Suppose now that $\dim J^2/J^3 \leq 1$. Then the dimension must be = 1, since otherwise X, Y would lie in soc $\Lambda$ and soc $\Lambda$ would not be simple. It follows that $\dim J^r/J^{r+1} \leq 1$ for all $r \geq 2$. Let m be the largest integer such that $J^m \neq 0$, then $J^m = \operatorname{soc} \Lambda$. If $m = 2$ then $\dim \Lambda = 4$, and $\Lambda$ is as in (b) or (b'), by III.3. Suppose now that $m > 2$. Since $\Lambda$ is symmetric, we know that $\operatorname{soc}_2\Lambda/\operatorname{soc} \Lambda$ is 2-dimensional. Therefore $J^{m-1} \subsetneq \operatorname{soc}_2\Lambda$. It follows that there is some $\omega \in J - J^2$ which lies in $\operatorname{soc}_2\Lambda$, and then $\omega$ is an arrow. Without loss of generality, $\omega = Y$.

We have $YJ = \operatorname{soc} \Lambda = JY$. Let X be a generator of J which is independent of Y. Since XY lies in soc $\Lambda$, we have that $X\Lambda$ is spanned by X, $X^2$, ..., $X^m$, and $\operatorname{soc} \Lambda = \langle X^m \rangle$. Then $XY = cX^m$ for some $c \in K$. We may assume hat $c = 0$, otherwise we replace Y by $Y - cX^{m-1}$. It follows from III.2 that $YX = 0$. Now $Y^2 = aX^m$, and $0 \neq a \in K$. We may assume that $a = 1$; and $\Lambda$ is as in part (a) of III.1, with $n = 2$.

III.5 LEMMA *Let $\Lambda = KQ/I$ be a local algebra with $\dim J/J^2 = 2$ and $\dim J^2/J^3 \leq 2$. Then there are $a, b \in J - J^2$ such that ab belongs to $J^3$.*

Proof: We work in $A = \Lambda/J^3$ and assume that $J^3 = 0$. Therefore $J^2 \cong (N \otimes N)/U$ with $N = \operatorname{span}\{X, Y\}$, where U is some subspace of dimension $\geq 2$, and where the multiplication is given by the tensor product $\otimes$.

We may assume that U intersects both $N \otimes X$ and $N \otimes Y$ trivially and $\dim U = 2$. Then

U is the graph of an isomorphism $\phi$: N $\to$ N with U = $\{ v \otimes X + \phi(v) \otimes Y: v \in N \}$. Let a be an eigenvector of $\phi$ with eigenvalue $\lambda$, then $0 \neq a \otimes (X + \lambda Y)$ belongs to U.

III.6 LEMMA *Let* $\Lambda$ *be a tame local algebra with quiver Q, of dimension 5, with* $J^3 = 0$. *Then* $\Lambda \cong KQ/L$ *where L is one of the following ideals:*

*(i)* $(XY, YX)$.

*(ii)* $(YX - X^2, XY)$.

*(iii)* $(YX - X^2, XY - aY^2)$ *where* $a \in K$ *and* $0 \neq a \neq 1$.

*(iv)* $(X^2, Y^2)$.

*(v)* $(YX - X^2, Y^2)$.

Proof: The hypothesis implies that $J^2$ has dimension 2. First, assume there is some Y $\in J - J^2$ with $Y^2 = 0$. Let X, Y be a basis of N where $N \oplus J^2 = J$. There must be a relation involving $X^2$; otherwise $\Lambda$ would be wild, as $\Lambda$ has some algebra in I.10.10(d) as a quotient. Hence there are elements a, b $\in$ K such that $aXY + bYX + X^2 = 0$. Replace X by X' where X' = X + aY. If a - b = 0 then $\Lambda$ is as in (iv). Otherwise, replace also Y by (a-b)Y, and $\Lambda$ is as in (v).

Next, assume $X^2 \neq 0$ for all $X \in J - J^2$. By III.5, there is a basis X, Y of J modulo $J^2$ with YX = 0. There must be a relation involving XY, for otherwise $\Lambda$ would have a quotient isomorphic to the wild algebra I.10.10(c). Hence there are a, b $\in$ K with $aX^2 + bY^2 + XY = 0$. If ab = 0 we see that $\Lambda$ is an algebra as in (i) or (ii). Suppose now that $ab \neq 0$. Without loss of generality, a = 1; then $b \neq 1$ (otherwise there would be a generator whose square is 0.) Put u = X and v = X + Y; then $vu = u^2$ and $uv = -bY^2 = (-b)/(-b+1)v^2$. Therefore $\Lambda$ is an algebra as in (iii).

We will now classify symmetric algebras where $\Lambda/J^3$ satisfies one of the above conditions. We will not use the property "tame" for the rest of the calssification in this chapter.

III.7 LEMMA *There is no symmetric algebra* $\Lambda$ *such that* $\Lambda/J^3$ *satisfies III.6 (ii).*

Proof: Suppose such $\Lambda$ exists.

(1) <u>J does not have generators X', Y' with (X'Y') and (Y'X') lying in $J^3$:</u> Let X' = cX + dY and Y' = eX + fY (mod $J^2$). Then X'Y' = e(c+d)YX + $dfY^2$ and Y'X' = c(e+f)YX + $dfY^2$ (moulo $J^3$). Suppose that X'Y' and Y'X' ly in $J^3$, then e(c+d) 0 = c(e+f). Hence det $\begin{bmatrix} c & d \\ e & f \end{bmatrix}$ = cf - de = cf + ce = 0, and X', Y' are dependent (modulo $J^2$).

We have that $X^3 \equiv XYX \in J^4$, therefore $J^3$ is generated by $Y^3$. So we may write XY = $uY^k$ where u is a unit in K[Y], and $k \geq 3$. Let X' = X - $Y^{k-1}u$, then X'Y = 0. It follows from III 2 that YX' = 0, and we have a contradiction to (1).

<u>III.8</u> LEMMA *Let Λ be a symmetric local algebra such that $X^2$ - YX and XY - $aY^2$ lie in $J^3$ where $0 \neq a \neq 1$. Then dim Λ = 4, and Λ is one of the algebras in III.1 (b) or (b'). In particular, III.6(iii) is excluded for Λ symmetric.*

Proof: We have that $X^2$ - YX and XY - $aY^2$ lie in $J^3$. Consequently, calculating modulo $J^4$, $X^3 \equiv XYX \equiv aY^2X \equiv aYX^2 \equiv aX^3$ and $X^2Y \equiv aXY^2 \equiv a^2Y^3 \equiv aYXY \equiv aX^2Y$. Since $a \neq 1$, we have that $X^3 \in J^4$ and $X^2Y \in J^4$. Since $a \neq 0$, it follows that all the other monomials occuring lie in $J^4$. Consequently $J^3 \subseteq J^4$ and $J^3 = 0$, and then $J^2 \subseteq$ soc Λ. Since Λ is symmetric, it follows that dim Λ = 4.

<u>III.9</u> *Algebras Λ where Λ/$J^3$ satisfies III.6(i).*

Suppose Λ is a symmetric local algebra such XY and YX lie in $J^3$. By III.3 we may assume that dim Λ > 4. The algebra has a basis of the form

(*) $\{1, X, X^2, \ldots, X^s, Y, Y^2, \ldots, \}$.

If XY = 0 then YX = 0 as well, by III.2. Hence in this case, Λ is of the form given in III.1 (a).

We consider now the case when $XY \neq 0 \neq YX$. Let p be as large as possible such that XY and YX $\in J^p$. Then one of them does not lie in $J^{p+1}$ [and $J^p \neq 0$]. Write XY $\equiv$ $X^pu + Y^pv$ and YX $\equiv$ $X^pw + Y^pz$ (modulo $J^{p+1}$) where u, v, w, z $\in$ K; and we may replace X, Y by X', Y' where X' = X - $Y^{p-1}z$ and X' = X - $Y^{p-1}z$. Then we have new relations XY $\equiv cY^p$ and YX $\equiv dX^p$ for c, d $\in$ K; moreover one of them, c say, is non-zero. Now, $J^{p+1}$ is generated by $X^{p+1}$ and $Y^{p+1}$; and $cY^{p+1} \equiv YXY \equiv dX^pY \equiv dX^{p-1}Y^p$ $\in J^{p+2}$ and already $J^{p+1} = \langle X^{p+1} \rangle$.

If $d \neq 0$ then similarly $X^{p+1} \in J^{p+2}$, and $J^{p+1} = 0$, so $J^p \subseteq \text{soc}\, \Lambda$. We assumed that $0 \neq XY \in J^p$, and it follows that $J^p = \text{soc}\, \Lambda$. We have $X^p = aY^p$ for $0 \neq a \in K$ and also $XY = cY^p$ and $YX = dX^p$, with $cd \neq 0$. Replace now X, Y by $X' = X - cY^{p-1}$ and Y. Then $X'Y = 0$, and by III.2 it follows that $YX' = 0$ as well. With respect to these generators, $\Lambda$ is of the form III.1(a).

Suppose now that $d = 0$. Then $YX \in J^{p+1} = \langle X^{p+1} \rangle$. We assumed that $YX \neq 0$, consequently $YX = X^k u$ for $k \geq p+1$, where u is a unit of K[X]. Replace Y by $Y' = Y - X^{k-1}u$, then $Y'X = 0$, and by III.2, it follows that $XY' = 0$. We see that $\Lambda$ is an algebra as in III.1 (a). This completes the proof.

<u>III.10</u> *Algebras $\Lambda$ where $\Lambda/J^3$ satisfies III.6(iv).*

Let $\Lambda$ be a local symmetric algebra such that $X^2$ and $Y^2$ lie in $J^3$. We assume also that $\dim \Lambda > 4$; otherwise, $\Lambda$ has been determined in III.3. Since $\dim J^2/J^3 = 2$, we have then that XY and YX are generators of $J^2$ which are independent (modulo $J^3$).

<u>Case 1</u>: There exist such X, Y where $X^2$ and $Y^2$ lie in soc $\Lambda$.

Let k be the integer such that $(XY)^k \neq 0$ and $(XY)^{k+1} = 0$. We claim first that $\text{soc}\, \Lambda = \langle (XY)^k \rangle$, that is, $(XY)^k X = 0$: From the choice of k it is clear that $(XY)^k XJ = 0$, therefore $(XY)^k X \in \text{soc}\, \Lambda$. Moreover, $X(XY)^k \in X^2 J = 0$. Let $\psi$ be a symmetrizing form for $\Lambda$; then $\psi[(XY)^k X] = \psi[X(XY)^k] = 0$, and $(XY)^k X$ spans an ideal $\subseteq \ker \psi$, and we deduce that $(XY)^k X = 0$. Since $\Lambda$ is symmetric we have then also $(XY)^k = (YX)^k$. Note also that $k \geq 2$ since we assume $\dim \Lambda > 4$.

Let $X^2 = c(XY)^k$ and $Y^2 = d(XY)^k$ for $c, d \in K$. If char $K \neq 2$ then we replace X and Y by $X' = X - (c/2)Y(XY)^{k-1}$ and $Y' = Y - (d/2)(XY)^{k-1}X$. These are independent generators whose squares are zero, and we deduce that $\Lambda$ is isomorphic to an algebra in III.1(c).

Now assume char $K \neq 2$. If one of c or d is zero then we may assume that the other is 0 or 1 and we are done. Suppose therefore that c and d are both non-zero. Set $X = \lambda X'$ and $Y = \mu Y'$; we wish to find $0 \neq \lambda, \mu \in K$ such that $\lambda^2 = c(\lambda\mu)^k$ and $\mu^2 = d(\lambda\mu)^k$. This has a solution, since $k \neq 1$; and hence $\Lambda$ is an algebra as in III.1(c').

<u>Case 2</u>: Otherwise, choose X, Y such that $X^2 \in J^m - J^{m+1}$ and $Y^2 \in J^l$ where $l \geq m$. Take also m as large as possible with respect to these conditions.

(1) $J^{m+1} \subsetneq$ soc $\Lambda$: Assume first that m is even, m = 2k say. Then $J^m$ is generated by $(XY)^k$ and $(YX)^k$. There are elements c, d $\in$ K such that $X^2 + c(XY)^k + d(YX)^k \in J^{m+1}$. We may assume that c = 0 ; otherwise, we replace X by $X' = X + c(YX)^{k-1}Y$. We have that $(X')^2 = X^2 + c(YX)^k + c(XY)^k + c^2((YX)^{k-1}Y)^2 \equiv (c-d)(YX)^k \equiv (c-d)(YX')^k$ modulo $J^{m+1}$.] Then, by the choice of m, we have that d $\neq$ 0. Therefore $(XY)^kX \equiv (-d^{-1})X^3 \equiv (YX)^kX \equiv (YX)^{k-1}Y(-d)(YX)^k$ (modulo $J^{m+1}$). Since $Y^2X$ lies in $J^4$ we deduce that $(XY)^kX \in J^{m+2}$; hence $J^{m+2} = \langle [(XY)^kX]Y, Y[(XY)^kX] \rangle \subseteq J^{m+3}$ and then $J^{m+2} = 0$, as required.

Now assume that m = 2k + 1; then $J^m$ is generated by $(XY)^kX$ and $(YX)^kY$. There are c, d $\in$ K such that $X^2 + c(XY)^kX + d(YX)^kY \equiv 0 \pmod{J^{m+1}}$. We may assume that c = 0 [otherwise we replace X by X' where $X' = X + c(XY)^k$]. Then d $\neq$ 0, by the choice of m, and therefore $(YX)^kYX \equiv (-d^{-1})X^3 \equiv (XY)^{k+1}$ modulo $J^{m+1}$. It follows that $J^{m+2} = \langle (XY)^{k+1}X, (YX)^{k+1}Y \rangle = \langle (YX)^{k+1}X, (XY)^{k+1}Y \rangle \subseteq J^{m+3}$ and $J^{m+2} = 0$, as required.

(2) soc $\Lambda$ is an even power of J: Suppose that soc $\Lambda = J^{2r+1}$. Say $(XY)^rX \neq 0$, and let $\psi$ be a symmetrizing form of $\Lambda$. Then $\psi[(XY)^rX] = \psi[YX(XY)^{r-1}X] \subseteq \psi[J^{2r+2}] = \psi[0] = 0$. Consequently $(XY)^rX$ spans an ideal $\subseteq$ ker $\psi$, a contradiction.

By the hypothesis at the beginning, $J^m \not\subseteq$ soc $\Lambda$. So $J^{m+1} \neq 0$ and then $J^{m+1}$ = soc $\Lambda \subsetneq J^m$. By (2), we have that m+1 = 2k. Then clearly, soc $\Lambda = \langle (XY)^k \rangle$ and $(XY)^k = (YX)^k$. Then $X^2 = c(XY)^{k-1}X + d(YX)^{k-1}Y + f(XY)^k$ and (c,d) $\neq$ 0. Using the argument from above, we see that, without loss of generality c = 0.

Now consider $Y^2$; by the hypothesis, the element lies in $\text{soc}_2\Lambda$. If $Y^2$ lies in soc $\Lambda$ then we show similarly as in Case 1 that $\Lambda$ is an algebra in III.1(d) or (d'). Otherwise $Y^2 = a(YX)^{k-1}Y + b(XY)^{k-1}X + u(XY)^k$ with a, b, u $\in$ K and (a, b) $\neq$ (0, 0). We may assume a = 0; if not, replace Y by $Y' = Y + a(YX)^{k-1}$. Then b $\neq$ 0, and to see that $\Lambda$ is one of the algebras III.1 (e) or (e') we need a scalar transformation:

We have now $X^2 = a(YX)^{k-1}Y + f(XY)^k$ and $Y^2 = b(XY)^{k-1}X + u(XY)^k$ for a, f, b and u $\in$ K with a and b both $\neq$ 0. If char K $\neq$ 2 then we may assume that f = u = 0, by the argument in case 1. Let $X = \lambda X'$ and $Y = \mu Y'$ for $\lambda, \mu \in K^*$; then we require $\lambda$, $\mu$ satisfying the equations

(*) $a^{-1}\lambda^3 = (\lambda\mu)^k$ and $b^{-1}\mu^3 = (\lambda\mu)^k$;

then we may replace X, Y by X', Y' and obtain relations as in III.1(e'), with $c = f\lambda^{k-2}\mu^k$ and $d = u\lambda^k\mu^{k-2}$. Now, the equations (*) have solutions.

This completes the proof of III.10.

III.10.1 REMARK Consider an algebra $\Lambda$ in III.1(e'). If $\Lambda$ has socle scalars f, u as in (2) of the above proof then $\Lambda$ is isomorphic to an algebra in the same family of the same dimension with socle scalars f' and u' if and only if $f' = \omega^{k-2}\mu^{2k-2}$ and $u' = u\omega^k\mu^{2k-2}$ where $\omega^3 = 1$ and $\mu$ is a root of $t^{2k-3} - \omega^{2k}$. In particular, given f, u, there are finitely many solutions for f' and u' and hence there are infinitely many isomorphism classes and Morita equivalence classes of algebras belonging to III.1(e').

III.11 LEMMA *There is no symmetric algebra $\Lambda$ such that $\Lambda/J^3$ satisfies III.6(v).*

Proof: Suppose such algebra exists. We use the argument of III.10 again.
(1) $J^3 = \langle YXY \rangle \subseteq \operatorname{soc}\Lambda$: We have that $J^3$ is generated by XYX and YXY, by the given relations. Moreover, modulo $J^4$ we have that $XYX = X^3 = YXX = Y^2X = 0$ and therefore $J^3 = \langle YXY \rangle$. This implies $J^4 = \langle (YX)^2 \rangle \subseteq YJ^4 \subseteq J^5$ and $J^4 = 0$, as required.

Let $\psi$ be a symmetrizing form for $\Lambda$. Then $\psi[YXY] = \psi[XY^2] = 0$ since $XY^2 \in J^4 = 0$. Now, YXY lies in soc $\Lambda$ and spans therefore an ideal; and we deduce that YXY = 0. Consequently $J^3 = 0$, and $J^2 \subseteq \operatorname{soc}\Lambda$, therefore soc $\Lambda$ is not simple, a contradiction.

III.12 *Proof of III.1.1* If $\Lambda$ is an algebra in III.1(a) to (c') then evidently $\Lambda/\operatorname{soc}\Lambda$ is special biserial. For the algebras in (d) and (d') we obtain (ii). It remains to show that for the algebras (e) and (e'), the simple module K has $\Omega$-period 1. This can be done directly, using the identifications as in I.6.5.

III.13 *The quiver and relations for the group algebras of the dihedral, semidihedral and quaternion groups for char $K = 2$.*

(1) Assume first that D is the dihedral group of order $2^n$. Then the group algebra KD is isomorphic to the algebra in III.1(c) with $k = 2^{n-2}$ [ or as in III.1(b) if D is the Klein 4-group ]:

It is well-known that the group D may be generated by two involutions, for example y and xy, in the notation of III.17 below. Let $\Lambda$ be the algebra in III.1(c) with $k = 2^{n-2}$. Then it is easy to see that an algebra homomorphism is induced by taking $X \to (1 - y)$ and $Y \to (1 - xy)$. This is necessarily an isomorphism.

(2) Suppose now that $D$ is semidihedral of order $2^n$. We claim that $KD$ is isomorphic to one of the algebras in III.1(d) or (d') with $k = 2^{n-2}$:

Here it is not convenient to give an explicit isomorphism. Instead, with the notation of III.17, let $Y = (1 - y) \in KD$, then $Y^2 = 0$ and $Y$ is a generator of $J$. Let also $X'$ be the element $(1 - xy)$. Then $(X')^2 = (1 - j) = (1 - x)^{2^{n-2}} \in J^3$. Therefore the algebra $KD/J^3$ satisfies III.6.(iv), and we may apply III.10; and therefore $\Lambda$ is one of the algebras (c) to (d').

If $KD$ were of the form (c) or (c') then by II.7, all periodic $KD$-modules would have $\Omega$-period $\leq 2$. On the other hand, $D$ has a subgroup $Q$ which is quaternion and the induced module $K \otimes_{KQ} KD$ has $\Omega$-period 4 (by Green's Theorem V.1.5, the module is indecomposable). This excludes (c) and (c').

(3) Suppose $D$ is quaternion of order $2^n$. Then $KD$ is isomorphic o one of the algebras in III.1(e) or (e') with $k = 2^{n-2}$:

We use the notation of III.17. Let $X' = (1 - xy)$ and $Y' = (1 - y)$. Then $(X')^2 = (1 - j) = (1 - x)^{2^{n-2}} = (Y')^2 \in J^3$. Therefore the algebra $KD/J^3$ satisfies III.6(iv). By III.10, $KD$ is isomorphic to one of the algebras in III.1.(c) to (e'). Now, it is well-known that the simple $KD$-module $K$ has $\Omega$-period 4 (see for example [Be]). This shows that $KD$ is not one of the algebras (c) to (d'), by III.12.

III.13.1 COROLLARY [BD] *Let $D$ be the semidihedral group and char $K = 2$. Then $KD$/soc $KD$ is isomorphic to $\Lambda_m$/soc $\Lambda_m$, with $m = 2^{n-2}$.*

In [BD], the authors state an explicit formula.
Also, in [$Da_1$], Dade obtains a presentation of the group algebra of the quaternion group; he calculates the precise socle scalars.

Now we shall study the centers of the algebras in III.1. It is clear that the algebras in (a) or (b), (b') are commutative.

III.14 *Suppose $\Lambda$ is one of the algebras in III.1(c) to (e'). Then a general element* *of the center of rad Λ is of the form*

$$\omega = \sum_{i=1}^{k-1} r_i[(XY)^i + (YX)^i] + s(XY)^k + t(XY)^{k-1}X + u(YX)^{k-1}Y$$

*where the $r_i$, s, t and u are arbitrary elements of K. In particular, dim $Z(\Lambda) = k+3$.*

Proof: Since Λ is symmetric, we know that $soc_2\Lambda \subseteq Z(\Lambda)$, see I.3.9. Observe also that $X^2$ and $Y^2$ lie in $soc_2\Lambda$, hence in $Z(\Lambda)$. Suppose $\omega \in$ rad Λ where $\omega = \Sigma\, r_i(XY)^i + \Sigma\, s_i(YX)^i + \Sigma\, t_i(XY)^iX + \Sigma\, u_i(YX)^iY$, with $r_i$, $s_i$, $t_i$ and $u_i \in K$. By the above remarks, we have $X\omega - \omega X = \Sigma\,(-r_i+s_i)(XY)^iX + \Sigma\, u_i[(XY)^{i+1} - (YX)^{i+1}]$ and also $Y\omega - \omega Y = \Sigma\,(r_i - s_i)(YX)^iY + \Sigma\, t_i[(YX)^{i+1} - (XY)^{i+1}]$. This implies directly the statement.

<u>III.15</u> *Suppose Λ is a tame local symmetric algebra, let $\Lambda = K\langle X, Y\rangle/I$ where I is one of the ideals in III.1(c) to (e'); and let J = rad Λ. Suppose $\alpha \in J - J^2$; then there is some $\beta \in J$ such that α, β are independent modulo $J^2$ and moreover soc $\Lambda = \langle (\alpha\beta)^k\rangle = \langle (XY)^k\rangle$. In fact, one can take β = X or Y.*

Proof: We may write $\alpha = aX + bY + z$ where a, b ∈ K, not both zero, and $z \in J^2$. An arbitrary element β such that α, β are independent modulo $J^2$ is of the form $\beta = cX + dY + w$ with $w \in J^2$ and c, d ∈ K such that ad - bc ≠ 0. In all cases, $X^2$ and $Y^2$ belong to $soc_2\Lambda \subseteq J^3$, therefore we have, modulo $J^3$, that $\alpha\beta \equiv ad(XY) + bc(YX)$ and then, modulo $J^5$, $(\alpha\beta)^2 \equiv (ad)^2(XY)^2 + (bc)^2(YX)^2$, and so on. In particular, $(\alpha\beta)^r \neq 0$ for r < k since $(XY)^r$ and $(YX)^r$ are linearly independent. Moreover, $(\alpha\beta)^k = ((ad)^k + (bc)^k)(XY)^k$. If a ≠ 0 then take d = 1 and c = 0, similarly for b ≠ 0.

<u>III.16</u> Suppose Λ is a tame local symmetric algebra, we wish to recognize the isomorphism type. In the course of the proofs in this chapter we have derived criteria to do this; we will summarize the results for later use.

Suppose J = rad Λ. If dim $J^2/J^3 \leq 1$ then Λ is one of the algebras in III.1(b) or (b') or of (a), with n = 1; in particular, Λ is commutative (see III.4).

Otherwise, dim $J^2/J^3 = 2$; and then ne of the possibilities (i) or (iv) in III.6

holds (the others are excluded by III.7, 8, 11). If Λ satisfies III.6(i) then Λ is an algebra in III.1(a) and is commutative. Suppose now that III.6(iv) holds. If there are independent generators α, β of J whose squares lie in soc Λ then Λ is isomorphic to an algebra in III.(c) or (c'). Otherwise, either there is some $\alpha \in J - J^2$ such that $\alpha^2$ lies in soc Λ; then Λ is of the form III.1(d) or (d'), or else Λ belongs to III.1(e) or (e').

We will apply this in the following way: Suppose Λ is an arbitrary tame symmetric (basic) algebra; if e is a primitive idempotent of Λ then the algebra $\Lambda_e := e\Lambda e$ is ocal and symmetric (I.3.3) and also either tame or of finite type (I.4.7). Moreover, soc $\Lambda_e$ = soc eΛ, and we will use III.15 and III.16 to determine socle relations, and also, in case there is a loop at vertex e, to obtain generators of eJe whose squares lie in $\mathrm{soc}_2\Lambda$. The case when $\Lambda_e$ is of finite type is clear by I.4.3.1.

## III.17 *Presentations of some 2-groups*

In the following, we will always denote by j the central element of order 2 of the group considered.

(1) The dihedral 2-group of order $2^n$ ($n \geq 2$) has a presentation

$$D = \langle x, y \mid x^{2^{n-1}} = y^2 = 1,\ y^{-1}xy = x^{-1} \rangle$$

The conjugacy classes of D: The elements in D - < x > form two classes , $y^D$ and $xy^D$, each of size $2^{n-2}$, consisting of involutions. The elements in < x > of order ≥ 4 (if any) lie in classes of size 2, The number of such classes is $2^{n-1}-1$. The classes with central elements are {1} and {j}. Hence D has $2^{n-2} + 3$ conjugacy classes.

We note the following property: Suppose < a > and < b > are cyclic subgroups of D of the same order, > 2; then < a > = < b >.

(2) The semi-dihedral group of order $2^n$ ( n > 4) has a presentation

$$SD = \langle x, y \mid x^{2^{n-1}} = y^2 = 1,\ y^{-1}xy = x^{-1+2^{n-2}} \rangle.$$

Note that $y^{-1}xy = x^{-1}j$, this shows that $\langle x^2, y \rangle$ is dihedral of order $2^{n-1}$.

The conjugacy classes of SD The elements of SD - < x > form two conjugacy classes, $y^{SD}$ consisting of involutions, and $(xy)^{SD}$, elements of order 4, each of size $2^{n-2}$.

The elements in $\langle x \rangle$ of order $\geq 4$ lie in classes of size 2. The number of such classes is $2^{n-2}-1$. Finally, the central elements form classes $\{1\}$ and $\{j\}$. The number of conjugacy classes is therefore $2^{n-2}+3$.

(3) <u>The (generalized) quaternion group of order $2^n$ $(n > 3)$</u> has a presentation

$$Q = \langle x, y \mid x^{2^{n-2}} = y^2,\ y^4 = 1,\ y^{-1}xy = x^{-1} \rangle.$$

This group has a unique element of order 2.

<u>Conjugacy classes:</u> The elements of $Q - \langle x \rangle$ form two conjugacy classes, $y^Q$ and $(xy)^Q$, both of size $2^{n-2}$, consisting of elements of order 4. The elements of $\langle x \rangle$ of order $\geq 4$ lie in classes of size 2, and the number of such classes is $2^{n-2}-1$. The classes consisting of central elements are $\{1\}$ and $\{j\}$. The group $Q$ has $2^{n-2}+3$ conjugacy classes.

## IV *More on modules, quivers, Auslander-Reiten sequences*

This chapter contains material on periodic modules of symmetric algebras. Moreover, we determine all possible quivers of tame symmetric algebras with at most three simple modules. After that we investigate consequences of the graph structure of $\Gamma_s(\Lambda)$ for projective modules and other closely related modules.

### IV.1 *Modules and* $\Omega$, $\tau$

We assume throughout this section that $\Lambda$ is a symmetric algebra. We will collect some observations about $\Omega$-periodic modules of small period, especially simple modules, exploiting the non-singularity of the Cartan matrix, and consequences for the quiver of $\Lambda$. Moreover, we show that simple modules in 2-tubes must lie at the end.

IV.1.1 LEMMA *Assume that $M$ and $N$ are periodic modules of periods $m$, $n$ respectively. If $m$ and $n$ are coprime then for all $r, s \in \mathbb{Z}$ we have $\underline{Hom}_\Lambda(M, N) \cong \underline{Hom}_\Lambda(\Omega^r M, \Omega^s N)$.*

Proof: It is well-known that $\underline{\mathrm{Hom}}_\Lambda(M, N) \cong \underline{\mathrm{Hom}}_\Lambda(\Omega M, \Omega N)$. Thus the statement follows by induction, using the Chinese remainder theorem for $\mathbb{Z}$.

We shall apply this when $m = 4$ and $n = 3$. A typical example is as follows:

IV.1.2 LEMMA *Assume that $S_0$ and $S_1$ are simple periodic modules whose periods are coprime. Then $Ext^1(S_0, S_1) = 0 = Ext^1(S_1, S_0)$. In particular, in the quiver of $\Lambda$, there is no arrow joining the vertices $S_0$ and $S_1$.*

Proof: By IV.1.1, we have that $\mathrm{Ext}^1(S_0, S_1) \cong \underline{\mathrm{Hom}}_\Lambda(\Omega S_0, S_1) \cong \underline{\mathrm{Hom}}_\Lambda(S_0, S_1) = 0$.

IV.1.3 *Let $\Lambda$ be a symmetric algebra. If $M$ is a module satisfying $\Omega M \cong M$, then $soc\ M \cong top\ M$. In particular, if $\Lambda$ is basic and $M = \alpha\Lambda$ for an arrow $\alpha$ then $\alpha$ must be a loop.*

Proof: For modules of symmetric algebras having no projective summands, we have top $M \cong$ soc $\Omega M$.

<u>IV.1.4</u> LEMMA *Suppose $\Lambda$ is indecomposable and has a simple module $S$ such that $S \cong \tau S$. Then $\Lambda$ has only one simple module; moreover $\Lambda$ is uniserial and of finite type.*

Proof: Without loss of generality, $\Lambda$ is basic. Let $P = P(S)$. If $S \cong \tau S \cong \Omega^2 S$, then also $\Omega^{-1}S \cong \Omega S$, that is, $P/\mathrm{soc}\, P \cong \mathrm{rad}\, P$. Hence rad P has a simple top isomorphic to S, in other words, the quiver of $\Lambda$ consists of one vertex and a loop. Then $\Lambda \cong K[\alpha]/I$ for some ideal, and the statement follows from I.4.3.1.

We note that for IV.1.4 it is enough to assume that $\Lambda$ is self-injective.

<u>IV.1.5</u> LEMMA *Let $\Lambda$ be symmetric, and suppose $S$ is simple periodic, with projective cover $P$. Then $\mathrm{rad}\, P/\mathrm{rad}^2 P \cong \mathrm{soc}_2 P/\mathrm{soc}\, P$ in each of the following cases:*
*(a) $\Omega^3 S \cong S$;*
*(b) $\Omega^4 S \cong S$, and the Cartan matrix of $\Lambda$ is non-singular.*

Proof: (a) Assume that S has period 3. Then $\Omega^2 S \cong \Omega^{-1}S$, hence $\mathrm{rad}\, P/\mathrm{rad}^2 P \cong \mathrm{top}\, \Omega S \cong \mathrm{soc}\, \Omega^2 S \cong \mathrm{soc}\, \Omega^{-1} S \cong \mathrm{soc}_2 P/S$.
(b) Suppose $\Omega^4 S \cong S$, then there is an exact sequence
$0 \to S \to P \to Q_2 \to Q_1 \to P \to S \to 0$ where $Q_1$ and $Q_2$ are projective, with top $Q_1 \cong \mathrm{rad}\, P/\mathrm{rad}^2 P$ and soc $Q_2 \cong \mathrm{soc}_2 P/S$. By the exactness we have that $\underline{\dim}\, [P \oplus Q_1] = \underline{\dim}\, [P \oplus Q_2]$; and since the Cartan matrix is non-singular, it follows that $Q_1 \cong Q_2$.

<u>IV.1.5.1</u> (i) Let S correspond to vertex e of the quiver $Q$ of $\Lambda$ . Then the above observation shows that the number of arrows $f \to e$ is the same as the number of arrows $e \to f$, for any vertex f.
(ii) The proof of VI.1.5.(b) is a typical example of how we exploit that the Cartan matrix is non-singular.

<u>IV.1.6</u> LEMMA *Let $\Lambda$ be a symmetric algebra whose Cartan matrix is non-singular. Assume that $S_0$ is a simple periodic module of period 4, and that there is no arrow $0 \to i$ in the quiver $Q$ of $\Lambda$. Then there is no relation $\rho$ in $e_0 \Lambda e_i$ involving a path of*

*length $\leq 2$.*

Proof: Let $\alpha_1, \alpha_2, \ldots, \alpha_k$ be all arrows starting at 0, say $\alpha_j$ ends at vertex j where these vertices need not be distinct. We identify $\Omega^2 S_0$, that is $\Omega(\mathrm{rad}\ e_0\Lambda)$, with the module $\{ (x_1,x_2, \ldots,x_k) \in \bigoplus_{j=1}^{k} e_j\Lambda : \Sigma\, \alpha_j x_j = 0 \}$. Suppose there is a relation involving a path of length 2 between 0 and i involving a path of length 2, say starting with $\alpha_1$. Then there is an arrow $\beta$ ending at i and there are elements $x \in J^2$ and $y_j \in J$ such that $\alpha_1\beta + \alpha_1 x + \sum_{j\geq 2} \alpha_j y_j = 0$. It follows that $(\beta + x,\ y_2, \ldots y_k)$ lies in $\Omega^2 S_0$ but not in rad $\Omega^2 S_0$. Consequently $S_i$ occurs in top $\Omega^2 S_0$ and then also in soc $\Omega^3 S_0 \cong$ soc $\Omega^{-1} S_0 \cong \mathrm{soc}_2 P_0/S_0$. By IV.1.5, there is an arrow $0 \to i$, contrary to the hypothesis.

A similar argument, using IV.1.5(a), shows that also the following holds:

IV.1.7 LEMMA *Let $\Lambda$ be symmetric and $S_0$ a simple $\Lambda$-module of period 3. Assume that j is a vertex in the quiver of $\Lambda$ with $j \neq 0$. Then no path of length 2 between 0 and j occurs is any relation.*

IV.1.8 LEMMA *Let $\Lambda$ be basic and symmetric having a simple module $S_0$ of period 3.*

*(a) Suppose the quiver around the vertex 0 is of the form*

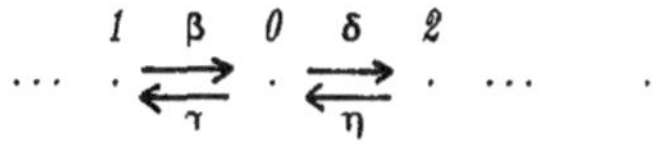


*Then we may assume that $\gamma\beta = \delta\eta$; and this is non-zero. Moreover, any relation of $\Lambda$ starting at vertex 0 is of the form $\gamma\beta\omega - \delta\eta\omega = 0$.*

*(b) Suppose the quiver around vertex 0 is of the form*

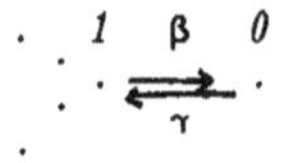


*Then we may assume that $\gamma\beta = 0$.*

Proof: (a) We identify $\Omega S_0$ with $\gamma\Lambda + \delta\Lambda$ and $\Omega^2 S_0$ with ker $\pi$ where $\pi$ is the projective cover $\pi$: $P_1 \oplus P_2 \to \gamma\Lambda + \delta\Lambda$ defined by $\pi(x, y) = \gamma x - \delta y$ (see I.6.5). By the hypothesis, ker $\pi \cong \Omega^{-1} S_0$; recall that $\Omega^{-1} S_0 \cong (\beta, \eta)\Lambda$. Consequently for some choice of $\beta, \eta$ we have that $(\beta, \eta)\Lambda$ = ker $\pi$, and the statements in (a) follow. Suppose

$\gamma\beta = 0$; then $(\beta, 0)$ lies in $(\beta,\eta)\Lambda$, which is not possible.

(b) Here we identify $\Omega S_0$ with $\gamma\Lambda$ and $\Omega^2 S_0$ with ker $\pi$ where $\pi$ is left multiplication with $\gamma$, $\pi$: $P_1 \to \gamma\Lambda$. Then ker $\pi \cong \Omega^{-1}S_0$, and this is isomorphic to $\beta\Lambda$, see I.6.5.

<u>IV.1.9</u> *Suppose $\Lambda$ is a basic algebra and $M$ is some $\Lambda$-module such that some simple module, $S = S(e_0)$ say, occurs only once in top $M$. Let $M = \Sigma\ \omega_i\Lambda$ where $\omega_i \notin$ rad $M$ and $\omega_i = \omega_i e_i$ for primitive idempotents $e_i$ (not necessarily distinct for $i \neq 0$). Now suppose $\omega_0' \in M$ - rad $M$ and $\omega_0' e_0 = \omega_0'$. Then $\omega_0'$ is of the form*

$$\omega_0' = \omega_0 z_0 + \sum_{i\neq 0} \omega_i z_i$$

*where $z_0$ is a unit in $e_0\Lambda e_0$ and for $i \neq 0$, $z_i \in$ rad $\Lambda$.*

Proof: We may write $\omega_0' = \Sigma\ \omega_i z_i$ such that $z_i \in e_i\Lambda e_0 \subseteq J$ for $0 \neq i$. Moreover, since also $M = \omega_0'\Lambda + \sum_{i\neq 0} \omega_i\Lambda$ we have $\omega_0 = \Sigma\ \omega_i u_i + \omega_0' u$ with $u \in e_0\Lambda e_0$. Substituting this gives an expression $\omega_0' = \omega_0' u z_0 + \sum_{i\neq 0} \omega_i r_i$; and we see that $uz_0$ is not nilpotent and hence is a unit since $e_0\Lambda e_0$ is local.

<u>IV.1.9.1</u> We note an elementary fact. Suppose $\Lambda$ is symmetric and basic, and let $\omega_1$ and $\omega_2 \in e_i\Lambda e_j$ where $e_i$ and $e_j$ are primitive idempotents of $\Lambda$. Any isomorphism $\omega_1\Lambda \to \omega_2\Lambda$ is induced by left multiplication by a unit of $e_i\Lambda e_i$.

<u>IV.1.10</u> LEMMA *Suppose $\Lambda$ is a symmetric indecomposable algebra of infinite type. Then any simple $\Lambda$-module with $\tau^2 S \cong S$ must lie at the end of a 2-tube.*

Proof: Suppose S is simple, and let $\Theta$ be the component containing S. Then by II.1.4, we have that $\tau S \neq S$, and therefore $\Theta$ is a 2-tube (I.8.6.1), and S must have $\Omega$-period 4 (I.8.4.1); in particular $\Omega\Theta \neq \Theta$.

(1) <u>There is a simple module in the first row of $\Theta \cup \Omega\Theta$:</u> Suppose not; then there is no projective attached to the first row of $\Theta$ or $\Omega\Theta$, and the length of any module M in the second row ( = $|X| + |\tau X|$ for some X) is > 1. In particular, there is no simple module in the second row of $\Theta \cup \Omega\Theta$ and then also no projective attached to the second row. This implies that the lengths of the modules in the third row are greater again

and hence > 1, and so on. By induction, there is no simple module in $\Theta \cup \Omega\Theta$, contrary to our hypothesis.

Suppose T is simple in the first row; we will show that this is the only simple module in $\Theta \cup \Omega\Theta$.

(2) <u>There is no other simple module in the first row</u>: The modules in the first row ar T, $\tau T$ and $\Omega T$, $\Omega^{-1}T$. Clearly $\Omega^{\pm 1}T$ are not simple (see I.3.8). If (2) does not hold then $\tau T$ is simple, say = V. Then it follows that $\Omega^{-1}T \cong \Omega V$ and $\Omega T \cong \Omega^{-1}V$, and therefore top $\Omega T \cong V$ and top $\Omega V \cong T$. It follows that the quiver of $\Lambda$ has a connected component $\overset{T}{\cdot} \rightleftarrows \overset{V}{\cdot}$ which is an algebra of finite type, by II.8.2; a contradiction to the hypothesis.

(3) <u>There is no other simple module in the second row of $\Theta \cup \Omega\Theta$</u>: The lengths of the modules in the second row of $\Theta$ are $|T| + |\tau T| \geq 3$, by (2); and for $\Omega\Theta$ the lengths are $|\mathrm{rad}\, P_T/T|$ and $|\Omega^{-1}T| + |\Omega T|$. Suppose (3) is false, then we must have that rad $P_T/T$ is simple, $\cong$ V say. Moreover top $\Omega^2 T \cong V$ and top $\Omega^{-2}T \cong V$, and $\tau V \cong \Omega^2 T \cong \Omega^{-1}T$ and this module has a simple socle and top isomorphic to V. Now, there is an almost split sequence $0 \to \tau T \to \Omega V \to T \to 0$. Since $\tau T$ is not simple, it follows from the AR-property that top $\Omega V \cong \mathrm{top}(\tau T) \oplus T \cong V \oplus T$. Hence the quiver of $\Lambda$ has a connected component

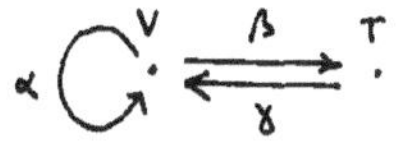


Since $P_T \cong \mathcal{U}(T, V, T)$ we deduce $\gamma\alpha = 0$. Moreover, since $\Lambda$ is symmetric, the multiplicity of T as a composition factor in $P_V$ is 1 and hence, in the basic algebra, T does not occur in $\alpha\Lambda$. We deduce that $\alpha\beta = 0$; and $\Lambda$ is a special biserial algebra. Then $\Lambda$ does not satisfy the hypothesis on $\Gamma_s(\Lambda)$ (II.7.1 and 2), a contradiction.

It follows that there is no projective in the second row of $\Theta \cup \Omega\Theta$; and the lengths of the modules in the third rows are determined and > 1. Inductively, there is no other simple module in $\Theta \cup \Omega\Theta$, and S = T.

<u>IV.1.11</u> LEMMA *Suppose $\Lambda$ is symmetric and basic, and assume that the quiver of $\Lambda$ contains*

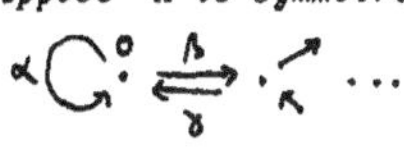

*Then* $\alpha\beta = 0$ *if and only if* $\gamma\alpha = 0$. *Moreover, if this happens then rad* $e_0\Lambda/soc\ e_0\Lambda$ *is decomposable.*

Proof: If $\gamma\alpha = 0$ then $\Lambda\alpha$ is spanned by $\alpha$, $\alpha^2$, ... .Consider the map $.\beta : \Lambda\alpha \to \Lambda\beta$. Since $soc(\Lambda\beta) \cong S_1$ but $e_1(\Lambda\alpha) = 0$ it follows that $\Lambda\alpha\beta = 0$. Similarly, $\alpha\beta = 0$ implies $\gamma\alpha = 0$.

Let $x \in \alpha\Lambda \cap \gamma\Lambda$, then $\beta x = \beta\alpha x_1 = 0$ and $\alpha x = \alpha\gamma x_2 = 0$ and consequently $Jx = 0$, and $\alpha\Lambda \cap \gamma\Lambda \subseteq soc(e_0\Lambda)$. Then rad $e_0\Lambda/S_0$ is a direct sum.

## IV.2 *Quivers of tame symmetric algebras with at most three simple modules*

Let $\Lambda$ be a symmetric algebra which is basic, with quiver $Q$ and radical J. We assume that in addition $\Lambda$ is tame and connected and has at most three simple modules. We will determine the possibilities for $Q$.

IV.2.1 Assume that $\Lambda$ is tame, with one simple module. Then the quiver of $\Lambda$ is of the form  , see I.4.3(a) and I.10.10.(a) (and I.4.6).

IV.2.2 *Assume that $\Lambda$ is tame, symmetric and connected, with two simple modules. Then the quiver of $\Lambda$ is one of the following:*

(2*A*) 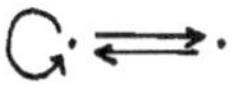 (2*B*) 

(2*C*)  (2*D*) 

Proof: The quiver $Q$ is connected and has two vertices. Moreover, $Q$ does not have a 'sink' or 'source', by I.6.1. Any algebra is wild if its quiver contains

 or  or  or a dual of these

(I.10.8).If there is no loop attached to some vertex of $Q$ then  $\subseteq$ $Q$ $\subseteq$ 

An algebra with quiver  is of finite type (II.8.2). This leaves the possibilities (2*C*) and (2*D*) for $Q$.

If $Q$ has one loop then it can only have the form (2*A*). Otherwise, $Q$ has two loops, and

 $\subseteq$ $Q$ $\subseteq$ 

The smaller quiver is excluded by I.6.1.

IV.2.3 LEMMA *Let $\Lambda$ be tame, symmetric and connected with three simple modules, with quiver $Q$. Assume that not all vertices of $Q$ are joined by arrows. Then $Q$ is one of* *the following quivers:*

(3$\mathcal{A}$) 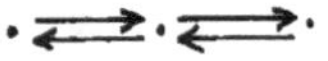

(3$\mathcal{B}$) 

(3$\mathcal{C}$) 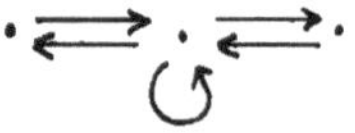

(3$\mathcal{D}$) 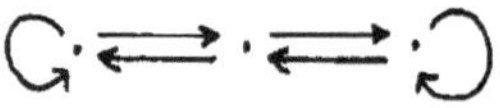

Proof: The quiver is connected, with three vertices and satisfies I.6.1. Consequently, by our hypothesis, it must contain a quiver of the form (3$\mathcal{A}$). Moreover, any other arrow of $Q$ can only be a loop, since $Q$ does not contain

or its dual (I.10.8). Similarly, $Q$ does not contain a double loop. This leaves the stated possibilities.

<u>IV.2.4</u> LEMMA *Let* $\Lambda$ *be tame, symmetric and connected with three simple modules, with quiver* $Q$. *Assume that any two vertices of* $Q$ *are joined by some arrow. Then* $Q$ *is one of the following quivers or their duals:*

(3$\mathcal{E}$) 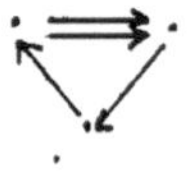

(3$\mathcal{F}$) 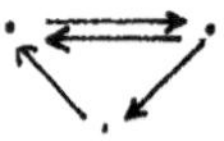

(3$\mathcal{G}$) 

(3$\mathcal{H}$) 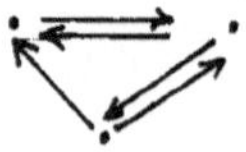

(3$\mathcal{I}$) 

(3$\mathcal{K}$) 

(3$\mathcal{L}$) 

(3$\mathcal{M}$) 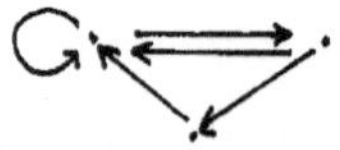

(3$\mathcal{N}$) 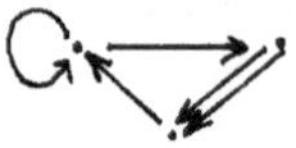

(3$\mathcal{O}$) 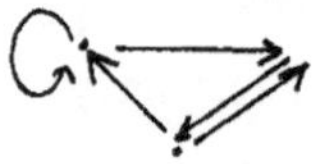

(3$\mathcal{P}$) 

(3$\mathcal{Q}$) 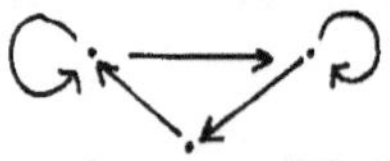

(3$\mathcal{R}$) 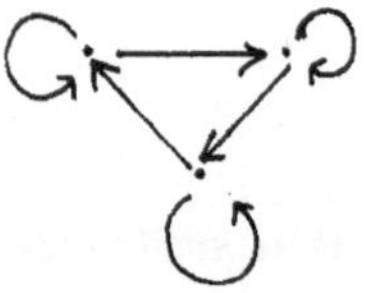

Proof: By I.6.1 and the hypothesis, $\mathcal{Q}$ must contain 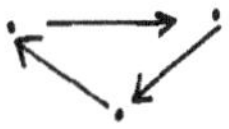
The number of arrows of $\mathcal{Q}$ is bounded since $\mathcal{Q}$ cannot contain a double loop, or a triple arrow.

We consider first quivers without loops. Here we use the fact that $\mathcal{Q}$ does not contain 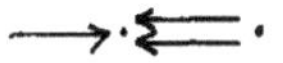 or its dual (I.10.8).

If $\mathcal{Q}$ has a double arrow then  $\subseteq$ $\mathcal{Q}$ $\subseteq$ ; consequently $\mathcal{Q}$ is one of ($3\mathcal{E}$, $\mathcal{G}$, $\mathcal{I}$). Now assume that $\mathcal{Q}$ has no double arrows. Then

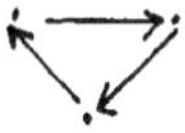 $\subseteq$ $\mathcal{Q}$ $\subseteq$ 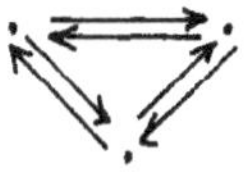

The smaller possibility is excluded since it only occurs for algebras of finite type (II.8.2). Hence $\mathcal{Q}$ is one of te quivers ($3\mathcal{F}$, $\mathcal{H}$, $\mathcal{K}$).

Now assume that $\mathcal{Q}$ has exactly one loop, say  $\subseteq$ $\mathcal{Q}$.
Then we exploit the fact that $\mathcal{Q}$ does not contain 
or its dual (I.10.8). If $\mathcal{Q}$ has $\leq$ five arrows then it is one of ($3\mathcal{L}$, $\mathcal{M}$, $\mathcal{N}$, $\mathcal{O}$) or a dual. The only quiver with six or more arrows including just one loop which is not wild is ($3\mathcal{P}$).

Now suppose that $\mathcal{Q}$ has precisely two loops. Then 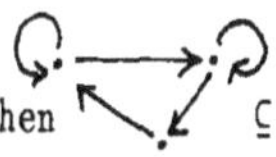 $\subseteq$ $\mathcal{Q}$. Any such quiver with two loops but more than two arrows is wild, by I.10.8; this leaves ($3\mathcal{Q}$).

Now assume that $\mathcal{Q}$ has three loops; then $\mathcal{Q}$ must be of the form ($3\mathcal{R}$). This is now the complete list.

## IV.3 *Exploiting the AR-quiver*

We assume here that $\Lambda$ is a symmetric algebra.

Our main object of interest are the indecomposable modules P, or equally well the indecomposable summands of rad P/ soc P. Let

$$S = \text{soc } P \quad \text{and} \quad H = \text{rad } P/S.$$

By I.7.7, the indecomposable summands of H are precisely the predecessors of P/S in the stable AR-quiver $\Gamma_s(\Lambda)$.

In order to study the structure of H we will now investigate the possibilities for the position of P/S in the stable AR-quiver $\Gamma_s(\Lambda)$. Equivalently, we may look at S; since $S = \Omega(P/S)$ and $\Omega$ induces a graph isomorphism of $\Gamma_s(\Lambda)$.

Let $\Theta$ be the component containing P/S. Then we have by I.7.7 the following:

IV.3.1 *If P/S has precisely two predecessors in $\Theta$ then rad $P/S \simeq U \oplus V$ where U and V are indecomposable and non-zero. Moreover, U and V are non-isomorphic if $\Theta$ does not have multiple arrows.*

We shall deal with algebras where $\Gamma_s(\Lambda)$ has multiple arrows at the end of this chapter in IV.3.8; this leads to a very special case. Therefore we will assume everywhere else that $\Gamma_s(\Lambda)$ does not have multiple arrows.

IV.3.2 LEMMA *Assume that $\Theta \simeq \mathbb{Z}D_\infty$ and that S has three predecessors in $\Theta$. Then there are two other simple modules $S_1$ and $S_2$ such that $P(S_i) = U(S_i, S, S_i)$ for $i = 1, 2$. In particular, $\Lambda$ has three simple modules.*

Proof: By the hypothesis, there are two AR-sequences

$$0 \to \tau X_i \to S \oplus Q_i \to X_i \to 0 \quad (i = 1, 2)$$

where $Q_i$ is either zero or indecomposable projective, and where $X_1$ and $X_2$ are non-isomorphic. Since the sequences are exact and S is simple, the modules $Q_i$ must be non-zero. Consequently the AR-sequences are standard (I.7.7). Let $S_i = \text{soc } Q_i$; then $S \cong \text{rad } Q_i/\text{soc } Q_i$ and $Q_i$ is determined. Moreover, since $X_1 \neq X_2$ and $S_i \cong \Omega^{-1}X_i$ we know that $S_1$ and $S_2$ are non-isomorphic. Furthermore, $S_1$ and $S_2$ each have only one predecessor in $\Gamma_s(\Lambda)$ but S does not, hence $S \neq S_1$ or $S_2$.

<u>IV.3.3</u> *Assume that* $\Theta \simeq \mathbb{Z}D_\infty$ *and that P/S has one predecessor in* $\Theta$. *Then one of the following holds for H = rad P/S:*

*(a) There is an AR-sequence of the form* $0 \to V \to H \to U \to 0$.

*(b) There is a simple module* $T \neq S$ *such that* $H \cong \operatorname{rad} P_T/T$.

Proof: The hypothesis implies that there is an AR-sequence

$$0 \to \tau U \to H \oplus Q \to U \to 0$$

where either Q is zero or Q is indecomposable projective; moreover $U \neq P/S$. In the first case, we obtain (a) with $V = \tau U$, and otherwise (b) follows, with $T = \operatorname{soc} Q$.

<u>IV.3.4</u> *Assume that* $0 \to V \xrightarrow{\lambda} H \xrightarrow{\kappa} U \to 0$ *is an AR-sequence as in IV.3.3(a). Then*

*(a) If U is not simple then* $\kappa$ *induces an isomorphism* $\operatorname{soc} H \cong \operatorname{soc} V \oplus \operatorname{soc} U$; *and if U is simple then* $\operatorname{soc} H \cong \operatorname{soc} V$.

*(b) If V is not simple then* $\lambda$ *induces an isomorphism* $\operatorname{top} H \cong \operatorname{top} V \oplus \operatorname{top} U$; *and if V is simple then* $\operatorname{top} H \cong \operatorname{top} U$.

This is a consequence of the AR-property I.7.3. Similarly, we have:

<u>IV.3.4.1</u> *With the notation of IV.3.4, if T is a proper simple submodule of V then* $H/\lambda(T) \cong V/T \oplus U$.

Now we shall study the situation of IV.3.3(a) where U or V is simple.

<u>IV.3.5</u> LEMMA *Assume that the simple module* $S_0$ *lies at the end of a* $\mathbb{Z}D_\infty$ *-component, with one predecessor, and that* $U_0$ *and* $V_0$ *are defined as in IV.3.3(a).*

*(a) If* $U_0$ *is simple, with* $U_0 \cong S_i$, *then* $\Omega^{-1}H_0 \cong H_i$, *and there is an AR-sequence*

$$0 \to S_0 \to H_i \to \tau^{-1}S_0 \to 0, \text{ so that } S_0 \cong V_i.$$

*(b) If* $V_0$ *is simple, with* $V_0 \cong S_i$ *then* $\Omega H_0 \cong H_i$, *and there is an AR-sequence*

$$0 \to \tau S_0 \to H_i \to S_0 \to 0, \text{ so that } S_0 \cong U_i.$$

*In both cases,* $S_0 \neq S_i$.

Proof: (a) Since $H_0$ is the unique predecessor of $U_0$ in $\Gamma_s(\Lambda)$, we know that $\Omega^{-1}H_0$ is the unique predecessor of $\Omega^{-1}U_0$ in $\Gamma_s(\Lambda)$. Hence if $U_0 \cong S_i$, then it follows from I.8.4.1 that $\Omega^{-1}H_0 \cong H_i$. As a consequence, we obtain that $H_i$ is the unique predecessor of $S_0$ in $\Gamma_s(\Lambda)$. Hence there is an AR-sequence $0 \to S_0 \to H_i \oplus P \to \tau^{-1}S_0 \to 0$ where P is indecomposable projective or zero. If P were non-zero then we would have, by I.7.7, that $S_0 \cong \text{rad } P$ and $\tau^{-1}S_0 \cong P/\text{soc } P$ and consequently $S_0 \cong \tau^{-1}S_0$. This is not possible in a $\mathbb{Z}D_\infty$-component, and therefore $P = 0$.
The proof of (b) is similar; and the last statement follows from the fact that $S_0$, $U_0$ and $V_0$ represent distinct vertices of the AR-quiver.

We note that IV.3.5 is true more generally for self-injective algebras.

<u>IV.3.6</u> LEMMA *Assume that Λ is symmetric with at most 2 simple modules, and that the Cartan matrix of Λ is non-singular. Let $P_0$ be an indecomposable projective module such that $H_0$ satisfies IV.3.3(a). Then $U_0$ and $V_0$ are not simple.*

Proof: Suppose that $U_0$ is simple; then $U_0 \neq S_0$ since these modules correspond to distinct vertices of $\Gamma_s(\Lambda)$. Say $U_0 = S_1$, then $S_0$ and $S_1$ are all simple Λ-modules. Since $\Omega^2U_0 \cong V_0$, there is an exact sequence $0 \to V_0 \to Q \to P_1 \to S_1$ 0 where $P_1 = P(S_1)$, and Q is projective, with $\text{soc } Q \cong \text{soc } V_0$.

We have that $\text{soc } V_0 \cong \text{soc } H_0$ by IV.3.4, and the socle factors correspond to the arrows in the quiver of Λ terminating at vertex 0. So in any case, $|\text{soc } V_0| \leq 2$; and since the quiver contains an arrow $1 \to 0$, we know that $S_1 \subseteq \text{soc } V_0$, and therefore $P_1$ is a summand of Q.

Moreover, $\text{top } Q \cong \text{top } \Omega S_1$, so the top factors correspond to the arrows in the quiver starting at 1. There is an arrow $1 \to 0$, therefore $S_0 \subseteq \text{top } \Omega S_1$, and $P_0$ is a summand of Q. Thus $Q \cong P_1 \oplus P_0$, and from the exactness we deduce $0 = \underline{\dim}\, S_1 - \underline{\dim}\, P_1 + \underline{\dim}\, P_1 + \underline{\dim}\, P_0 - \underline{\dim}\, V_0 = \underline{\dim}\, S_1 + \underline{\dim}\, P_0 - \underline{\dim}\, V_0 = \underline{\dim}\, 2S_1 + \underline{\dim}\, 2S_0$, a contradiction.

<u>IV.3.7</u> LEMMA *Suppose Λ is a symmetric algebra such that the only ends of $\Gamma_s(\Lambda)$ are*

*at tubes and at $\mathbb{Z}D_\infty$-components. Assume the quiver of $\Lambda$ contains $\vdots\ 0 \rightleftarrows 1$ such that no other arrow starts or terminates at vertex 1. Then one of the following holds:*

*(a) $S_1$ is periodic.*

*(b) The quiver of $\Lambda$ is of the form $2 \rightleftarrows 0 \rightleftarrows 1$, and moreover $H_1 \cong H_2$.*

Proof: The module $H_1$ is indecomposable. So if $S_1$ is not periodic then $S_1$ must satisfy IV.3.3. If there were an AR-sequence $0 \to V_1 \to H_1 \to U_1 \to 0$ then we would have that $U_1 \cong S_1$ and $V_1 \cong S_1$, by IV.3.4. This is not possible since $U_1$ and $V_1$ correspond to distinct vertices of the AR-quiver.

Hence IV.3.3(b) must hold, and there is a projective module $P_2 \neq P_1$ with $(\mathrm{rad}\ P_2)/S_2 \cong H_1$, and then the quiver contains $2 \rightleftarrows 0\ \vdots$ and no other arrow starts or ends at vertex 2. Thus (b) holds.

IV.3.8 *Symmetric algebras with $A_{12}$-components.*

Suppose that $\Theta$ is a stable AR-component of a symmetric algebra of type $\mathbb{Z}\tilde{A}_{12}$ (see I.8.5).

IV.3.8.1 LEMMA *Let $W$, $X$ lie in $\Theta$ such there are irreducible maps $W \to X$.*

*(a) If $|W| > |X|$ then there is an AR-sequence*

$$0 \to \tau X \to W \oplus W \to X \to 0, \text{ and moreover } |\tau X| > |W|.$$

*(b) If $|W| < |X|$ then there is an AR-sequence*

$$0 \to W \to X \oplus X \to \tau^{-1} W \to 0, \text{ and moreover } |X| < |\tau^{-1} W|.$$

Proof: (a) The AR-sequence of X is of the form

$0 \to \tau X \to W \oplus W \oplus P \to X \to 0$ where either $P = 0$, or $P$ is indecomposable projective. Suppose that $P \neq 0$. Then $X \cong P/\mathrm{soc}\ P$ and $\tau X \cong \mathrm{rad}\ P$ (I.7.7); consequently $|X| = |\tau X| = |P| - 1$. By the exactness we have that $2|W| = |X| - 1$ and $|W| < |X|$, a contradiction. Hence $P = 0$, and $|\tau X| = 2|W| - |X| = |W| + (|W| - |X|) > |W|$.

Part (b) is proved similarly.

IV.3.8.2 LEMMA *Let* $\Theta \simeq \mathbb{Z}\tilde{A}_{12}$. *Then*

*(a) The modules in* $\Theta$ *can be indexed by integers such that there are irreducible maps* $X_{k-1} \to X_k$, *and* $|X_k| < |X_m|$ *if and only if* $k < m$.

*(b) There is a unique projective attached to* $\Theta$ *which lies between* $X_{-1}$ *and* $X_1$.

Proof: Choose some $X_0$ in $\Theta$ of minimal length. Now label the modules in $\Theta$ according to the first condition in (a). There are irreducible maps $X_{-1} \to X_0$ and $X_0 \to X_1$. These must be epi and mono respectively, by the minimality of $|X_0|$. Now (a) follows from IV.3.8.1.

It follows also that there are no projectives $P \to X_{-k}$ or $X_k \to P$ for any $k \geq 1$. Now, there must be some projective, P say, attached to $\Theta$ (see I.8.13), consequently there must be an AR-sequence $0 \to X_{-1} \to P \oplus 2X_0 \to X_1 \to 0$. This completes the proof of the Lemma.

IV.3.8.3 THEOREM *Suppose* $\Lambda$ *is symmetric and connected such that* $\Gamma_s(\Lambda)$ *has a non-periodic component* $\Theta$ *with multiple arrows in which every module has two predecessors. Then* $\Theta \simeq \mathbb{Z}\tilde{A}_{12}$ *and moreover:*

*(a)* $\Lambda$ *has one or two simple modules,* $S$ *or* $S_1$, $S_2$ *say, with projective covers* $P$ *or* $P_1$, $P_2$ *respectively, and one of the following holds:*

*(i)* $rad\, P_i/S_i \simeq S_j \oplus S_j$ *where* $\{i, j\} = \{1, 2\}$ *and* $\Theta \neq \Omega(\Theta)$.

*(ii)* $rad\, P/S \simeq S \oplus S$; *and* $\Theta = \Omega(\Theta)$.

*Moreover, in (i), the Cartan matrix of* $\Lambda$ *is singular.*

*(b)* $\Theta$ *and* $\Omega(\Theta)$ *are the only non-periodic components of* $\Lambda$.

Proof: Clearly $\Theta \cong \mathbb{Z}\tilde{A}_{1,2}$. It follows from IV.3.8.2 that there is a projective attached to $\Theta$. Hence there must be a simple module, S say, lying in the component $\Delta := \Omega(\Theta)$. Now, $\Delta$ is again of type $\tilde{A}_{12}$, so it has the properties described in IV.3.8.2 above. With the notation there, it follows that $S = X_0$, and we have astandard AR-sequence $0 \to X_1 = rad\, P \to S \oplus S \oplus P \to P/soc\, P \to 0$

where P is some indecomposable projective $\Lambda$-module. Let $T = P/rad\, P$; since there are two irreducible maps $S \to \Omega^{-1}T$, we deduce from I.8.4.1 that there are also two irreducible maps $\Omega S \to T$. Now, by the hypothesis, the AR-sequence of $\Omega S$ has only two

indecomposable middle terms, therefore there is an AR-sequence
$0 \to \Omega(S) \to T \oplus T \oplus P_S \to \Omega^{-1}(S) \to 0$. If $S \neq T$ we obtain (i) and otherwise (ii).

(b) From the structure of the projectives we see that $\Lambda/\mathrm{soc}\ \Lambda$ is special biserial. III s easy to see that there is a special biserial symmetric algebra $\Lambda_1$ with $\Lambda_1/\mathrm{SOC}\ \Lambda_1 \cong \Lambda/\mathrm{soc}\ \Lambda$ Then $\Gamma_s(\Lambda) = \Gamma_s(\Lambda_1)$ (see I.8.11), and (b) follows using II.7.1.

IV.3.9 Suppose $\Lambda$ is an algebra as in the theorem; then trivially $\Lambda/\mathrm{soc}\ \Lambda$ is special biserial. It follows therefore from II.7.1 (and I.8.11) that $\Gamma_s(\Lambda)$ consists of one component $\mathbb{Z}\tilde{A}_{12}$ and otherwise only 1-tubes. (Note that in the notation used in II.7.1, a component isomorphic to $\mathbb{Z}\tilde{A}_{1,1}$ is the same as a component with tree class $\tilde{A}_{12}$ as is used in IV.3.8.3, I.8.5, which appears to be introduced by [DR]).

## IV.4 *Ends of components*

The problem we will study now is to show that certain modules lie "at the end" of some component. We recall that the module $M$ lies at the end of the component $\Theta$ of $\Gamma_s(\Lambda)$ if $M$ has either one or three predecessors in $\Theta$.

In this section we assume that $\Lambda$ is an arbitrary finite-dimensional algebra. The next result is due to [BR].

IV.4.1 DEFINITION [BR] *Let $\gamma \in eJf$ where $e$, $f$ are primitive idempotents, and let $\pi$: $e\Lambda/\gamma J \to e\Lambda/\gamma\Lambda$ be the canonical epimorphism. The element $\gamma$ is non-supportive if for any non-zero summand $C$ of $eJ/\gamma\Lambda$ we have that $C \cap \pi(\mathrm{soc}\ e\Lambda/\gamma J) \neq 0$.*

Non-supportive elements will provide us with modules lying at ends of AR-components.

IV.4.2 THEOREM *Let $\Lambda$ be a finite-dimensional algebra. If $0 \neq \gamma$ is non-supportive, $\gamma \in eJf$, then $e\Lambda/\gamma\Lambda$ has a unique predecessor in $\Gamma_s(\Lambda)$.*

Proof: Put $X = e\Lambda/\gamma\Lambda$, and let $0 \to Y \xrightarrow{\mu} N \xrightarrow{\epsilon} X \to 0$ be the AR-sequence ending in X; we have to show that N is indecomposable.

Then Y has a simple socle (in general, top $\Omega X \cong$ soc $\tau X$, see e. g. [$G_2$]3.1). If X is simple then soc N ≅ soc Y, and then N is indecomposable. So we assume that X is not simple.

Let $N = \bigoplus_{i=1}^{t} N_i$ with all $N_i$ indecomposable, $\mu = [\mu_1, \ldots, \mu_t]^T$ and $\epsilon = [\epsilon_1, \ldots, \epsilon_t]$ where $\mu_i: Y \to N_i$ and $\epsilon_i: N_i \to X$. Since soc Y is simple, one of the $\mu_i$ must be mono, say $\mu_1$. Let $C = \bigoplus_{i=2}^{t} N_i$, and set $p = [\mu_2, \ldots \mu_t]^T: Y \to C$ and $q = [\epsilon_2, \ldots \epsilon_t]: C \to X$. With $\mu_1$ also q is mono. (We have that q is irreducible, and $|C| < |X|$.) Since $\epsilon$ is epi and X has a simple top, it follows that $\epsilon_1$ must be epi; and then also p is epi. Since soc Y is simple, p vanishes on soc Y.

Let $\pi : e\Lambda/\gamma J \to X$ be the canonical projection. This is not a split epi (since $e\Lambda/\gamma J$ is indecomposable), therefore by the AR-property there is a map $[\pi_1, \pi_1']^T: e\Lambda/\gamma J \to N_1 \oplus C$ with $\pi = \epsilon_1\pi_1 + q\pi_1'$. Since $\pi$ is surjective and q maps into rad X, we have that $\epsilon_1\pi_1$ must be surjective.

(1) <u>We may assume that $\pi = \epsilon_1\pi_1$:</u> Let $\iota$: ker $\pi = \gamma\Lambda/\gamma J \to e\Lambda/\gamma J$ be the inclusion. Since $0 = \pi\iota = (\epsilon_1\pi_1 + q\pi_1')\iota$ it follows that ker $\pi$ maps into ker $[\epsilon_1, q]$. Hence there is $\pi'$: $e\Lambda/\gamma J \to Y$ with $[\mu_1, p]^T\pi' = [\pi_1, \pi_1']^T\iota$. Now, ker $\pi$ is simple, and p vanishes on soc Y, so $\pi_1'\iota = p\pi' = 0$ and $\epsilon_1\pi_1\iota = (\pi - q\pi_1')\iota = 0$ and ker $\pi \subseteq \ker(\epsilon_1\pi_1)$. Then we get equality, by the lengths, and the sequence

$0 \to \gamma\Lambda/\gamma J \to e\Lambda/\gamma J \to X \to 0$ is exact. Then there is an automorphism $\eta$ of X such that $\pi = \eta\epsilon_1\pi_1$. (Define $\eta(x) = \pi(a)$ if $a \in (\epsilon_1\pi_1)^{-1}(x)$.) Now let $\epsilon_1' = \eta\epsilon_1$ and $q' = \eta q$. Since $\eta$ is an automorphism, the sequence $0 \to Y \to N_1 \oplus C \to X \to 0$ is also an AR-sequence. Note that $\epsilon_1'$ is epi, and q' is mono. Moreover, $\pi = \epsilon_1'\pi_1$. This proves (1).

If $f : M \to M'$ is a map we denote by soc f the restriction of $f$ : soc M → soc M'. Taking socles is always left exact, so there is an exact sequence

$$0 \to \text{soc } Y \xrightarrow{\begin{bmatrix}\text{soc } \mu_1 \\ 0\end{bmatrix}} \text{soc } N_1 \oplus \text{soc } C \xrightarrow{[\text{soc } \epsilon_1,\ \text{soc } q]} \text{soc } X.$$

Since we started with an AR-sequence and soc X $\subsetneq$ X it follows that [soc $\epsilon_1$, soc q'] is also epi. This shows that soc X = $\epsilon_1$(soc $N_1$) ⊕ q(soc C). Since $\pi = \epsilon_1\pi_1$ we see that $\pi(\text{soc } e\Lambda/\gamma J) \subseteq \epsilon_1(\text{soc } N_1)$ and hence for any indecomposable summand $C'$ of $qC$ we have that $C' \cap \pi(\text{soc } e\Lambda/\gamma J) = 0$. On the other hand, q is an irreducible mono, with image $\subseteq$ rad $X = eJ/\gamma\Lambda$, and therefore $qC$ is a direct summand of $eJ/\gamma\Lambda$. We assumed that $\gamma$ is non-supportive. So it follows that $qC = 0$ and $C = 0$.

<u>IV.4.3</u> LEMMA *Any element $\gamma \in eJf - eJ^2f$ is non-supportive.*

Proof: Let $\pi: e\Lambda/\gamma J \to e\Lambda/\gamma\Lambda$ be the canonical epi. We claim that $eJ/\gamma J \cong eJ/\gamma\Lambda \oplus \gamma\Lambda/\gamma J$, induced by $\pi$: Let $\pi'$ be the restriction of $\pi$ to $eJ/\gamma J$. Since $\gamma\Lambda \subseteq eJ$, we get an exact sequence $0 \to \gamma\Lambda/\gamma J \to eJ/\gamma J \xrightarrow{\pi'} eJ/\gamma\Lambda \to 0$. The simple module $\gamma\Lambda/\gamma J$ is not contained in $J(eJ/\gamma J) = \text{rad}(eJ/\gamma J)$ since $\gamma \notin J^2$. Hence the sequence splits. It follows that with this identification, $\pi(\text{soc } eJ/\gamma J) = \text{soc}(eJ/\gamma\Lambda)$ which has non-zero intersection with any non-zero summand $C$ of $eJ/\gamma\Lambda$.

## IV.5 *Ends of components II*

In this section, $\Lambda$ is a self-injective algebra.

We shall now define a class of modules which are important for the study of indecomposable projective modules. Then we will see that they lie at ends of AR-components.

IV.5.1 *Definition of modules $\tilde{U}$, $\underset{\sim}{V}$, [and $\tilde{V}$, $\underset{\sim}{U}$ ].*

Let P be indecomposable projective, with soc P ≅ S.

(a) Suppose rad P/S ≅ U ⊕ V as in IV.3.1. Then we have an embedding U ⊕ V → P. We define $\tilde{U}$ := (P/S)/V and $\underset{\sim}{V}$ := $\Omega\tilde{U}$. Similarly, let $\tilde{V}$ := (P/S)/U and $\underset{\sim}{U}$ := $\Omega\tilde{V}$.

(b) Suppose P/S satisfies VI.3.3 (a); then we have an embedding

V → P/S. We now take $\tilde{U}$ := (P/S)/V and $\underset{\sim}{V}$ := $\Omega\tilde{U}$. Here $\underset{\sim}{U}$ and $\tilde{V}$ are not defined.

(c) Assume that U and V are as in IV.3.3 (a). There is an extension of U by $\mathrm{soc}_2(\underset{\sim}{V})$. We denote such a module by < U >.

We note that these modules are not unique in general.

We shall now prove that the modules $\tilde{U}$, $\underset{\sim}{V}$ (and $\underset{\sim}{U}$, $\tilde{V}$) lie at ends of components. In the special case when $\underset{\sim}{V}$ or $\underset{\sim}{U}$ is cyclic we know this already from IV.4.2 since in this case the module generated by a non-supportive element.

IV.5.2 DEFINITON (a) We say that the module M is "*tall*" if it has the following property: Whenever there is an inclusion j: M → M' where M' is indecomposable and non-projective then m ∉ rad M implies j(m) ∉ rad M'.

(b) We say that the module M is "*deep*" provided any epimorphism $\pi$: X → M with X indecomposable non-projective has the property that $\pi$(soc X) = soc M.

This definition is motivated by the following observations:

IV.5.3 LEMMA Let 0 → X → M → *W* → *0 be an AR-sequence . Then M is not isomorphic to rad Q/soc Q for Q indecomposable projective* if *either X is tall*, *or W is deep.*

Proof: Suppose, for example, that there is some indecomposable projective $\Lambda$-module $Q$ with $\operatorname{soc} Q = T$ such that $M \cong \operatorname{rad} Q/T$. Then there is a monomorphism $j: X \to Q/T$. Now, $Q/T$ has a simple top and $X$ is tall. Consequently, $Q/T$ is projective, a contradiction.

<u>IV.5.4</u> LEMMA *(i) Let $U$, $V$ be as in IV.3.3(a). If $\operatorname{soc} H \cong \operatorname{soc} U \oplus \operatorname{soc} V$ then $U$ is deep. If $\operatorname{top} H \cong \operatorname{top} U \oplus \operatorname{top} V$ then $V$ is tall.*
*(ii) If $\alpha$ is an arrow then $\alpha\Lambda$ is tall.*
*(iii) A tall module with a simple top must be generated by an arrow.*

Proof: We will prove (ii); for (i), one uses the same idea (see p.119 in [$E_5$]). Suppose $\alpha\Lambda$ is not tall. Then there is an inclusion $j: \alpha\Lambda \to M'$ where $M'$ is indecomposable non-projective and $j(\alpha)$ lies in rad $M'$; say $j(\alpha) + u = \omega x$ for $\omega$, $u \in M'$ and $x \in J$. Since $\alpha$ is an arrow, we have an inclusion $\iota: \alpha\Lambda \to P$ where $P$ is indecomposable projective (= injective). Thus there is some $\varphi \in \operatorname{Hom}(M', P)$ such that $\varphi j = \iota$. Let $\varphi(\omega) = ez$ for $z \in \Lambda$. We have $ezx = \varphi(\omega x) = \varphi[j(\alpha) + u] = \varphi j(\alpha) + \varphi(u) = \iota(\alpha) + \varphi(u) = \alpha + \varphi(u)$. Since $\alpha$ is an arrow we deduce that $z$ does not lie in $J$. Thus $\varphi$ is onto, and since $P$ is projective and $M'$ is indecomposable, it follows that $M' \cong P$, a contradiction.

(iii) Let $X = \omega\Lambda$ where $\omega \in e\Lambda f$ and where $e$, $f$ are primitive idempotents of $\Lambda$. Suppose that $\omega$ is not an arrow; then $\omega = \alpha\omega_1$ for some arrow $\alpha$ and $\omega_1 \in J$. But then $X \subseteq \alpha J$, and $X$ is not tall.

Now we will obtain more modules lying at ends of components. Let $P$ be indecomposable projective; then soc $P$ and top $P$ are simple, related by the Nakayama permutation, see I.3.4, but $\operatorname{soc} P \neq \operatorname{top} P$ in general. Suppose that Let $H = \operatorname{rad} P/\operatorname{soc} P$; then it may happen that $H$ is a direct sum of two indecomposable modules, or else that there is an AR-seuence $0 \to V \to H \to U \to 0$ with $U \not\cong P/\operatorname{soc} P$. This occurs at ends of $\mathbb{Z}D_\infty$ components but also at ends of $\mathbb{Z}\tilde{D}_n$-components, or of $\mathbb{Z}D_n/\langle \tau^q \rangle$-components for algebras of finite type, in the situation studied by Riedtmann in [$Ri_2$].

IV.5.5 **PROPOSITION** *Assume that* $\Lambda$ *is a self-injective algebra. Let* $P$ *be an indecomosable projective* $\Lambda$ *module satisfying one of the following conditions:*

*(i)* $rad\ P/soc\ P = U \oplus V$ *where* $U$, $V$ *are both non-zero.*

*(ii) There is an AR-sequence* $0 \to V \to radP/soc\ P \to U \quad 0$ *as in IV.3.3(a); moreover* $U$ *and* $V$ *are not simple.*

*Then* $\tilde{U}$ *has one predecessor in* $\Gamma(\Lambda)$.

Proof: Let $0 \to \tau\tilde{U} \to M \to \tilde{U} \to 0$ be the AR-sequence ending in $\tilde{U}$. We have to show that $M$ is indecomposable. Both hypotheses imply that $\tilde{U} \neq Q/soc\ Q$ for $Q$ indecomposable projective, therefore $M$ does not have a projective sommand.

(1) soc $M \cong$ soc $\tilde{U} \oplus$ soc $\tau\tilde{U}$: In both cases, $\tilde{U}$ is not simple. By the AR-property, the inclusion soc $\tilde{U} \to \tilde{U}$ factors through $\epsilon$.

Now let $M = \bigoplus_{i=1}^{k} M_i$ with indecomposable summands $M_i$, and let $\epsilon_i = \epsilon_{|M_i}$. Then $\epsilon_i$ is irreducible, and hence is either 1-1 or onto.

(2) If (i) holds then $\epsilon_i$ is not 1-1: Suppose $\epsilon_i$ is 1-1. Then $M_i$ is isomorphic to a summand of U. With the hypothesis (i) we have that the inclusion map $\epsilon_i : M_i \to \tilde{U}$ has a non-trivial factorization through P/soc P; on the other hand the map should be irreducible; this is not possible.

(3) If $\epsilon_i$ is 1-1 then $M_i$ = rad $\tilde{U}$, and hence soc $M_i$ = soc U:

If $\epsilon_i$ is 1-1 then $\epsilon_i$ factors through rad $\tilde{U}$ which is the unique maximal submodule of $\tilde{U}$. Moreover, we are in case (ii), and rad $\tilde{U}$ ( = U) is indecomposable, so (3) follows.

(4) If $\epsilon_i$ is surjective then $\epsilon_i$(soc $M_i$) = soc $\tilde{U}$ and $|$soc $M_i| > |$soc $U|$: The first statement follows from the fact that U is tall. If $\epsilon_i|$soc $M_i$ were an isomorphism then $\epsilon_i$ would be 1-1 and an isomorphism. This is not possible since $\epsilon_i$ is irreducible.

Since $\tilde{U}$ is cyclic, one of the $\epsilon_i$ must be surjective. Hence (3) and (4) imply that

(5) soc $M \cong k$[soc U] $\oplus$ X where X $\neq$ 0.

Now, soc $\tau U \cong$ top $\Omega U$ = top V $\cong$ top V. We deduce therefore from (1) and (5) that

(6) top V $\cong$ (k-1)[top U] $\oplus$ X.

Now let the AR-sequence ending with $\underset{\sim}{V}$ be $0 \to \underset{\sim}{V} \to W \to \tau^{-1}\underset{\sim}{V} \to 0$. Recall that $\underset{\sim}{V} \cong \Omega\tilde{U}$, and that $\Omega$ induces a graph isomorphism of $\Gamma_s(\Lambda)$. Hence W also has precisely k non-projective indecomposable summands. We may dualize the above arguments; note that

$\underset{\sim}{V}$ is deep, by the hypothesis. We obtain

(5*) <u>top $W \cong k[\text{top } V] \oplus Y$ where $Y \neq 0$;</u>

and since $\text{top}(\tau^{-1}\underset{\sim}{V}) \cong \text{soc } \Omega^{-1}\underset{\sim}{V} \cong \text{soc } \tilde{U} \cong \text{soc } U$, we deduce

(6*) <u>$\text{soc } U \cong (k-1)[\text{top } V] \oplus Y$.</u>

By (6) and (6*) we calculate lengths and have $|\text{top } V| = (k-1)|\text{soc } U| + |X| = (k-1)^2|\text{top } V| + (k-1)|Y| + |X|$. It follows that $k-1 = 0$ as required.

It remains to study the position of $\tilde{U}$ and $\underset{\sim}{V}$ in $\Gamma(\Lambda)$ in the case when IV.3.3(a) holds and one of U, V is simple.

<u>IV.5.6</u> PROPOSITION *Assume $\Lambda$ is self-injective and that P is indecomposable projective with H = rad P/soc P. Suppose that U, V are as in IV.3.3(a) and suppose one of U, V is simple. If the AR-component of U is not fixed by $\Omega$ then $\tilde{U}$ has one predecessor in $\Gamma(\Lambda)$.*

Proof: (1) <u>It suffices to proof the statement for the case when U is simple:</u> Say $P = P(S_0)$; then $U = U_0$ and $V = V_0$. Assume $V_0$ is simple, $\cong S_i$ say. Then by IV.3.5 it is also true that $U_i$ is defined and is simple (and $V_i \neq 0$), actually $U_i = \nu^{-1}S_0$. Then $\underset{\sim}{V}_0 \cong \mathcal{U}(S_i, \nu^{-1}S_0) \cong \tilde{U}_i$. Since $\Omega$ induces a graph isomorphism, $\tilde{U}_i$ and $\tilde{U}_0$ have the same number of predecessors in $\Gamma_s(\Lambda)$; and they do not have projective predecessors since $V_0$ and $V_i$ are non-zero.

Suppose $U_0 = S_i$, then $V_{\nu i} \cong S_0$. We consider the AR-sequence ending in X where $X = \Omega^{-1}\tilde{U}_0$ ( $= \tilde{U}_{\nu i}$), it is of the form $0 \to Y \xrightarrow{\mu} N \xrightarrow{\epsilon} X \to 0$. We shall prove that X is indecomposable.

(2) <u>N does not have a projective summand:</u> Otherwise $X = P/\text{soc } P$ for some projective P. Then top $P = S_{\nu i}$, and $P = P_{\nu i}$ and $\Omega X$ is simple, so that $V_{\nu i} = 0$, a contradiction.

The module Y has a simple socle ($\text{soc } \tau X \cong \text{top } \Omega X \cong \text{top } V_{\nu i}$); and X is not simple. Let $N = \bigoplus_{i=1}^{t} N_i$ with all $N_i$ indecomposable, $\mu = [\mu_1, \ldots, \mu_t]^T$ and $\epsilon = [\epsilon_1, \ldots, \epsilon_t]$. Since soc Y is simple, one of the $\mu_i$ must be mono, say $\mu_1$. Let $C = \bigoplus_{i=2}^{t} N_i$ and $q = [\epsilon_2, \ldots, \epsilon_t]$. With $\mu_1$ also q is mono.

(3) <u>$C \cong U_{\nu i}$:</u> We have that q is not onto and rad X is the unique maximal submodule of

X, therefore q factorizes through the inclusion j, say $\begin{array}{ccc} C & \xrightarrow{q} & X \\ & {\scriptstyle q_1}\searrow \quad \nearrow{\scriptstyle j} & \\ & \text{rad } X & \end{array}$

Since q is irreducible and j is non-split, it follows that $q_1$ is a split mono. Now rad $X \cong U_{\nu i}$, which is indecomposable, and (3) follows.

The module $U_{\nu i}$ has only one successor in $\Gamma_s(\Lambda)$ [it satisfies IV.3.3(a)]. Hence there is an AR-sequence $0 \to C \overset{\epsilon'}{\to} X \oplus P \to \tau^{-1}C \to 0$ where P is projective or is zero.

(4) P = 0: Suppose not; then $C = U_{\nu i} \cong \text{rad } P$. Let j be the component of $\epsilon'$ from C to X; it is one-to-one. Now, P is also the injective hull of C, hence there is a

commutative diagram $\begin{array}{ccc} & P & \\ \nearrow & & \nwarrow{\scriptstyle \psi} \\ C & \overset{\epsilon'}{\to} & X \end{array}$. The map $\psi$ is 1-1 and then an isomorphism, by the lengths; and consequently X is projective, a contradiction.

It follows now that $\tau^{-1}C \cong S_{\nu i}$, and $U_{\nu i}$, $S_{\nu i}$ lie in the same AR-component, contrary to the hypothesis.

We note that in the case $\Lambda$ is symmetric and $\Theta = \mathbb{Z}D_\infty$ then $\Theta$ is not fixed by $\Omega$. This is true because $\Theta$ has only one end, and $\Omega^2 \cong \tau$.

## IV.6 *Ends of components III*

Assume first that $\Lambda$ is a self-injective algebra; and suppose $\Lambda$ has components of tree class $D_\infty$ and also tubes, or may be, $D_n$ or $\tilde{D}_n$. Then there are two types of modules X having only one predecessor in $\Gamma_s(\Lambda)$. Let

$$(*) \quad 0 \to \tau X \xrightarrow{\lambda} M \xrightarrow{\kappa} X \to 0$$

be the AR-sequence ending at X, and assume M is indecomposable. We say that *X is of type d* if M has three predecessors in $\Gamma_s(\Lambda)$. If $M \neq \text{rad } Q/\text{soc } Q$ for Q indecomposable projective, this happens if and only if there is some module $R \neq X$ and an AR-sequence

$$(**) \quad 0 \to \tau R \xrightarrow{\lambda_1} M \xrightarrow{\kappa_1} R \to 0$$

We shall now study now a component $\Theta$ which contains AR-sequences (*) and (**). We will need to know whether there are projectives attached to $\Theta$ or $\Omega\Theta$ Therefore we should investigate whether end terms in (*) or (**) (or in their $\Omega$-translates) are simple.

<u>IV.6.1</u> LEMMA *Suppose there are AR-sequences (*) and (**), with M indecomposable. Assume that $\tau X$ is tall and that $T$ = soc X is simple and does not have three predecessors in $\Gamma_s(\Lambda)$. Assume also that $\tau X/\text{soc } \tau X$ is indecomposable. Then*

*(a) X is not simple.*

*(b) If $\tau X/\text{soc } \tau X \neq 0$ then R is not simple.*

Proof: Suppose that X is simple. Then soc $M \cong$ soc $\tau X$ since $\kappa$ is non-split, hence soc M is simple.

(1) <u>R is simple</u>: If not then by the AR-property, the inclusion soc R $\to$ R factors through $\kappa_1$, and soc $M \cong$ soc R $\oplus$ soc $\tau R$, which is not possible since soc M is simple.

We consider X and $\tau R$ as submodules of M. Since soc M is simple, it follows that $T \subseteq \tau X \cap \tau R$; and by the AR-property we have therefore $M/T \cong \tau X/T \oplus X$ and $M/T \cong \tau R/T \oplus R$. These are two indecomposable summands, by the hypothesis.

Now $R \neq X$ since they represent different vertices of the AR-quiver, hence $X \cong \tau R/T$ and $R \cong \tau X/T$. Recall that R is simple, so it follows that $\tau X$ is cyclic, hence it is generated by an arrow, by IV.5.4(iii). Consequently $R = \tau X/T$ is isomorphic to a direct summand of H(S). But then R lies between $\Omega^{-1}T$ (= P/T) and rad P in the AR-quiver and we have that M ≅ rad P. Now, M has 3 predecessors, and hence P/T and T as well, contrary to the hypothesis.

(b) Assume for contradiction that R is simple. If τX ∩ τR were zero then $\kappa_1|_{\tau X}$ would be a monomorphism, and thus τX ≅ R and τX would be simple, contrary to the hypothesis that τX is tall. Hence τX ∩ τR ≠ 0. Since T = soc τX is simple it follows that T ⊆ τR. The AR-property implies that M/T ≅ τX/T ⊕ X and M/T ≅ τR/T ⊕ R. We know that R ≠ X and that τX/T is indecomposable. It follows that R ≅ τX/T and then τX is cyclic. By IV.5.4(iii), τX may be generated by an arrow. It follows as in (a) that R is a direct summand of H(S), and we get a similar contradiction.

This completes the proof.

Now let $\tilde{U}$, $\underset{\sim}{V}$ be as in IV.5.1. These modules lie always at the end of some component, by IV.4.2 or IV.5.5 or IV.V.6. One might expect that in the case when U, V satisfy IV.3.1 or IV.3.3(a) with U and V both not simple the modules $\tilde{U}$, $\underset{\sim}{V}$ should lie in tubes. However, the best we are able to prove (in more generality) is the following. Recall that a module at the end of a $\mathbb{Z}D_\infty$- or $\mathbb{Z}\tilde{D}_n$-component is said to be "of type d":

<u>IV.6.2</u> PROPOSITION *Assume* Λ *is a self-injective algebra. Suppose U and V are defined as in IV.3.1 or IV.3.3(a) and moreover that top H ≅ top U ⊕ top V and also soc H ≅ soc U ⊕ soc V. Assume also that soc U and top V are simple. Then* $\tilde{U}$ *and* $\underset{\sim}{V}$ *are not of type d.*

Proof: The hypothesis implies that $\tilde{U}$ is deep and $\underset{\sim}{V}$ is tall (see IV.5.4). It suffices to consider $\underset{\sim}{V}$, since Ω induces a graph isomorphism of $\Gamma_s(\Lambda)$. Let

$0 \to \underset{\sim}{V} \to M \to X \to 0$ be the AR-sequence starting with $\underset{\sim}{V}$; then M is indecomposable by IV.5.5.

(1) <u>top X ≅ soc U:</u> We have top X = top $\tau^{-1}\underset{\sim}{V}$ ≅ soc $\Omega^{-1}\underset{\sim}{V}$ ≅ soc $\tilde{U}$ ≅ soc U. Suppose that $\underset{\sim}{V}$ is of type d. By IV.5.5 we know that M ≠ rad Q/soc Q for Q indecomposable projective. Hence there is another AR-sequence

(*) $0 \to \tau R \to M \to R \to 0$ . Recall that, by IV.6.1, R and X are not simple.

We consider $\underset{\sim}{V}$ and $\tau R$ as submodules of $M$.

<u>Case 1</u> $\underset{\sim}{V} \cap \tau R \neq 0$. Then $T = \mathrm{soc}\,\underset{\sim}{V} \subseteq \tau R$, and by the AR-property we have $M/T \cong V \oplus X$ and also $M/T \cong \tau R/T \oplus R$. In particular, $M/T$ has only two indecomposable summands. Since $X$ and $R$ represent distinct vertices of $\Gamma(\Lambda)$, we have $R \neq X$, and it follows that $X \cong \tau R/T$ and that $V \cong R$. In particular, $V$ lies at the end of a $\mathbb{Z}D_\infty$ -component with one predecessor, and so $S$ must satisfy IV.3.3(a). We deduce $\tau^{-1}M \cong H(S)$ and $X \cong \Omega S$. Consequently top $\Omega S \cong$ top $X \cong$ soc $U$ and is simple. On the other hand, top $\Omega S \cong$ top $H$ and this not simple, by the hypothesis, a contradiction.

<u>Case 2</u> $\underset{\sim}{V} \cap \tau R = 0$. Then we have the following:

(2) <u>top $M \cong$ top $V \oplus$ top $X$, hence is of length 2:</u> The hypothesis implies that $\underset{\sim}{V}$ is not simple. By the AR-property the inclusion map rad $\underset{\sim}{V} \to \underset{\sim}{V}$ factorizes through $\lambda$ and the first statement follows. We assume $|\mathrm{top}\ V| = 1$; and by (1) we know $|\mathrm{top}\ X| = 1$.

(3) <u>If $\tau R$ is not simple then top $R$ is simple</u>: Suppose $\tau R$ is not simple. By the AR-property, the inclusion rad $\tau R \to \tau R$ factors through $\lambda_1$ and it follows that top $M \cong$ top $R \oplus$ top $\tau R$, and (3) follows using (1).

By the hypothesis of case 2, we have that $\kappa_{1|\underset{\sim}{V}}$ is a monomorphism and that $\underset{\sim}{V} \subsetneq R$. Now, $R$ has a simple top and hence $\underset{\sim}{V} \subseteq$ rad $R$. Since $R$ is not projective we deduce that $\underset{\sim}{V}$ is not tall, a contradiction (IV.5.4).

This shows that $\tau R$ must be simple, call it $E$; and apply $\Omega^{-1}$ to the appropriate part of $\Gamma_s(\Lambda)$. Since there is an irreducible map $\tau M \to E$, there is also one $\Omega^{-1}(\tau M) \to \Omega^{-1}E = P(\nu E)/E$. It follows that $\Omega^{-1}(\tau M) = H(\nu E)$. On the other hand, $\Omega^{-1}(\tau M)$ is the middle term of $\mathcal{A}(U)$, and therefore $\mathfrak{U} \cong U(\nu E)$. Let $Q = P(\nu E)$, then there is an epimorphism $\pi$: rad $Q \to \mathfrak{U}$ with $\pi[\mathrm{soc}(Q)] = 0$ and $\mathfrak{U}$ is not deep. We have therefore obtained a contradiction to IV.54. This completes the proof of the Proposition.

We also have some information about the position of $U$ and $V$ in the case when $U$, $V$ are as in IV.3.3(a) and when one of $U$ or $V$ is simple.

<u>IV.6.3</u> LEMMA *Suppose $\Lambda$ is a symmetric algebra, $S_0$ is a simple $\Lambda$-module such that $U_0$ and $V_0$ are defined and satisfy IV.3.3(a). Assume that $V_0$ is simple, say $V_0 \cong S_1$. Then*
*(a) $U_0$, $\underset{\sim}{V}_0$ and $\underset{\sim}{V}_1$ are not of type d.*

*(b) If the Cartan matrix of $\Lambda$ is non-singular then $U_0$, $\underset{\sim}{V}_0$ and $\underset{\sim}{V}_1$ do not have $\Omega$-period dividing 4.*

We will apply these results in VIII, in order to study algebras of semidihedral type, see VIII.1. In that situation we deduce more, namely:

<u>IV.6.4</u> COROLLARY *With the hypothesis of IV.6.3, if $\Lambda$ is of semidihedral type then $U_0$, $\underset{\sim}{V}_0$ and $\underset{\sim}{V}_1$ lie at the end of the 3-tube.*

Proof of IV.6.3: By IV.3.6 we have that $S_1$ also satisfies IV.3.3(a), and moreover $U_1 \cong S_0$. Therefore we take $\underset{\sim}{V}_0 \cong U_1$; and then $\Omega^2 U_0 \cong \underset{\sim}{V}_1$. Let

(*) $0 \to \underset{\sim}{V}_1 \to M \to U_0 \to 0$ be the AR-sequence ending in $U_0$; then $M$ is indecomposable, by IV.5.6. Suppose that $\underset{\sim}{V}_1$ is of type d. Since $\underset{\sim}{V}_1$ is tall, $M \neq \mathrm{rad}\, Q/\mathrm{soc}\, Q$ for $Q$ indecomposable projective (see IV.5.3), and hence there is also an AR-sequence

(**) $0 \to \tau R \to M \to R \to 0.$

We consider $\underset{\sim}{V}_1$ and $\tau R$ as submodules of $M$.

<u>Case 1</u> $\underset{\sim}{V}_1 \cap \tau R \neq 0$. Then $S_1 = \mathrm{soc}\, \underset{\sim}{V}_1 \subseteq \tau R$; and using the AR-property twice we have $M/S_1 \cong V_1 \oplus U_0 \cong \tau R/S_1 \oplus R$. Since $V_1$ and $U_0$ are both indecomposable, we see that $M/S_1$ has only two summands. Moreover, $U_0$ and $R$ represent distinct vertices of the AR-quiver, therefore $U_0 \neq R$; and it follows that $V_1 \cong R$ and $U_0 \cong \tau R/S_1$.

(1) <u>$\tau^{-1}R \cong S_0$:</u> We have $\tau^{-1}R \cong \tau^{-1}V_1 \cong U_1 \cong S_0$.

(2) <u>$\tau R \cong \Omega^{-1}S_0$:</u> We calculate composition factors. We obtain $\underline{\dim}\, \tau R = \underline{\dim}\, U_0 + \underline{\dim}\, S_1 = \underline{\dim}\, U_0 + \underline{\dim}\, V_0 = \underline{\dim}\, \Omega^{-1}S_0$. Moreover, $\tau R$ is indecomposable and $S_1$ is simple, and $\tau R/S_1$ ($\cong \tilde{U}_0$) has a simple top $\cong S_0$. Hence $\mathrm{top}(\tau R) \cong S_0$, and the claim follows.

This shows that $\tau^2 S_0 \cong \Omega^{-1}S_0$, which is not possible for $S_0$ in a $\mathbb{Z}D_\infty$-component.

<u>Case 2</u> $\underset{\sim}{V}_1 \cap \tau R = 0$: Then $\kappa$ and $\kappa_1$ induce embeddings $\tau R \to U_0$ and $\underset{\sim}{V}_1 \to R$ respectively. Let $M_0 = \kappa^{-1}(U_0)$; then $M_0$ is a maximal submodule of $M$ and $\kappa(M_0)$ is a maximal submodule of $U_0$. We claim that also $R_0 = \kappa_1(M_0)$ is maximal in $R$. This holds because $\ker\, \kappa_1 = \tau R \subsetneq U_0$ and so $\ker\, \kappa_1 \subseteq M_0 \subset M$. Using the AR-property we deduce now that $M_0 \cong \underset{\sim}{V}_1 \oplus U_0 \cong \tau R \oplus R_0$. This is again a direct sum of two

indecomposable modules. We also have that $\underset{\sim}{V}_1 \neq \tau R$ since they correspond to distinct vertices of $\Gamma_s(\Lambda)$. Therefore we must have that $U_0 \cong \tau R$ and $\underset{\sim}{V}_1 \cong R_0$.

(1) $\underline{\tau^2 R \cong S_1}$: We have $\tau^2 R \cong \tau U_0 \cong V_0 \cong S_1$, by the hypothesis.

(2) $\underline{R \cong \Omega S_1}$: Comparing composition factors shows that $\underline{\dim}\, R = \underline{\dim}\, \underset{\sim}{V}_1 + \underline{\dim}\, S_0 = \underline{\dim}\, \underset{\sim}{V}_1 + \underline{\dim}\, U_1 = \underline{\dim}\, \Omega S_1$. Moreover, R is indecomposable and $R_0$ is a maximal submodule,and $R_0$ has a simple socle $\cong S_1$. Hence soc $R \cong S_1$, and (2) follows.

This leads again to the contradiction that $S_1$, $\Omega S_1$ lie in the same $\tau$-orbit and in a $\mathbb{Z}D_\infty$-component.

(b) Now assume that $\Omega^4 \tilde{U}_0 \cong \tilde{U}_0$. Then there is an exact sequence

$0 \to \tilde{U}_0 \to P \to P' \to P_1 \to P_0 \to \tilde{U}_0 \to 0$ where P and P' are projective, and soc $\tilde{U}_0 \cong$ soc P. Suppose the Cartan matrix of $\Lambda$ is non-singular. Then it follows from the exactness that $P_0 \oplus P' \cong P_1 \oplus P$. Hence $P_0$ is a summand of P, and $S_0$ occurs in soc $U_0$. This shows that there is a loop at vertex 0, and then $S_0 \subsetneq$ top $H_0$. On the other hand, since $V_0$ is simple we must have that top $H_0$ is simple. So top $H_0 \cong S_0$; and the algebra $\Lambda$ is decomposable, contrary to our hypothesis.

<u>IV.6.5</u> PROPOSITION *Assume that the algebra $\Lambda$ is symmetric. Suppose that $X$ is uniserial of length 2, say $X = U(S_0, S_1)$ such that*

*(i) $\Omega X$ is generated by an arrow, and*

*(ii) $\Omega^{-1}X$ has a simple socle.*

*Let $0 \to \Omega X \to M \to \Omega^{-1}X \to 0$ be the AR-sequence; then $M$ is indecomposable. If $X$ is of type d then there is an AR-sequence $0 \to \tau R \to M \to R \to 0$ and moreover either $R \cong \Omega S_0$, or $\tau R \cong \Omega^{-1} S_1$.*

Proof: Since $\Omega X$ is generated by an arrow which is non-supportive (IV.4.1) it follows from IV.4.2 that $\Omega X$ lies at the end of some stable component. Moreover, $\Omega X \neq$ rad P for P projective, and therefore M is indecomposable.

Assume that X is of type d. Since $\Omega X$ is tall (IV.5.4), we have that $M \neq$ rad Q/soc Q for Q projective (IV.5.3); hence there is another AR-sequence

$0 \to \tau R \to M \to R \to 0$. It remains to prove the last part. Note that soc $\Omega X \cong S_0$ and top $\Omega^{-1}X \cong S_1$ which are both simple. We consider $\Omega X$ and $\tau R$ as submodules of M.

<u>Case 1</u> $\Omega X \cap \tau R \neq 0$. Then $M/S_0 \cong \Omega X/S_0 \oplus \Omega^{-1}X \cong \tau R/S_0 \oplus R$. Since $\Omega X$ has a simple

top, we know that $\Omega X/S_0$ is indecomposable; hence $M/S_0$ has only two summands. Moreover R and $\Omega^{-1}X$ are non-isomorphic since they belong to different vertices of the AR-quiver. Therefore $\tau R/S_0 \cong \Omega^{-1}X$. We see now that $\tau R$ has a simple top $\cong S_1$; moreover $\underline{\dim}\ \tau R = \underline{\dim}\ \Omega^{-1}S_1$ and hence $\tau R \cong \Omega^{-1}S_1$.

<u>Case 2</u> $\Omega X \cap \tau R = 0$. Then M has a submodule $M_0 = \Omega X \oplus \tau R$. Let $M_1 = \lambda^{-1}(\text{rad}\ \Omega^{-1}X)$; then clearly $\lambda(M_1) \subset \Omega^{-1}X$. Moreover since $\ker \lambda_1 \subsetneq M_0 \subsetneq M_1$ we deduce that $R_0 := \lambda_1(M_1)$ is a maximal submodule of R. It follows from the AR-property that $M_1 \cong \Omega X \oplus \text{rad}\ \Omega^{-1}X \cong \tau R \oplus R_0$; and the hypothesis implies that $\Omega X \cong R_0$ and consequently, since soc R is simple and $\underline{\dim}\ R = \underline{\dim}\ \Omega S_0$ we deduce $R \cong \Omega S_0$.

For convenience we include a further application; this is essentially (4.1) in $[E_5]$II, with revised hypotheses.

<u>IV.6.6</u> **LEMMA** *Let $S = S_f$ be simple and not periodic, such that there is no loop at vertex $f$, and two arrows start and two arrows end at vertex $f$. Assume also that $H_f \neq$ rad $Q$/soc $Q$ for $Q$ projective $\neq P_f$. Then*

*(a) S lies at the end of a $\mathbb{Z}D_\infty$-component with one predecessor.*

*(b) $\mho$ lies at the end of the 3-tube.*

Proof: The hypotheses imply that either IV.3.1 or IV.3.3(a) holds for $P = P_f$. In any case, U and V are defined, and moreover, U and V have simple socles and tops, by IV.3.4. Hence we know by IV.6.2 that $\mho$ and $\underset{\sim}{V}$ lie at ends of tubes.

(1) $\underline{\Omega^4\mho \neq \mho}$: Suppose $\Omega^4\mho \cong \mho$, then there is an exact sequence

$0 \to \mho \to Q_3 \to Q_2 \to Q_1 \to P \to \mho \to 0$ where $P = P_f$, and where the $Q_i$ are projective; moreover top $Q_1 \cong$ top V and soc $Q_3 \cong$ soc U. By exactness, $\underline{\dim}\ P \oplus Q_2 = \underline{\dim}\ Q_1 \oplus Q_3$, and since the Cartan matrix is non-singular, we deduce $P \oplus Q_2 \cong P \oplus Q_3$. Hence S occurs either in top V or in soc U and there is a loop at the vertex corresponding to S, contrary to the hypothesis.

Hence $\mho$ and then also $\underset{\sim}{V}$ lie at the end of the 3-tube, and (b) follows.

Assume for contradiction that IV.3.1 holds. Then $\underset{\sim}{U}$ and $\mathcal{V}$ are defined as well. The arguments in (1) apply to $\mathcal{V}$, therefore $\mathcal{V}$ and also $\underset{\sim}{U}$ lie at the end of the 3-tube. So either $\mathcal{V} \cong \underset{\sim}{V}$, or $\underset{\sim}{U} \cong \mho$, and there is a loop at the vertex corresponding to S, a contradiction.

## *V Stable Auslander-Reiten components for tame blocks*

We start with a summary of some general modular representation theory. Then we will study a small number of tame blocks for 2-local groups in detail: We determine their basic algebras and their stable AR-quiver. It is possible to determine the basic algebras for 2-local groups fairly directly since one can exploit appropriate skew group rings.

Then we give a summary of more general principles which are available now to determine $\Gamma_s(B)$ for arbitrary blocks B, provided $\Gamma_s(b)$ is known for blocks b of appropriate p-local blocks. These include powerful new results by Kawata. We will then use these and our earlier results to determine $\Gamma_s(B)$ for arbitrary tame blocks.

Moreover, we will determine B-subpairs and k(B) for arbitrary tame blocks. This forms part of original work by Brauer and Olsson ([$B_2$], [O]), we present the material in the terminology of [AB].

### V.1 *Modular group representations*

We assume that K is an algebraically closed field of characteristic $p > 0$. Details for the following may be found in various sources ([A], [CR], [F], [L], [$M_2$])

V.1.1 Suppose G is an arbitrary finite group and H is a subgroup of G. We consider KH as a subalgebra of KG; and KG is a (KH,KG)-bimodule.

If $W \in$ mod KH then we write $W^G := W \otimes_{KH} KG$, the KG-module $W^G$ is *induced*.

Any KG-module M may be considered as a KH-module, the *restriction* of M to H, denoted by $M_H$.

*The Mackey decomposition:* Suppose H and L are subgroups of G and $W \in$ mod KH. Then

$$(M^G)_L \cong \bigoplus_{x\in\mathcal{T}} [(M \otimes x)_{H^x \cap L}]^L$$

where $\mathcal{T}$ is a system of representatives for the (H,L)-double cosets in G.

A KG-module M is *H-projective* if there is a KH-module W such that M is isomorphic to a direct summand of $W^G$.

V.1.2 Suppose M is an indecomposable KG-module. Then a *vertex* of M is a minimal subgroup V of G such that M is V-projective. The vertices of M form a G-conjugacy class of p-subgroups of G (see [F] or [$L_3$]). I V is a vertex of M we write V = vx(M) (or V $=_G$ vx(M)). Hence M is H-projective for H $\leq$ G if and and only if V is G-conjugate to a subgroup of H. For this one uses the notation "$\leq_G$".

Suppose M is an indecomposable KG-module and V is a vertex of M. Then a *source* of M is an indecomposable KV-module S such that M is isomorphic to a summand of $S^G$. Fixing V, the sources of M form one $N_G(V)$-orbit of V-modules. (If g $\in$ G and W $\in$ mod KV then $W^g$ = W $\otimes$ g is a KV-module. This defines an action of the group $N_G(V)$ on mod KV.)

Suppose B is a block of KG. Then a *defect group* of B may be defined as a minimal subgroup D of H such that all M $\in$ mod B are D-projective; or equivalently, a maximal element in the set of vertices of indecomposable B-modules. The defect groups of G form one conjugacy class of p-subgroups of G. (For other equivalent definitions, see the standard text books.) We note that a normal p-subgroup of G is contained in the defect group of every block of KG.
We often denote a defect group of B by $\delta(B)$.

The principal block of KG is the block to which the simple KG-module K belongs. Its defect group is a p-Sylow subgroup of G; and this is also a vertex of K.

<u>V.1.3</u> *The Green correspondence* Suppose G is a finite group, V $\leq$ G is a p-subgroup and H is a subgroup of G containing $N_G(V)$. Let

$\mathcal{X}$ = { $gVg^{-1} \cap V$: g $\in$ G and g $\notin$ H },

$\mathcal{Y}$ = { $gVg^{-1} \cap H$: g $\in$ G and g $\notin$ H },

$\mathcal{Z}$ = { R $\leq$ V: R is not G-conjugate to any X $\in \mathcal{X}$ }.

THEOREM(J.A. Green) *There is a one-to-one correspondence between isomorphism classes of indecomposable KG-modules with vertex in $\mathcal{Z}$ and isomorphism classes of indecomposable KH-modules with vertex in $\mathcal{Z}$. If M and W are such modules for G and H respectively which correspond then M and W have the same vertex, and*

$$M_H \cong W \oplus Y, \qquad W^G \cong M \oplus X$$

*where Y is a $\mathcal{Y}$-projective KH-module, and X is an $\mathcal{X}$-projective KG-module.*

One writes W = fM and M = gW.

Suppose B and b are the blocks of KH and KG containing M, fM respectively, and assume $H = N_G(V)$. Then the block idempotents of B, b correspond to each other under the Brauer map with respect to V, and moreover $\delta(b) \leq_G \delta(B)$ (see [$Gr_1$] or [F], also V.5.)

V.1.4 The operators $\Omega$, $\Omega^{-1}$ (see I.1) preserve vertices; and also, $\Omega$ commutes with the Green correspondence. Recall that for group algebras, $\Omega^2 \cong \tau$, and hence also $\tau$ commutes with the Green correspondence. It is in general not clear how AR-sequences behave under Green correspondence. The technical problems arise since the modules occuring in the middle of an AR-sequence may have a different vertex. In V.3 we will study this question in detail.

To do this, we will need a few details about modules with particular vertices. Modules with cyclic or quaternion vertices are $\Omega$-periodic; in particular, they lie in tubes. Moreover, suppose M has period q then the rank of the tube is q if q is odd, and q/2 otherwise. In the special case when the prime p is 2 then all modules with cyclic vertices have $\tau$-period 1. If p = 2 and the vertex of M is quaternion then the $\tau$-period of M is 1 or 2 or 4; this may be proved using (V.2).

The number of indecomposable KG-modules having cyclic vertices is finite. Hence if such a module belongs to a block of infinite type then its component, being infinite, (I.8.3) contains modules with non-cyclic vertices.

V.1.5 Suppose H is a normal subgroup of G such that the index |G:H| is a power of p. If $M \in$ mod KH is indecomposable then so is $M \otimes_{KH} KG$. This is a special case of *Green's Theorem* (see for example [F]). More generally, the same conclusion holds when H is a subnormal subgroup of G whose index is a power of p.

V.1.6 *Reduction modulo p, decomposition numbers.* Suppose G is a finite roup. A splitting p-modular system for G is a triple (K, R, S) where R is a complete discrete valuation ring, $K = R/(\pi)$ where $(\pi)$ is the maximal ideal of R, and S is the field of fractions of R, and where K and S are splitting fields for all subgroups of G.
We start off with an algebraically closed field K of characteristic p; then there exist R, S such that (K, R, S) is a splitting p-modular system for G.

Denote by $\mathcal{L}(RG)$ the category of RG-lattices; an RG-lattice is by definition an

RG-module which is finitely generated and free as an R-module. We will now describe correspondences between

(i) $\mathcal{L}(RG)$ and mod KG,

(ii) mod SG and $\mathcal{L}(RG)$.

(i) If L is an RG-lattice then $\bar{L} := L/L\pi$ is a KG-module such that $\mathrm{rank}_R L = \dim_K \bar{L}$. Not every KG-module is of the form $\bar{L}$ for some $L \in \mathcal{L}(RG)$.

(ii) Suppose $L \in \mathcal{L}(RG)$, then $L \otimes_R S$ is an SG-module. Conversely, any SG-module M has an "R-form", that is, an RG-submodule T which contains an S-basis of M; and then $T \otimes_R S \cong M$.

Given $M \in$ mod SG, then a *reduction modulo p* of M is a KG-module $\bar{T}$ where T is some R-form of M. The composition factors of $\bar{T}$ as a KG-module are uniquely determined by M and do not depend on the choice of T. This is due to Brauer, see for exampe [F], [L].

*Decomposition numbers*: Let $\{\chi_i\}$ be the set of ordinary irreducible characters of G; and let $M_i$ be an SG-module whose character is $\chi_i$; and choose an R-form $T_i$ of $M_i$. Suppose $\{S_j\}$ is a full set of simple KG-modules. Then the decomposition number $d_{ij}$ is is defined to be the multiplicity of $S_j$ as a composition factor of $\bar{T}_i$.

<u>V.1.7</u> *The decomposition matrix and the Cartan matrix* It is well-known that RG is a semi-perfect ring, and idempotents of RG modulo J(RG) can be lifted; actually, one can lift orthogonal decompositions of $1_{RG}$ into primitive idempotents (see for example, [CR], [F], [L]). Correspondingly one can lift decompositions of projectives. Also, if e is a block idempotent of KG then there is a block idempotent $\hat{e}$ of RG with $\hat{e} \equiv e \bmod (\pi)G$.

Let B be a block of KG, with indecomposable projectives $\{e_i KG : i \in I\}$ and simple modules $\{S_i\}$ where $S_i = e_i KG/e_i J(KG)$, $i \in I$. The number of simple modules is denoted by $\ell(B)$.

Let $\{\hat{e}_i : i \in I\}$ be a set of idempotents of RG with $\hat{e}_i \equiv e_i \bmod (\pi)G$.

The *characters belonging to B* are the irreducible characters $\chi$ of G which occur as constituents of the character of the SG-modules $(\hat{e}_i RG) \otimes_R S$. Then k(B) is the number of these characters.

If $\chi_i$ belongs to B then all composition factors of $\bar{T}_i$ belong to B. The ***decomposition matrix of B*** is the matrix $D = [d_{ij}]$ where the $d_{ij}$ are the decomposition numbers where $\chi_i$ runs over the irreducible characters belonging to B, and $S_j$ over the simple B-modules. Then D has k(B) rows and l(B) columns.

Recall that the Cartan matrix C of the block B has entries $c_{ij}$ = the multiplicity of $S_j$ as a composition factor of $e_iKG$. Relating Cartan numbers and decomposition numbers, we have that $D^tD = C$. This is due to Brauer (see for example [F], [L]).

Suppose C is the Cartan matrix of a block. It is important that det C is a power of p; in particular, C is non-singular. This has been first proved by Brauer; there is a new proof due to Alperin, Collins, Sibley (see [$L_3$]). Moreover, the highest elementary divisor of C is equal to the order of a defect group of the block, and it occurs with multiplicity one (see [F]).

V.1.8 Let Z(B) be the centre of the block, considered as an algebra. It is not difficult to see that $dim_K Z(B) = k(B)$. We will use this to relate the structure of B as an algebra to the arithmetic of reduction modulo p.

*The height of an irreducible character:* If $\chi$ is an irreducible character belonging to B then it is true that the degree $\chi(1)$ is divisible by $|P : \delta(B)|$ where P is a Sylow p-subgroup of G. Write $\chi(1) = |P:\delta(B)|p^h m$ where p and m are coprime. Then h is said to be the height of $\chi$. It is a well-known fact that the block contains always some character of height zero.

Analogously, if M is a B-module then $|P:\delta(B)|$ divides $\dim_K M$; and one may similarly define the height of M. It is also true that there exists always some simple B-module of height zero (see [$M_1$]).

## V.2 *Some tame blocks of 2-local groups*

The main aim of this section is to determine the basic algebras of a number of tame blocks for 2-local groups. A group is said to be p-local it it has a non-trivial mormal p-subgroup.

With this hypothesis, there are naturally twisted group rings arising; and we wish to exploit more generally the structure of a number of particular algebras arising in this way. We first study some twisted group rings and then turn to blocks (from V.2.9).

There is much literature on blocks of p-local groups. We note that the structure of tame blocks of 2-local groups can also be deduced from more general results by Kulshammer and Puig [Ku], [KP], $[Pu_{1,2}]$.

V.2.1 Let A be an algebra and G a finite group acting on A as a group of algebra automorphisms.

(a) The skew group ring $A * G$ is the free A-module with basis $\{ \bar{g} : g \in G \}$ on which the multiplication is defined by the rules

(S1) $a \bar{g} = \bar{g} a^g$ $(g \in G, a \in A)$ and

(S2) $\bar{g}_1 \bar{g}_2 = \overline{g_1 g_2}$ $(g_1, g_2 \in G)$.

(b) More generally, let $\gamma: G \times G \to A^*$ be a factor set. We define the twisted group ring $A *_\gamma G$ to be the algebra with underlying space as for $A * G$ satisfying (S1), and where (S2) is replaced by

$(S2^\gamma)$ $\bar{g}_1 \bar{g}_2 = \gamma(g_1,g_2) \overline{g_1 g_2}$ $(g_1, g_2 \in G)$.

We usually write g instead of $\bar{g}$. Normally the values of $\gamma$ lie in the centre of $A^*$.

V.2.2 LEMMA *Suppose that A is a local symmetric algebra over a field of characteristic p, and G is a p'-subgroup of Aut A. Denote by Λ the associated crossed product. If Λ is symmetric then G must act trivially on soc A.*

Proof: Let $\chi$ be the character corresponding to the G-action on soc A. The primitive idempotent for the trivial KG-module is $\epsilon_0 = |G|^{-1}\sum_{g\in G} g$ ; and the module $\epsilon_0\Lambda$ is a direct summand of $\Lambda$ as a $\Lambda$-module. It is indecomposable since as an $A$-module it is isomorphic to $A$; and moreover $\epsilon_0\Lambda/\text{rad }\epsilon_0\Lambda \cong K$. This shows that $\epsilon_0\Lambda$ is the indecomposable projective module whose simple quotient is the trivial $\Lambda$-module. If $\Lambda$ is symmetric then soc $\epsilon_0\Lambda$ must be isomorphic to the trivial $\Lambda$-module as well. The socle of $\epsilon_0\Lambda$ is spanned by $\epsilon_0\omega$ where $\omega$ spans soc $\Lambda$; and if $g \in G$, then $\epsilon_0\omega = \epsilon_0\omega g = \epsilon_0 g(\omega^g) = \epsilon_0\omega^g = \epsilon_0\chi(g)\omega$ and hence $\chi(g) = 1$.

<u>V.2.3</u> *Suppose $A$ is a local algebra, $G$ a finite subgroup of Aut $A$ and $\Lambda = A * G$. If $S$ is a simple $\Lambda$-module then $S$ is semisimple as an $A$-module. In particular, $A$ acts trivially of $S$ in the case when $A/J(A) \cong K$. Therefore $\Lambda$ and $KG$ have the same simple modules.*

To see this, consider $A$ as a subalgebra of $\Lambda$ by identifying $a \in A$ with a.1. Now, $J(A)*G$ is a nilpotent ideal of $\Lambda$ and therefore $J(A)*G \subseteq J(\Lambda)$. We deduce that $SJ(A) \subseteq SJ(\Lambda) = 0$; and $S$ is an $A/J(A)$-module, and therefore semisimple.

<u>V.2.3.1</u> We note that this generalizes the fact that a normal p-subgroup of a group $H$ acts trivially on simple $KH$-modules when char $K = p$.

We will now study some skew group rings over the Kronecker algebra; let
$A = K[X, Y]/(X^2, Y^2)$.

<u>V.2.4</u> LEMMA *Assume $\Lambda = A * G$ where $G \leq$ Aut $A$ is cyclic of order 3, $A$ is the Kronecker algebra and char $K \neq 3$; and assume also that $\Lambda$ is symmetric. Then $\Lambda$ is isomorphic to $KQ/I$ where $Q$ is the quiver*

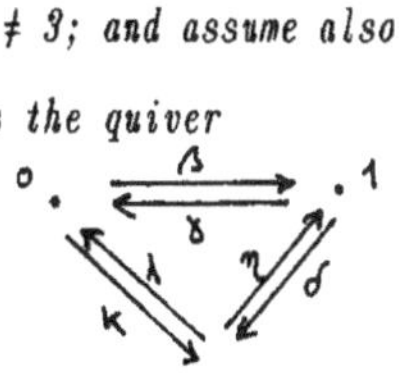


*and $I$ is the ideal generated by* $\beta\gamma = \kappa\lambda$, $\gamma\beta = \delta\eta$, $\eta\delta = \lambda\kappa$,

$0 = \beta\delta = \delta\lambda = \lambda\beta$ *and* $0 = \gamma\kappa = \kappa\delta = \delta\gamma$.

*In particular, there is a unique such algebra.*

Proof: Let $G = \langle g \rangle$. Since char $K \neq 3$ we have that every $KG$-module is semisimple. Now, rad $A$ and soc $A$ are $KG$-modules. Thus rad $A$ has a $K$-basis $v_0$, $v_1$, $v_2$ consisting of eigenvectors of $g$, such that $\langle v_0 \rangle$ = soc $A$ and moreover $v_0 = v_1 v_2$. Then $v_1$ and $v_2$ are automatically independent generators of rad $A$ (as $A$-module).

Thus a matrix of $g$ with respect to such a basis is $g = \mathrm{diag}[c_1 c_2, \ c_1, \ c_2]$ where $c_1$ and $c_2$ are roots of $t^3 - 1$ and at least one of $c_1$, $c_2$ is $\neq 1$. Conversely, any such matrix determines an algebra automorphism of $A$ of order 3. We see that there are in general essentially 3 possibilities for $g$, namely

(I) $c_1 = \omega$ and $c_2 = \omega^2$, (II) $c_1 = \omega = c_2$ and (III) $c_1 = \omega$ and $c_2 = 1$.

Here $\omega$ is a root of $t^2 + t + 1$.

By V.2.2, for a symmetric algebra we must have that $v_0{}^g = v_0$; this excludes (II) and (III).

We will now determine the structure of $A * G$ with action (I):

(1) <u>We may assume that $v_1 = X$ and $v_2 = Y$:</u> If char $K = 2$ then this is clear, since by the structure of the algebra, any two independent generators of rad $A$ satisfy the same relations. Now suppose char $K \neq 2$. Then $X^g$ and $Y^g$ are generators of $J$ whose square is 0, and $g$ has order 3. Using this one shows by an easy direct calculations that $X^g = cX + rXY$ and $Y^g = dY + sXY$ for some $c, d \in K^*$; actually $\{c, d\} = \{\omega, \omega^2\}$. It is then clear that $v_1$ and $v_2$ are, up to scalar multiples, of the form $X + aXY$ and $Y + bXY$ for $a, b \in K$; in particular $v_1{}^2 = 0$ and $v_2{}^2 = 0$; and we may take $X = v_1$ and $Y = v_2$.

By V.2.3, there are three simple modules and these are 1-dimensional; in particular, $\Lambda$ is basic. Let $e_i = (1/3)(1 + \omega^{-i}g + \omega^{-2i}g^2)$ for $i = 0, 1, 2$ ; then the $e_i$ are orthogonal idempotents with $\Sigma\, e_i = 1$, and $e_i g = \omega^i e_i$. The $e_i$ must be primitive; and we identify $e_i$ with vertex $i$ of the quiver.

Now, $e_i X = X e_{i+1}$ and $e_i Y = Y e_{i+2}$ where the index is taken modulo 3. We take as arrows

$$\beta = e_0 X e_1, \quad \delta = e_1 X e_2, \quad \lambda = e_2 \Lambda e_1,$$
$$\kappa = e_0 Y e_2, \quad \eta = e_2 Y e_1, \quad \gamma = e_1 Y e_0.$$

It is a straightforward calculation to show that these satisfy the relations. Therefore there is an epimorphism $KQ/I \to \Lambda$ which is an isomorphism; since the algebra $KQ/I$ also has dimension 12.

<u>V.2.4.1</u> COROLLARY *Let K be a field of characteristic 2 which contains a root of $t^2 + t + 1$. Then the group algebra $KA_4$ is isomorphic to the algebra in V.2.4.*

Proof: Since $A_4$ is a semidirect product of the Klein 4-group V and $C_3$, it follows that $KA_4 \cong KV * C_3$, and $KV \cong A$, see I.4.3. Moreover, in V.2.4 we only used that K contains roots of $t^2 + t + 1$.

<u>V.2.5</u> LEMMA *Suppose that A is the Kronecker algebra and char $K \neq 3$. Then there is a unique symmetric algebra $\Lambda = A * G$ where $G \leq Aut\ A$ is isomorphic to the symmetric group $S_3$. The basic algebra of $\Lambda$ is isomorphic to KQ/I where Q is the quiver*

*and I is the ideal generated by the relations*

$$\alpha^2 = 0, \quad \gamma\beta = 0, \quad \alpha\beta\gamma = \beta\gamma\alpha, \quad \eta^2 = \gamma\alpha\beta, \quad \eta\gamma = 0 = \beta\eta.$$

Proof: Let $S_3 = \langle g, h: g^3 = 1 = h^2, hgh = g^2 \rangle$. We have already seen that, without loss of generality, $X^g = \omega X$ and $Y^g = \omega^2 Y$ where $\omega$ is a root of $t^2 + t + 1$. It remains to determine the action of h. Now $\omega^2 X^h = (\omega^2 X)^h = X^{g^2 h} = (X^h)^g$. So $X^h$ is an eigenvector of g with eigenvalue $\omega^2$, and consequently $X^h = cY$ for $0 \neq c \in K$. Without loss of generality, c = 1; otherwise replace Y by cY. Similarly $Y^h = dX$ for $0 \neq d \in K$, and we deduce $X = X^{h^2} = Y^h = dX$ and d = 1. Thus the action is determined.

To find the basic algebra, we start off with the algebra $B = A * C_3$ and take it in the form V.2.4. Then $\Lambda \cong B * C_2$ where $C_2 = \langle h \rangle$. The action of h on B is given by $e_0 h = he_0$, $e_1 h = he_2$ and $e_2 h = he_1$. Moreover, $X^h = Y$ and $Y^h = X$. Hence the action of h on the arrows of the quiver of B is $(\beta\ \kappa)(\gamma\ \lambda)(\delta\ \eta)$.

*The basic algebra for* $\Lambda$ may be taken as $e\Lambda e$ where $e = e_0 + e_1$. We have $e_0\Lambda e_0 = \text{span}\{ e_0, \beta\gamma, e_0 h, \beta\gamma h \}$, $e_0\Lambda e_1 = \text{span}\{ \beta, \kappa h \}$, $e_1\Lambda e_0 = \text{span}\{\gamma, \gamma h\}$ and $e_1\Lambda e_1 = \text{span}\{ e_1, \delta h, \gamma\beta \}$. Since $\gamma + \gamma h = \gamma\alpha$ and $\beta + \kappa h = \alpha\beta$ and moreover $(\delta h)^2 = \delta\eta = \gamma\beta$, we may take the quiver of $\Lambda$ as follows:

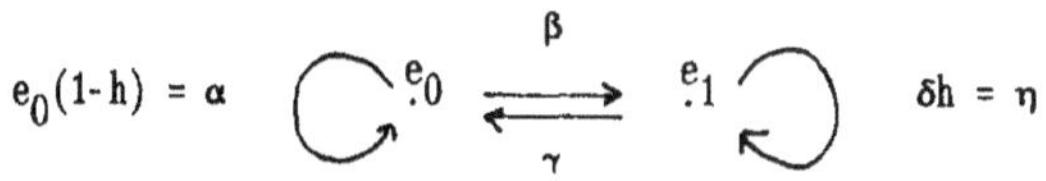

Then it is straightforward to derive the stated relations.

<u>V.2.5.1</u> COROLLARY *Suppose char K = 2, and K contains the roots of $t^2 + t + 1$. Then the group algebra $KS_4$ is isomorphic to the algebra Λ in V.2.5.*

<u>V.2.6</u> *Suppose Λ is a twisted group ring $\Lambda = A *_\gamma G$ where A is the Kronecker algebra and G is cyclic of order 2, and where char K = 2. Assume also that Λ is tame. Then Λ is isomorphic to one of the algebras in III.1(c), (c') or (d), (d').*

Let G = < τ >, then γ is determined by $\gamma(\tau, \tau) \in A^*$; and we may assume that $\gamma(\tau, \tau) = 1 + x$ for $x \in J(A)$. Since Λ is tame and local, its radical must have two generators (see I.10.10 and I.4.3.1); we deduce that its radical is generated by X and $\alpha = (1 - \tau)$. Since $X^2 = 0$ and Λ is not commutative, the statement follows from III.16.

We will now study a skew group ring over the "quaternion algebra". That is, let A be the 8-dimensional local algebra A = K< X, Y >/ L where L is the ideal

$$L = ( (XY)^2 - (YX)^2, X^2 - YXY - r(XY)^2, Y^2 - XYX - s(XY)^2 ) \text{ for } r, s \in K.$$

(as in III.1(e), (e')). If char K = 2 then the group algebra of the quaternion group of order 8 is isomorphic to A, where r = s = 0 (see III.17 and V.2.7.1 below).

<u>V.2.7</u> LEMMA *Suppose Λ = A * G where $G \leq Aut\ A$ is cyclic of order 3 and let char K ≠ 3. Assume also that Λ is symmetric. Then Λ ≅ KQ/I where Q is the quiver in V.2.4 and I is the ideal generated by the relations*

$$\beta\delta = \kappa\lambda\kappa, \quad \delta\lambda = \gamma\beta\gamma, \quad \lambda\beta = \eta\delta\eta,$$

$$\kappa\eta = \beta\gamma\beta, \quad \gamma\kappa = \delta\eta\delta, \quad \eta\gamma = \lambda\kappa\lambda, \text{ and } 0 = \beta\delta\eta = \delta\lambda\kappa = \lambda\beta\gamma$$

*There is a unique such algebra.*

<u>V.2.7.1</u> With the hypothesis as above, if such G exists then A has generators and relations as above with r = s = 0; in particular this is the case when A is the group algebra of the quaternion group with char K = 2.

Proof of V.2.7: Let $G = \langle g \rangle$. Since char $K \neq 3$ we have that every KG-module is semisimple. We will study eigenvectors of g on $J/J^2$.

(1) We must have $X^g = aX + x_1$ and $Y^g = bY + y_1$ with $a, b \in K$ roots of $t^3 - 1$ and $x_1, y_1 \in J^2$: Since $X^g$ and $Y^g$ are again independent generators of J modulo $J^2$, we have (*) $X^g = aX + cY + x_1$ and $Y^g = dX + bY + y_1$ with $ab - cd \neq 0$ $(a, b, c, d \in K)$ and $x_1, y_1 \in J^2$.

Moreover, we require that g is an algebra automorphism, therefore $(X^g)^2$ and $(Y^g)^2$ must lie in $soc_2(A)$. On the other hand, by (*) we have $(X^g)^2 \equiv ac(XY + YX)$ modulo $soc_2 A$ and therefore $ac = 0$; similarly $bd = 0$. Since $ab - cd \neq 0$, either $a = b = 0$ or $c = d = 0$. Since $g^3 = 1$, the second possibility is excluded; and it also follows that $a^3 = b^3 = 1$.

Let $v_1$ and $v_2$ be independent eigenvectors of g such that $v_1$ and $v_2$ generate J modulo $J^2$. By (1), we may assume that $v_1 = X + w_1$ and $v_2 = Y + w_2$ where $w_1$ and $w_2 \in J^2$. Then soc $A = \langle (v_1v_2)^2 \rangle$. We require that A is symmetric; by V.2.2 we must have that $[(v_1v_2)^2]^g = (v_1v_2)^2$ hence $(ab)^2 = 1$. Now a, b are 3-rd roots of 1, hence $ab = 1$. Since g is non-trivial we deduce $a = \omega$ and $b = \omega^2$ where $\omega$ is a root of $t^2 + t + 1$.

(2) We may assume that $v_1^2 = v_2v_1v_2$ and $v_2^2 = v_1v_2v_1$: The eigenspace of g with eigenvalue $\omega^2$ is spanned by $v_2$ and $v_2v_1v_2$. Since $v_1^2$ lies also in this space and in $J^2$, we deduce that $v_1^2 = c(v_2v_1v_2)$ with $c \in K$. Moreover, $v_1^2 \neq 0$ since otherwise XYX would lie in soc A. Similarly $v_2^2 = d(v_1v_2v_1)$ with $0 \neq d \in K$. After scalar transformations (see VII) we obtain $c = d = 1$. Actually, we may take now $X = v_1$ and $Y = v_2$; then we simplify the relations at the beginning by having $r = s = 0$. (Since A is symmetric, the socle relation is unchanged.)

(3) The quiver and the relations for A: By V.2.3 we know that there are three simple modules; these are 1-dimensional; and in particular, A is basic. Let $e_i$ be as in V.2.4; we identify these with the vertices of the quiver. As in V.2.3 we take as arrows $\beta = e_0Xe_1$, $\delta = e_1Xe_2$, $\lambda = e_2Xe_1$, $\kappa = e_0Ye_2$, $\eta = e_2Ye_1$ and $\gamma = e_1Ye_0$. It is straightforward to show that $A \cong KQ/I$ where I is as stated.

V.2.7.2 COROLLARY *Let char $K = 2$; then the group algebra of $SL_2(3)$ over K is isomorphic to the algebra A in V.2.7.*

In the attempt to show that certain blocks are Morita equivalent to $KS_4$, unfortunately, non-trivial factor sets could appear. However, one does not have to go into detailed calculation since there is a different way to recgnize the Morita-equivalence class, at least modulo the socle. This uses the following characterization:

V.2.8 LEMMA *Suppose* $\Lambda$ *is a symmetric basic algebra with quiver* 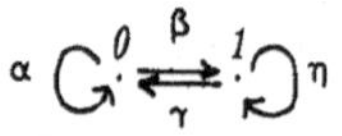



*and Cartan matrix* $C = \begin{bmatrix} 4 & 2 \\ 2 & 3 \end{bmatrix}$. *Then* $\Lambda \cong KQ/I$ *where* $I$ *is one of the following:*

*(i)* $I = [\,\beta\eta,\ \eta\gamma,\ \gamma\beta,\ \eta^2 - \gamma\alpha\beta,\ \alpha^2 - c(\alpha\beta\gamma),\ \alpha\beta\gamma - \beta\gamma\alpha\,]$ *for* $c \in K$.

*(ii)* $I = [\,\beta\eta,\ \eta\gamma,\ \gamma\beta,\ \eta^2 - \gamma\alpha\beta,\ \alpha^2 - \beta\gamma - c(\alpha\beta\gamma),\ \alpha\beta\gamma - \beta\gamma\alpha\,]$ *for* $c \in K$.

Proof: We study first the projective module $e_1\Lambda$, let $H_1 = \text{rad}\ e_1\Lambda/S_1$. By the hypotheses, $\dim H_1e_1 = 1$; moreover $0 \neq \text{top}\ H_1e_1$ and $0 \neq \text{soc}\ H_1e_1$. This implies that $\eta \in \text{soc}_2\ e_1\Lambda$ and consequently $\eta\gamma \in \text{soc}\ e_1\Lambda \cap e_1\Lambda e_0 = 0$, and $\eta\Lambda$ is spanned by $\eta$ and $\eta^2$, with $\text{soc}\ e_1\Lambda = \langle \eta^2 \rangle$. Consider a projective cover $\pi$: $e_1\Lambda \to \beta\Lambda$ given by $\pi(x) = \beta x$; since $\text{soc}\ \beta\Lambda \cong S_0$ but $\eta\Lambda e_1 = 0$, we deduce $\pi(\eta) = 0$ and $\beta\eta = 0$.

The hypothesis on the Cartan matrix implies $\gamma\Lambda = \text{span}\ \{\gamma, \gamma\alpha, \gamma\alpha\beta\}$; therefore the element $\gamma\beta$ which lies in $e_1J^2e_1$, can be written in the form $\gamma\beta = c\gamma\alpha\beta$ for some $c \in K$. We may assume $\gamma\beta = 0$; otherwise replace $\beta$ by $\beta' = (1 - c\alpha)\beta$. Considering the above projective cover again, we see that $\beta\Lambda = \text{span}\ \{\beta, \beta\gamma, \beta\gamma\alpha\}$, and in particular $\langle \beta\gamma\alpha \rangle = \text{soc}\ e_0\Lambda$. It follows that $e_0\Lambda/\beta\Lambda$ is spanned by the cosets of $e_0$, $\alpha$, $\alpha\beta$, and we have a basis of $e_0\Lambda$. Therefore $\alpha^2 = a(\beta\gamma) + c(\beta\gamma\alpha)$ for $a, c \in K$. If $a \neq 0$ then without loss of generality $a = 1$ and $\Lambda$ is an algebra as in (ii); otherwise, $\Lambda$ is as in (i).

This result could also be applied for a shortcut in the proof of V.2.5.

V.2.8.1 Suppose that $\Lambda$ is as in V.2.8(i); then $\Lambda/\text{soc}\ \Lambda$ is visibly special biserial. Now consider an algebra $\Lambda$ given in (ii); it has the following properties:

(1) $H_0 = \text{rad}\ e_0\Lambda/S_0$ is indecomposable; in particular, $\Lambda$ is not isomorphic to an algebra in (i). In fact, $H_0$ has $\geq$ three predecessors in $\Gamma_s(\Lambda)$. Firstly, $H_0$ occurs in

the middle of the standard AR-sequence I.7.7, moreover there is an AR-sequence $0 \to \beta\Lambda/S_0 \to H_0 \to \alpha\Lambda/\alpha\Lambda \cap \beta\Lambda \to 0$, and then $H_0$ must have another predecessor in $\Gamma_s(\Lambda)$, by the composition lengths.

(2) The algebra $\Lambda$ in (ii) has modules of $\Omega$-period 4; an example is the module given by

$$K \xleftarrow{\gamma=1} K \xrightarrow{\eta=1} K \xleftarrow{\beta=1} K$$

We will now investigate Morita equivalence for some blocks of p-local groups.

V.2.9 Suppose G is a finite groups and H is a normal subgroup of G. Let also K be a field of characteristic p. The situation we have to deal with is as follows:

There are blocks b and B of KH and KG respecively, and moreover $b \sim_M \mathrm{End}_{KH}(P_0)$ for a projective module $P_0$ of b, and $B \sim_M \mathrm{End}_{KG}(P_0{}^G)$. It is a well-known fact that in case $P_0$ is G-stable, $\mathrm{End}_{KG}(P_0{}^G)$ is isomorphic to a twisted group ring of G/H over $\mathrm{End}_{KH}(P_0)$, see $[C_2]$. We wish to use this, in special cases, therefore we give some details.

Suppose $M \in \mathrm{mod}\ KH$; then there are well-known vector space isomorphisms

$$(*)\quad \mathrm{Hom}_{KG}(M^G, M^G) \underset{F}{\overset{E}{\rightleftarrows}} \mathrm{Hom}_{KH}(M, (M^G)_H),$$

given by $E(f) = f_{|M}$ and $F(h) = \mathrm{Tr}_H^G(\tilde{h})$ where $\mathrm{Tr}_H^G(\phi) = \sum_{G\setminus H} \phi^g$ and

$$\tilde{h}(m \otimes g) = \begin{cases} 0 & \text{if } g \notin H, \\ h(mg) \otimes 1 & \text{if } g \in H. \end{cases}$$

Now assume that $M \otimes g \cong M$, as a KH-module, for all $g \in G$; and let $A = \mathrm{End}_{KH}M$. Choose a set $\mathcal{T}$ of representatives for the cosets of H in G, and for each $g \in \mathcal{T}$, fix an isomorphism $\epsilon_g : M \otimes 1 \to M \otimes g$ of KH-modules. Then as a vecor space, $\mathrm{Hom}_{KH}(M, M^G{}_H) \cong \bigoplus_{g \in \mathcal{T}} \epsilon_g A$. Now, using the isomorphisms in (*), one defines a ring structure on $\oplus\, \epsilon_g A$; and one obtains a twisted group ring of G/H over A. This is particularly useful in cases when elements $\epsilon_g$ exist with nice properties.

V.2.10 LEMMA *Let b be a block of KG such that the defect group D of b is central in G. Then b is Morita equivalent to KD.*

Proof: (1) $\ell(b) = 1$: Let S be a simple b-module; then D acts trivially on S, by V.2.3.1, and S has vertex D (this is well-known and easy to see). Therefore S is a projective G/D-module, and $\mathrm{Ext}^1_{G/D}(S, T) = 0$ for any simple H-module T.

Suppose now that T is an G-module with $\mathrm{Ext}^1_G(S, T) \neq 0$. We have to show that $S \cong T$. Take a non-split exact sequence $0 \to T \to E \to S \to 0$. If E is indecomposable then there is some $1 \neq g \in Z(D)$ which does not act trivially on E. Then multiplication by $1 - g$ induces an endomorphism $\mu_g$, say, of E which is non-zero and not an isomorphism. It follows that $T = \mathrm{im}\, \mu_g \cong E/T \cong S$. (This argument is due to P. Landrock.)

Let now P be the indecomposable projective b-module.

(2) $\mathrm{End}_G(P) \cong KD$: The Cartan number of b is just $|D|$ which is equal to $\dim \mathrm{End}_G(P)$. For each $g \in D$, let $\rho_g$ be the endomorphism of P given by multiplication with g. Define a map $KD \to \mathrm{End}(P)$ by $g \to \rho_g$ and linear extension. This is a ring homomorphism, and we have to show that it is one-to-one. Let $0 = \Sigma\, c_g\rho_g$ where $c_g \in K$. Since $P_{|KD}$ is projective, hence free, there is some $m \in P$ such that $\{ mg: g \in D \}$ is linearly independent in P. We have that $0 = \Sigma\, c_g\rho_g(m) = \Sigma\, c_g mg$ and then $c_g = 0$ for all $g \in D$.

Generalizing this, we have the following:

V.2.11 LEMMA *Let B be a block with defect group D, and assume that $G = DC_G(D)$. Then B is Morita equivalent to KD.*

Proof: Let $C = C_G(D)$, and let b be the block of C corresponding to B; then $C \cap D = Z(D)$ is the defect group of b (this is well-known, see for example [F], [$M_1$] or [AB]). Therefore the results of V.2.10 can be used for b; in particular there is a unique indecomposable projective b-module, $P_0$ say. Now, every indcomposable pojective B-module occurs as a summand of $P_0^G$ that is, $P_0^G$ is a projective generator of mod B; and consequently $B \sim_M \mathrm{End}_{KG}(P_0^G)$, by I.2.1.

We will now show that $\Lambda := \mathrm{End}_{KG}(P_0^G) \cong KD$. Let $A = \mathrm{End}_{KC}(P_0)$; by V.2.9, we have that $\Lambda$ is a twisted group ring of G/C over A; and by V.2.10 there is a isomorphism of

KZ and A with $z \to \rho_z$, where z is right multiplication by $z \in Z$.

We choose a transversal $\mathcal{T}$ of Z in D; then $\mathcal{T}$ is also a transversal of C in G; and moreover, every element g of $\mathcal{T}$ commutes with every element of C. Therefore the map $m \otimes 1 \to m \otimes g : P \otimes 1 \to P \otimes g$ is a C-isomorphism, and we take this for the element $\epsilon_g$ in V.2.9. With the notation of V.2.9, we have a map $\Psi$: $KD \to \Lambda$ defined as follows: Write $d \in D$ in the form $d = gz$ for $g \in \mathcal{T}$ and $z \in Z$, and let $\Psi(d) = \epsilon_g\rho_z$ and extend this linearly; and this is an isomorphism of K-algebras.

<u>V.2.12</u> PROPOSITION *Let H be a subgroup of some group G. Suppose $e_1$ is a block idempotent of KH, and e is a block idempotent of KG which is of the form*

*$e = \Sigma\ e_1^{g_i}$ where the sum is taken over a transversal of H in G, and assume also that $\{ e_1^{g_i} \}$ is a set of orthogonal idempotents of KN where $N = \cap H^{g_i}$. Let b and $b_1$ be the blocks corresponding to e and $e_1$, then b and $b_1$ are Morita equivalent.*

We shall prove this in a number of steps.

(1) <u>Let $M \in \mathrm{mod}(b)$; then there is some $M_1 \in \mathrm{mod}(b_1)$ with $M_1^G \cong M$. If M is simple then so is $M_1$:</u> We have, as N-modules, that $M = Me = \Sigma \oplus Me_1^{g_i}$, by the orthogonality. Let $M_1 = Me_1$; then $M_1$ is also an N-module. We claim now that $M \cong M_1^G$. Let $\Psi$: $M_1^G \to M$ be the linear transformation defined by $\Psi\ [\ me_1 \otimes g_j] = (mg_j)e_1^{g_j}$. It is straightforward to check that this is a G-homomorphism; and it is also bijective. Suppose M is simple, then so is $M_1$; since the functor $-\otimes_{KH}KG$ is exact.

(2) <u>Let X be a $b_1$-module. Then $(X^G)e_1 = X \otimes 1 = (X \otimes 1)e$</u>: Since $X^G = X \otimes_{KH}KG$ and also $X = Xe_1$, we have that $X \otimes g_ie_1 = X \otimes e_1g_ie_1 = X \otimes g_i(e_1^{g_i})e_1 = 0$ if $g_i$ does not lie in H. So $(X^G)e_1 = (X \otimes 1)e_1 = X \otimes 1$, which is equal to $(X \otimes 1)e$ since $(X \otimes 1)e_1^{g_i} = X \otimes e_1(e_1^{g_i}) = 0$ for $g_i$ not in H.

(3) <u>Let S be a simple $b_1$-module, then $S^G$ is a simple b-module</u>: Let $0 \neq m = \Sigma\ s_i \otimes g_i$, and, say, $s_k \neq 0$; then $mg_k^{-1} = \Sigma\ s_i' \otimes g_i \in mKG$ with $s_1' = s_k \neq 0$. By (2) we have $(mg_k^{-1})e_1 = s_k \otimes 1$, and it follows now that $mKG = S^G$, and hence $S^G$ is simple. Moreover, $(S^G)e = (S \otimes 1)e_1 \neq 0$.

Let S be a simple $b_1$-module, then $S^G$ is a simple b-module; and every simple b-module is of this form. Moreover, if S and T are simple $b_1$-modules such that $S^G \cong T^G$ then $(S^G)e_1 \cong (T^G)e_1$, and by (2) we deduce $S \cong T$. Hence the map $S \to S^G$ gives a one-one correspondence between the simple modules for $b_1$ and those for b. Consequently we also have a bijection $P \to P^G$ between the indecomposable projective modules for $b_1$ and those for b. Moreover, b and $b_1$ have the same Cartan matrix; in particular the basic algebras of b and $b_1$ have equal dimensions.

Let $P = \bigoplus_{i=1}^{k} P_i$ where $P_1, \ldots, P_k$ are representatives for the indecomposable projective $b_1$-modules; then $A_1 = \mathrm{End}_H(P)$ is the basic algebra of $b_1$ and $A = \mathrm{End}_G(P^G)$ is the basic algebra of b (I.2.1); and our aim is to show that A and $A_1$ are isomorphic as algebras.

Define $\psi: A_1 \to A$ by $\psi(f) = f \otimes 1$. This is an algebra homomorphism and is also one-to-one; and by the dimensions, $\psi$ is an isomorphism.

<u>V.2.12.1</u> Suppose G is a finite group, V is a normal p-subgroup of G and char K = p, and assume also that b is a block of KG, with idempotent e. Now let $H = VC(V)$; then also $H \trianglelefteq G$. It is known that e lies already in KH (see for example [AB]). Write e as a sum of block idempotents of KH, and fix one of the summands, call it $e_V$. If we wish to determine the Morita equivalence class of b then we may assume that $e_V$ is G-stable:

Let $N = N(e_V)$ be the stabilizer of $e_V$; then $e_V$ is a primitve idempotent of N and moreover $e = \sum_{G/N} e_V^{g_i}$, see [AB]. Moreover, we have that $H = \cap N^{g_i}$, and the $\{e_V^{g_i}\}$ are orthogonal idempotents of KH. Therefore, by V.2.12, the blocks b and $e_VKH$ are Morita equivalent; and we may assume that $e = e_V$, that is, $e_V$ is G-stable.

<u>V.2.13</u> *Suppose B is a tame block whose defect group D is normal in G, and $M \in$ mod B is periodic, with $\Omega$-period m.*

*(i) If D is a Klein 4-group then m = 1 or 3.*

*(ii) If D is dihedral of order $\geq 8$ then $m \leq 3$.*

*(iii) If D is semidihedral then $m \leq 4$.*

*(iv) If D is a quaternion group then m = 1 or 2 or 4.*

Proof: The module $M$ is $H$-projective where $H = DC_G(D)$, let $W \in \mathrm{mod}\ KH$ be indecomposable such that $M$ is a direct summand of $W^G$ (see V.1.2); and moreover $W$ must periodic as well. Then $W$ belongs to a block b of KH, and D must have defect group D. Without loss of generality, b is G-stable (V.2.12). By V.2.11 we have that $b \sim_M KD$, hence the possible periods for $W$ are known, recall that $\Omega^2 \cong \tau$.

(a) Assume first that G/H is a 2-group; then by Green's Theorem, the module $W^G$ is indecomposable, and $W$ and $M$ have the same period. Now the statements follow from II.7.6 (and II.7.3) and II.10.1 in the case D is dihedral or semidihedral. Suppose D is quaternion, then we know that $\Omega^4 K \cong K$. If $M$ is arbitrary then a projective resolution of $M$ is obtained by taking $M \otimes_K \mathcal{P}$ where $\mathcal{P}$ is projective resolution for K. Consequently $\Omega^4(M) \cong M$ and m divides 4.

(b) Now assume that G/H contains elements of odd order; then either D is a Klein 4-group, or else a quaternion group of order 8. We claim first that

(*) If D is a Klein 4-group then the period is $\leq 3$.

As a module for the Kronecker algebra, $W$ has $\Omega$-period 1 (this can be seen directly from I.4.3) and the summands of $W^G$ are a union of $\Omega$-orbits. Moreover, it is easy to see that $W^G$ has either one or three summands.

The fact (*) is used for V.2.14.

The remaining details for case (b) are consequences of V.2.14 and V.2.15: If D is a Klein 4-group then we are in the situation of V.2.14 (c) or (d), and we apply II.7.4 and II.7.5. Otherwise, by V.2.15, the block B containing $M$ is Morita equivalent to the algebra $\Lambda$ in V.2.7 with char K = 2. One can write down explicitly minimal projective resolutions of length 4 for the simple modules of the algebra. Then one observes that $\Lambda$ is Morita equivalent to the group algebra of $SL_2(3)$ (see V.2.7.1); and for the general case, one works in this group algebra by using tensor product, as in (a).

<u>V.2.14</u> PROPOSITION *Let V be a Klein 4-group which is normal in G and let B be a tame block of G with defect group D. Then one of the following holds:*

*(a) $B \sim KV$ [ and $\ell(B) = 1$].*

*(b) $B \sim KD$ [ and $\ell(B) = 1$].*

*(c) $B \sim K\Lambda$, [ and $\ell(B) = 3$]*

*(d) $B \sim KS_4$, modulo the socle [ and $\ell(B) = 2$].*

*In particular, a defect group of B is either V [ in (a) and (c)], or otherwise a dihedral group of order 8.*

Proof: Let $H = C(V)$; then H is a normal subgroup of G; and let b be the block of KH corresponding to B (see V.2.11). Suppose D is a defect group of B.

(1) Either $D = V$, or $V \subset D$ and D is dihedral of order 8: By V.1.2, $V \subseteq D$; and since b is tame, D must be dihedral or semidihedral or quaternion. The only such group which also contains a normal Klein 4-group are V and a dihedral group of order 8.

(2) V is a defect group of b: From [AB(2.9)] we have that $V \leq \delta_N(b)$, and moreover, V is central in $\delta_N(b)$. On the other hand, $\delta_N(b) \leq D \leq$ a dihedral group of order 8, by (1), and the statement follows.

Therefore we may apply V.2.10. In particular, there is just one indecomposable projective b-module, $P_0$ say, and $A := End_{KN}(P_0) \cong KV$. Moreover every projective B-module occurs as a summand of $P_0^G$ and by I.2.1, B is Morita equivalent to $End_{KG}(P_0^G)$.

Let $\Lambda = End_{KG}(P_0^G)$, then $\Lambda$ is isomorphic to a twisted group ring $A *_\gamma G/N$, and the group G/N is isomorhic to a subgroup of Aut V which is $S_3$.

If $G = N$, then (a) holds. Now suppose G/N is cyclic of order 2; then $\delta(B)$ has order 8, and by (2) there is some element $\tau$ of order 2 in $G - N$, and by V.2.8, the algebra $\Lambda$ is one of III.1(c) to (d'). The algebras (d) and (d') have modules of period 4 (see II.10), on the other hand, the periods of B-modules are $\leq 2$, by V.2.13, and therefore B must satisfy (b).

Now consider the case when G/N is cyclic of order 3. Fix $g \in G - N$, and choose some $\epsilon_g$ as in V.2.9; let $\epsilon_g(m \otimes 1) = \varphi_g(m) \otimes g$. Then we have that $\epsilon_g^3(m \otimes 1) = \varphi_g^3(m) \otimes 1$, and $\varphi_g^3 =: u$ is a unit in A.

(3) We may assume that $u = 1$: Without loss of generality, $u = 1 + x$ for $x \in J(A)$. Now, $x^2 = 0$ and therefore $u^2 = 1$, and $(\varphi_g u)^3 = uu^3 = 1$ since $\varphi_g$ cummutes with u. We replace $\varphi_g$ by $\varphi_g u$.

This shows that $\Lambda$ is a skew group ring, and B satisfies (c), by V.2.4.

Now assume that $G/N \cong S_3$. Then the basic algebra of B satisfies the hypothesis of

V.2.8. Moreover, by V.2.13, condition (*), we know that B does not have modules of $\Omega$-period 4. Consequently we deduce from V.2.8.1 that the basic algebra of B is of the form V.2.8(i), and then $\Lambda/\mathrm{soc}\ \Lambda \cong KS_4/\mathrm{soc}\ KS_4$, by V.2.5.1.
This completes the proof of V.2.14.

<u>V.2.15</u> *Let D be a quaternion group of order 8 which is normal in the group G, and let B be a tame block of KG whose defect group is D. Then one of the following holds:*
*(i) $B \sim_M KQ$ and $\ell(B) = 1$.*
*(ii) $B \sim_M KG$ where $G = SL_2(3)$ and $\ell(B) = 3$.*

Proof: Let $H = DC_G(D)$, then H is normal in g. Let also b be the block of KH corresonding to B (see V.2.12). We may assume that b is G-stable, by V.2.11. Since D is a defect group of B, we have that the index $|G:H|$ is odd (see V.5). On the other hand, G/H is isomorphic to a subgroup of Aut D, and hence either G = H, or G/H is cyclic of order 3. In the first case, we deduce that B = b, and (i) holds by V.2.11. Now assume G/H is cyclic of order 3. For b we apply V.2.11; let $P_0$ be the indecomposable projective b-module, then $\Lambda = \mathrm{End}_{KH}(P_0) \cong KD$ and this is isomorphic to the algebra in III.1.(e') with k = 2. Fix $g \in G - H$ and choose some $\epsilon_g$ as in V.2.9; let $\epsilon_g(m \otimes 1) = \varphi_g m \otimes g$; then we have $\epsilon_g^3(m \otimes 1) = \varphi_g^3 m \otimes 1$, and moreover $\varphi_g^3 = u$ is a KH-isomorphism of $P_0$.
(1) <u>We may assume that $\varphi_g^3 = 1$:</u> As we just have seen, u is a unit of $\Lambda$, a which is a local ring. We may assume that $u = 1 + x$ where $x \in J$; otherwise we replace $\varphi_g$ by a scalar multiple. Note also that $\varphi_g$ commutes with u. Now, $J(\Lambda)^4 = 0$ and therefore $(1 + x)^4 = 1 + x^4 = 1$ and therefore $u^3 = u^{-1}$. We deduce that $(\varphi_g u)^3 = 1$, ad we may replace $\varphi_g$ by $\varphi_g u$.
This shows that $\Lambda$ is a skew group ring of the cyclic group of order 3 over $\Lambda$, and it follows from V.2.7.1 that B satisfies (ii).

<u>V.2.16</u> *Suppose B is a block whose defect group D is normal in G, and suppose D is dihedral or semidihedral of order $\geq 8$, or quaternion of order $\geq 16$. Then $B \sim_M KD$.*

Proof: Let $H = DC_G(D)$, then H is normal in G and the index of H in G is a power of 2 since Aut D is a 2-group. Let b be the block of KH corresponding to B, by V.2.12 we may assume that b is G-stable. On the other hand, the defect group of B is D, therefore the index is odd, and we deduce H = G; and the statement follows now from V.2.10.

## V.3 *On the AR-quiver of group algebras*

In this chapter we shall discuss results based on restriction and induction of modules between groups and subgroups, in order to determine the graph structure of components of $\Gamma_s(KG)$ where G is a arbitrary finite group. Details may be found in [Be], $[E_{4,9}]$, $[K_{1,2}]$.

V.3.1 LEMMA *Let $A(M)$ be the AR-sequence of some KG-module M, and let $H \leq G$. Then the sequence $A(M)_{|H}$ splits if and only if $vx(M) \subsetneq H$.*

V.3.2 LEMMA *Suppose that $0 \to \tau M \to E \to M \to 0$ is an AR-sequence of KG-modules, and let X be an indecomposable summand of E. Then either $vx(X) \leq_G vx(M)$, or $vx(M) <_G vx(X)$.*

V.3.3 LEMMA *Let $H \leq G$, and assume that C is an indecomposable KH-module such that $(C^G)_H$ has a unique summand isomorphic to C. Suppose $C^G = M \oplus C'$ where M is indecomposable and $M_H \cong C \oplus *$. Then $A(C)^G \cong A(M) \oplus \mathcal{E}$ where $\mathcal{E}$ is a split sequence.*

In particular, the hypothesis is satisfied if $N_G(V) \leq H$ where V is a vertex of C; by V.1.3, then M is the Green correspondent gC of C in G.

Conversely, given an indecomposable KG-module M one may apply the above result by choosing H to be some subgroup of G containing $N_G(V)$ where $V = vx(M)$, and by taking for C the Green correspondent fM.

After having seen that the AR-sequence $\mathcal{A}(M)$ splits off from an induced AR-sequence $\mathcal{A}(C)^G$ one needs to study the middle term of $\mathcal{A}(M)$; important would be to know the number of indecomposable summands. This is to some extent answered by the following:

<u>V.3.4</u> PROPOSITION (Kawata) *Suppose $M$ is a KG-module and $fM$ is the Green correspondent of $M$ in $N_G(V)$ where $V = vx(M)$. Suppose the AR-sequences are $0 \to \tau M \to E \to M \to 0$ and $0 \to \tau fM \to E' \to fM \to 0$ respectively. Then there is a one-one-correspondence between*

*(i) the indecomposable summands $Y'$ of $E'$ with $vx(Y')_N\geq V$, and*

*(ii) the indecomposable summands $Y$ of $E$ with $vx(Y)\ {}_G\geq V$.*

*In fact, $(Y')^G$ has a unique summand $Y$ with $vx(Y) =_G vx(Y')$ ; and $Y' \to Y$ gives the correspondence.*

Concerning the graph structure of components, there is the following result, also due to Kawata:

<u>V.3.5</u> THEOREM (Kawata) *Suppose $\Theta$ is a connected component of $\Gamma_s(KG)$. Define $Vx(\Theta) = \{\ vx(M):\ M \in \Theta\ \}$, and set $Q$ be a minimal element in $Vx(\Theta)$. Choose $M_0$ in $\Theta$ with $vx(M_0) = Q$, let $N = N_G(Q)$ and $\Delta$ be the component of $\Gamma_s(KN)$ containing $fM_0$.*
*Then there is a subquiver $\Delta_0$ of $\Delta$ and a graph isomorphism $\Psi$: $\Delta_0 \to \Theta$; here $\Psi$ arises from taking summands of induced modules. Moreover, $Q \leq H \leq_G N_G(Q)$ for all $H \in Vx(\Theta)$.*

We need to know this subquiver explicitly: It consists of all modules L such that there is a walk in $\Delta$ : $fM_0 = L_0 - L_1 - \dots - L_t = L$ such that $Q \leq_G vx(L_i)$.

We will also need some results on vertices of modules in stable components of the form $\mathbb{Z}A_\infty^\infty$.

<u>V.3.6</u> PROPOSITION *Suppose $\Theta$ is a component of tree class $A_\infty^\infty$ of $\Gamma_s(KG)$ where $G$ is a p-group and char $K = p$. Then one of the following holds:*

*(i) All modules in $\Theta$ have the same vertex.*

*Otherwise, only two distinct vertices $Q$, $P$ occur, with $Q \lneq P$. Moreover, $p = 2$, and either*

*(ii) $|P:Q| = 2$ (and $|Q| > 4$). The modules with vertex $Q$ lie in a subquiver $\Theta_Q$ such that both $\Theta_Q$ and the complement $\Theta - \Theta_Q$ are connected and as graphs isomorphic to $\mathbb{Z}A_\infty$.*

*(iii) $Q$ is a Klein 4-group and $P$ is dihedral of order 8. The modules with vertex $Q$ lie in two adjacent $\tau$-orbits.*

V.3.7 LEMMA *Suppose $\Lambda = KD$ where $D$ is dihedral of order 8 and char $K = 2$. Then the non-periodic $\Lambda$-modules whose vertex is a fixed Klein 4-group lie in one component of $\Gamma_s(\Lambda)$ which has tree class $A_\infty^\infty$.*

That is, condition (iii) in V.3.6 does occur. For the proof, fix a Klein 4-subgroup V of D. Let $M \in$ mod KD have vertex V and source S; then actually $M \cong S \otimes_{KV} KD$, by Green's Theorem (V.1.5); and with M also S is non-periodic.

The non-periodic KV-modules are the modules $\{\Omega^n K: n \in \mathbb{Z}\}$, see II.7.3, and they form therefore two $\tau$-orbits only. Hence we are done if we show:

(*) *There is an irreducible map $(\Omega K)^D \to K^D$.*

Recall that $KD \cong K\langle X, Y\rangle/(X^2, Y^2, (XY)^2 - (YX)^2)$, see III.13, which is special biserial; hence we know the irreducible maps between non-periodic modules by II.5. With the parametrization of modules as in II.3, we have that the non-periodic indecomposable KD-modules are strings. One finds here by an easy direct calculation that $(K_V)^D \cong M(C)$ where $C = X$, the word of length 1, say. For convenience of notation, we use left modules here; then by II.5, there is an irreducible map $M({}_cC) \to M(C)$ where ${}_cC = (XYX)^{-1}YX$, and one sees directly that $M((XYX)^{-1}YX)) \cong \Omega(M(C))$.

V.3.8 LEMMA *Suppose $\Lambda = KG$ where $G$ is the symmetric group $S_4$ and char $K = 2$. Then the non-periodic modules whose vertex is a normal Klein 4-group of $G$ lie in one component of $\Gamma_s(KG)$, and they form four adjacent $\tau$-orbits.*

Proof: Let $A_4$ be the alternating group, and let V be the normal Klein 4-group $\leq G$; then $V < A_4$. By V.1.2, every module M with vertex V is $A_4$-projective, and by Green's Theorem (V.1.5) there is an indecomposable $A_4$-module $M_0$ with $(M_0)^G \cong M$. Moreover,

$\mathcal{A}(M_0)^G \cong \mathcal{A}(M)$, by V.3.3. In II.7.4 we have proved that the non-periodic $KA_4$-modules lie in a component of tree class $\tilde{A}_5$, that is, there are just six (adjacent) $\tau$-orbits. Now, two of the simple modules of $A_4$ are conjugate in G; and their $\Omega$-orbits become the $\Omega$-orbit of the induced module. Consequently the induced non-periodic modules with vertex V form four $\tau$-orbits of G, and they are adjacent.

Explicitly, let $S_0$ and $S_1$, $S_2$ be the simple $A_4$-modules and $S_1{}^G \cong S_2{}^G = E$; then $S_0{}^G = \mathcal{U}(K, K)$. We work in the basic algebra of KG. With the notation of II.7.5, the quiver of KG is of the form

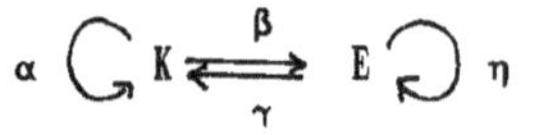


With the parametrization of modules as in II.3 ( using left modules for the moment), we have $E = M(1_E)$ and $(S_0)^G = M(\alpha)$. By II.5, there are irreducible maps

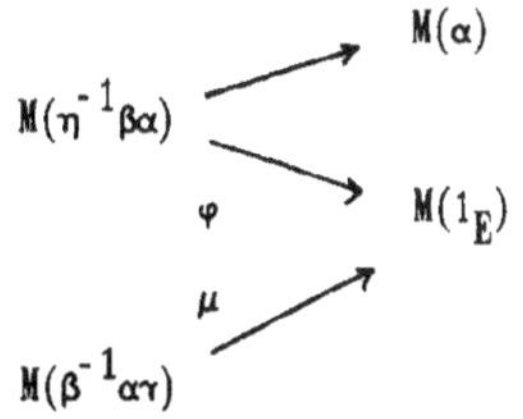


where ker $\varphi = M(\beta\alpha)$ and ker $\mu = M(\alpha\gamma)$. On the other hand, $M(\eta^{-1}\beta\alpha) = \Omega M(1_E)$ and $M(\beta^{-1}\alpha\gamma) = \Omega M(\alpha)$.

## V.4 *The stable AR-quiver for tame blocks*

First we will determine the graph structure of the stable AR-components for dihedral blocks. The result is as follows:

<u>V.4.1</u> **THEOREM** *Let B be a block with dihedral defect groups; then $\Gamma_s(B)$ has only the following components:*

*(a) Tubes of rank 1 or 3. There are infinitely many 1-tubes and at most two 3-tubes.*

*(b) Non-periodic components on which two arrows end at each vertex.*

<u>V.4.1</u> *The following details hold as well:*

In (a) we have that each 3-tube corresponds to a 3-tube of a local block $\sim_M KA_4$ or $KS_4$, and the modules in it have vertex $V_4$. In particular, the number of such tubes is 0 if each local $N(V_4)$-B-subpair (see V.5) has only one simple module, it is 2 if there are two non-conjugate $N(V_4)$-B-subpairs each of which has > 1 simple module, and 1 otherwise. We note that this is the same division into cases as in [$B_2$].

In (b): Assume first that $\delta(B)$ is a Klein 4-group. If $\ell(B) = 1$ then there is just one non-periodic component; it has tree class $\tilde{A}_{12}$. Otherwise, $\ell(B) = 3$, and there is one non-periodic component, and it has tree class $\tilde{A}_5$. Now let $\delta(B)$ be of order $\geq 8$; then there are infinitely many non-periodic components, and they are all isomorphic to $\mathbb{Z}A_\infty^\infty$.

<u>Proof of V.4.1</u>: We shall first study some local blocks.

<u>Case 1</u> $\delta(B) = V_4 \trianglelefteq G$ and $\ell(B) = 1$. By V.2.14 we know that $B \sim_M KV_4$. Also, $KV_4$ is isomorphic to the Kronecker algebra, and we obtain the statement from II.7.3.

<u>Case 2</u> $\delta(B) = V_4 \trianglelefteq G$ and $\ell(B) > 1$. In V.2.14 we have seen that $B \sim_M KA_4$; the statement follows now from II.7.4.

<u>Case 3</u> $V_4 \subsetneqq \delta(B)$ and $V_4 \triangleleft G$. By V.2.14 we have that $B \sim_M KS_4$, with two simple modules. By II.7.5 there is just one tube of rank 3, and otherwise there are only 1-tubes. The non-periodic components are isomorphic to $\mathbb{Z}A_\infty^\infty$.

<u>Case 4</u> $D = \delta(B) \trianglelefteq G$ and $|D| \geq 8$. Then by V.2.16 we have that $B \sim_M KD$. Now we obtain from II.7.6 that there are only 1-tubes and $\mathbb{Z}A_\infty^\infty$-components; as stated.

Now we consider the general case. We shall apply the results from V.3.

*Tubes:* Green correspondence preserves periodicity and the period (V.1.4); hence the tubes of B have rank 1 or 3 (see V.2.13).

*There are $\leq$ two 3-tubes:* It suffices to show that there are at most six modules at the end of 3-tubes. Let M be a module at the end of a 3-tube, let $Q = vx(M)$ and $N = N_G(Q)$. Then the Green correspondent fM lies in a local block b and also has period 3. From cases 1 to 4 we deduce that Q must be a Klein 4-group and that b has one or two 3-tubes. Moreover,

(*) all modules in such a tube have the same vertex.

It suffices now to show the following: If X belongs to a 3-tube $\Theta$ and X does not lie at the end then gX does not lie at the end.

By (*) we know that all modules occuring in the AR-sequence of X have vertex Q; Now it follows from V.3.4 that $\mathcal{A}(gX)$ has at least two middle terms.

To finish the statements on 3-tubes we need that there are in total at most two 3-tubes.

If $\delta(B)$ has order 4 then there is only one such local block; which has $\leq$ two 3-tubes. Suppose $|\delta(B)| \geq 8$. Then there are two conjugacy classes of local $V_4$-blocks to be taken into account, see V.5.13 below. They both have defect groups of order 8; they are blocks of the form studied in case 3 and 4. We have seen that each of then has at most one 3-tube, as required. Using II.4 one sees that there are infinitely many 1-tubes.

*Non-periodic components:* It suffices to show that $\beta(M) \geq 2$ for any non-periodic module M. Let V be a vertex of such M; then V is not cyclic (V.1.4), hence must be dihedral here. Let b be the block of $N_G(V)$ containing fM, and let $\Delta$ be the component of fM. We know that b is Morita equivalent to KD or $KA_4$ or $KS_4$, and also, the Morita equivalence in V.2.14 preserves vertices. We have studied vertices of modules in non-periodic components of these algebras; applying V.3.5 to V.3.8 shows that either the vertices in a component are constant, or else only two vertices occur. Therefore, if $\Delta_0$ is the subquiver in V.3.5 then $\Delta_0 = \Delta$. Moreover, we have determined the tree class of $\Delta$ above in cases 1 to 4; and we obtain the result for the component of M.

Now we shall determine the graph structure of the stable AR-components for semidihedral blocks. The result is as follows:

<u>V.4.2</u> THEOREM *Let B be a block with semidihedral defect groups; then $\Gamma_s(B)$ has only the following components:*

*(a) Tubes of rank $\leq$ 3. There are infinitely many 1-tubes and 2-tubes, but $\Gamma_s(B)$ has at most one 3-tube.*

*(b) Non-periodic components $\cong \mathbb{Z}A_\infty^\infty$ and $\mathbb{Z}D_\infty$.*

<u>V.4.2.1</u> *The following details are true:* In (a), a 3-tube corresponds to a 3-tube of a

local block with dihedral defect groups $\sim_M KS_4$, and the modules in it have vertex $V_4$. In particular, there is a 3-tube if and only if the local $N_G(V_4)$-block has more than one simple module.

Proof of V.4.2 We shall study first some blocks of 2-local groups.
Assume first that $1 \neq O_2(G)$ (the largest normal 2-subgroup of $G$), and $O_2(G)$ is dihedral (and then of order $\geq 8$) or semidihedral. Recall that $O_2(G)$ is contained in a defect group of B (V.1.2). Therefore we deduce from V.2.16 that $B \sim KD$ where D is a defect group. Now the statement follows from II.10.1.
Now let G be arbitrary.
(a) *Tubes:* The Green correspondence preserves periodicity and the period (V.1.4), and Green correspondents lie in tame blocks again, hence by V.2.13, the tubes of B have rank $\leq$ 3. The arguments in the proof of (1) in V.4.1 show that the 3-tubes of B correspond to 3-tubes in local $N_G(V_4)$-blocks. There is only one conjugacy class of Klein 4-groups $\leq$ D and only one such block has to be considered. Its defect group must be dihedral of order 8, and so by V.2.14 there is $\leq$ one 3-tube.

There are infinitely many 1-tubes and 2-tubes of modules with dihedral and quaternion vertices respectively, see II.10.9.
(b) *Non-periodic components*: Suppose M lies in a non-periodic component, $\Theta$ say. Let V be a vertex of M; then V is not cyclic or quaternion (V.1.4), hence is dihedral or semidihedral. Let b be the block of $N_G(V)$ containing fM, and let $\Delta$ be the component of fM. Then b is tame and its defect group is $\leq_G D$, so $\delta(b)$ is dihedral or semidihedral, and then by V.2.14, either $b \sim_M KS_4$ or $b \sim_M KD$. Continue as in V.3.1.

The result on the graph structure of $\Gamma_s(B)$ for quaternion blocks is essentially well-known. It is for group algebras usually proved using cohomology, and this appears frequently in the literature. Here we give a different proof.

V.4.3 THEOREM *Suppose B is a block with quaternion defect groups. Then the stable AR-quiver $\Gamma_s(B)$ consists only of tubes of rank 1 or 2.*

Proof: Let $\Theta$ be a component of $\Gamma_s(B)$. Then $\Theta$ is infinite (I.3.8); hence it must contain some module $M$ with $vx(M) = Q$ non-cyclic. Then $Q$ is some quaternion group. Moreover, the Green correspondent of $M$ lies also in some block with quaternion defect group (V.5.3). The periodicity is preserved under Green correspondence; so we may assume that $Q$ is normal in $G$.

We have that $C(Q) \Delta G$. Let $b_Q$ be a block of $C(Q)$ appearing in a decomposition of $B$ as a central idempotent of $C(Q)$.

(1) <u>We may assume that $b_Q$ is G-stable:</u> Let $N = N(b_Q)$; then $b_Q$ is aso a block of $N$ [ and also of any subgroup of $N$ containing $C(Q)$ ], by [AB;2.9]. Say, this block of $N$ is $b_1$. Any B-module $M$ is isomorphic to $M_1{}^G$ where $M_1$ lies in $b_1$ (see V.2.12). So we may replace $G$ by $N$ and $B$ by $b_1$ and get (1).

Then $B = b_Q$, and $B$ is also a block of $C := C(Q)$. There is a summand $M_0$ of $M_{|C}$ ( with $M_0$ in $b_Q$) and $M \mid M_0{}^C$.

<u>Case 1</u> $G/QC(Q)$ is a 2-group. Then by Green's theorem (V.1.5), the module $M_0{}^G$ is indecomposable, and therefore $M$ and $M_0{}^G$ have the same period. Now, $b_Q$ is Morita equivalent to the group algebra $KQ$ (see V.2.6). By V.3.7 below, all $KQ$-modules have period 1, 2 or 4.

<u>Case 2</u> $G/QC(Q)$ is not a 2-group. Then $|Q| = 8$, and $C_3 \leq G/QC(Q) \leq S_3$. Hence $G$ as a normal subgroup $S$ of index 2 or 1 with $QC(Q) \Delta S \leq G$ and $S/QC(Q)$ cyclic of order 3. Every B-module is S-projective; and by Green's theorem there is an indecomposable S-module $M_0$ in $b_Q$( as a block of S ) with $M \cong M_0{}^G$. Now, the block $b_Q$ of $S$ is Morita equivalent to the group algebra of $SL_2(3)$ (V.2.16). By V.2.13, all modules in this block have period 1, 2 or 4.

<u>V.4.4</u> LEMMA *Let $\Lambda$ be the local algebra of quaternion type. Then*

*(a) The simple module $K$ has $\tau$-period 2.*

*(b) If $\dim \Lambda$ is a power of 2 and char $K = 2$ then all $\Lambda$-modules have $\tau$-period 1 or 2.*

Proof: By I.8.11 we may assume that $\Lambda = K\langle X, Y\rangle/I$ where $I = ( X^2 - (YX)^{k-1}Y,\ Y^2 - (XY)^{k-1}X,\ (XY)^k - (YX)^k,\ (XY)^kX,\ (YX)^kY )$. We determine first the $\Omega$-period of $K$; and we identify $\Omega K$ with the module $X\Lambda + Y\Lambda$ and $\Omega^{-1}(K)$ with the submodule $(X, Y)\Lambda$ of $\Lambda \oplus \Lambda$ (see I.6.4).

*A projective cover of* $\Omega K$: We define $\pi_1 : \Lambda \oplus \Lambda \to \Omega K$ by $\pi_1(u, v) = Xu - Yv$; then $\pi_1$ is a surjective $\Lambda$-homomorphism. Let $L = \ker \pi_1$; then by the exactness, $\dim L = 2|\Lambda| - |\Omega K| = |\Lambda| + 1$, and consequently $L \cong \Omega^2 K$.

*Generators for* $L$: Define elements $\omega_0$ and $\omega_1$ of $\Lambda \oplus \Lambda$ by $\omega_0 = (X, (XY)^{k-1})$ and $\omega_1 = ((YX)^{k-1}, Y)$. Then $\omega_0\Lambda + \omega_1\Lambda$ is a submodule of $\ker \pi_1$. Moreover, it is not difficult to see that they have the same dimension: The intersection of $\omega_0\Lambda$ and $\omega_1\Lambda$ is generated by $\omega_0 X$ which is equal to $\omega_1 Y$ and has dimension 3. Hence $\omega_0\Lambda + \omega_1\Lambda \cong \Omega^2 K$.

*A projective cover of* $\Omega^2 K$: Define $\pi : \Lambda \oplus \Lambda \to \omega_0\Lambda + \omega_1\Lambda$ to be the map $\pi(u, v) = \omega_0 u - \omega_1 v$. Let $L = \ker \pi$; then $\dim L = |\Lambda| - 1$. We have alredy seen that $(X, Y)\Lambda$ is contained in $\ker \pi$; and by dimension, equalitiy holds. Therefore $\Omega^3 K \cong \Omega^{-1} K$, as stated.

(b) Suppose now that char $K = 2$ and $\dim \Lambda$ is a power of 2. Then $\Lambda/\mathrm{soc}\,\Lambda \cong KQ/\mathrm{soc}\,KQ$ where $Q$ is a quaternion 2-group. Then tensor products of $\Lambda$-modules are defined. In particular, if $M$ is any indecomposable non-projective $\Lambda$-module then we obtain a projective resolution for $M$ by tensoring $M$ with a projective resolution of $K$. Now, $\Omega^4 K \cong K$, consequently $\Omega^4 K \otimes M \cong \Omega^4 M \oplus$ projectives, so $M \cong \Omega^4 M$, and the period of $M$ divides 4, as required.

## V.5 *The number of irreducible characters for a tame block*

The aim of this chapter is to determine the number k(B) of irreducible characters of B when B is a tame block. Moreover, at the end we will determine the Klein 4-group subpairs for dihedral blocks.

We start by giving a summary of the relevant block theory. This is based on the approach in [AB]. In this chapter, " block" means usually "block idempotent".

V.5.1 *Brauer's formula* Let B be an arbitrary p-block of some finite group G. Then

$$k(B) = \ell(B) + \Sigma\ \ell(b_u)$$

where the sum is taken over the subsections $(u, b_u)$ of B in G $(u \neq 1)$.

A subsection of B in G is a G-conjugacy class of pairs $(u, b_u)$ where u is a p-element of G and $b_u$ is a block of $C_G(u)$ such that $Bb_u \neq 0$. Recall that $\ell(b)$ is the number of simple modules of the algebra bKG.

In order to use this formula, one should first determine the subsections of B in G. This may be done locally.

V.5.2 Suppose G is a finite group. A convenient analogue for blocks of "p-subgroups" of G is the notion of "subpairs". A subpair of G is, by definition, a pair $(P, b_P)$ where P is a p-subgroup of G, and $b_P$ is a block of $C_G(P)$. (When $P = \langle u \rangle$ one writes $(u, b_u)$.)

If $(P, b_P)$ and $(Q, b_Q)$ are two subpairs of G then $(Q, b_Q)$ is normal in $(P, b_P)$, provided

(i) $Q \vartriangle P$,

(ii) $b_Q$ is stable under P-conjugation,

(iii) $b_P Br_P(b_Q) = b_P$.

Here $Br_P$ is the Brauer map (see for example [AB]; we will not use it here.)

Then inclusion of subpairs is the transitive closure of normality. One writes $(Q, b_Q) \subset (P, b_P)$ and $(Q, b_Q) \vartriangle (P, b_P)$. The following is of great importance:

**THEOREM** [3.4, AB ] *If $(P, b_P)$ is a subpair of G and Q is a subgroup of G then there is a unique block $b_Q$ of $C_G(Q)$ such that $(Q, b_Q) \subset (P, b_P)$. Moreover, if $Q \vartriangle P$ then*

*$(Q, b_Q) \vartriangle (P, b_P)$.*

Now we fix a block B of G. A B-subpair is a subpair $(P, b_P)$ such that $(1, B) \subset (P, b_P)$. (It can be shown that this is the same as requiring that $b_P B \neq 0$.)

Following [AB], a Sylow B-subpair is a B-subpair $(P, b_P)$ such that P is a defect group of B. This definition is motivated by:

THEOREM [3.10, AB] (a) *The Sylow B-subpairs are the maximal B-subpairs.*
*(b) G acts transitively by conjugation on the Sylow B-subpairs,*

In group theory, if one wants to determine the G-conjugacy classes of p-elements of G, then by Sylow's theorem, these are the G-conjugacy classes of elements u where u belongs to a fixed Sylow p-subgroup of G. These may then be determined locally, if necessary, by Alperin's theorem on fusion (see for example [Go]). This has an analogue in block theory: We consider the problem to determine the G-conjugacy classes of B-subpairs $(u, b_u)$; that is, the subsections of B (returning to the starting point, see V.5.1).

By [AB; 3.10] we may fix a Sylow B-subpair $(D, b_D)$ and determine the G-conjugacy classes of $(u, b_u)$ where $(u, b_u) \subset (D, b_D)$. There is an analogue of Alperin's fusion theorem, namely:

COROLLARY [4.13, AB] *Suppose $(u, b_u)$ and $(v, b_v)$ are B-subpairs contained in the Sylow B-subpair $(D, b_D)$. Then $(u, b_u)$ and $(v, b_v)$ are G-conjugate if and only if they are locally conjugate.*

We should define local conjugation. First, $(u, b_u)$ is locally conjugate to $(u, b_u)^g$ if g belongs to the stabilizer $N(b_P)$ of some $(P, b_P)$ where $(u, b_u) \subset (P, b_P) \subset (D, b_D)$. In general, local conjugation is the transitive closure of this relation.
(Here $(D, b_D)$ is fixed.)

Hence, for the original problem to determine the subsections of B (and k(B), see V.5.1), we fix $(D, b_D)$ where D is a defect group of B, and then determine G-conjugacy classes of $(u, b_u) \subset (D, b_D)$; it is enough to do this locally if necessary. By [AB; 3.4] we only have to study the corresponding group elements $u \in D$; since the block $b_u$ is then uniquely determined.

V.5.3 *Local information on defect groups* We will use the following facts:

V.5.3.1 [4.5; AB] *Let $(1, B) \subset (P, b_P)$. Then there is a defect group $D_1$ of $B$ such that* (1) $P \subset D_1$,

(2) $N_{D_1}(P)$ *is a defect group of* $b_P$ *in* $N(b_P)$, *and*

(3) $PC_{D_1}(P)$ *is a defect group of* $b_P$ *in* $PC(P)$.

V.5.3.2 Let $(P, b_P)$ be any subpair; and let $P_1$ be a defect group of $b_P$ in $N(b_P)$. Then $|N(b_P) : P_1C(P)|$ is not divisible by p. (This is well-known.)

V.5.3.3 Suppose B is a 2-block, $(D, b_D)$ is a fixed B-subpair and $J \in Z(D)$ has order 2, and let $(J, b_J) \subset (D, b_D)$. Then

(a) The defect groups of $b_J$ are defect groups of B.

(b) Suppose $|Z(D)| = 2$. Let $\tau \in D$ have order 2 such that $(\tau, b_\tau) \neq (J, b_J)$. Then the defect groups of $b_\tau$ are not the defect groups of B.

V.5.4 EXAMPLE Suppose B is a dihedral block with defect group D, and let $\tau \in D$ be a non-central involution such that $C_G(\tau)$ is equal to a G-conjuate of D.

[Explicitly, one may take $G = A_6$, $B = B_0$; moreover $D = \langle (1\ 2\ 3\ 4)(5\ 6), \tau \rangle$ where $\tau = (1\ 2)(3\ 4)$. Then $C_G(\tau)$ is the group $\langle (1\ 3\ 2\ 4)(5\ 6), (1\ 4)(2\ 3) \rangle$.]

Then $KC_G(\tau)$ is a local algebra, hence $b_\tau = 1$ and trivially $(1, B) \subset (\tau, b_\tau)$. For the group $D_1$ in V.5.3.1 one must take $C_G(\tau)$ (since $C(\tau) = N(\tau) = N(b_\tau)$ ). On the other han, also $(\tau, b_\tau) \subset (D, b_D)$ where $(D, b_D)$ is a Sylow B-subpair. But D does not have the properties to satisfy V.5.3.1.

V.5.5 HYPOTHESIS In the following we assume that B is a tame block of G with defect group D of order $2^n$. Moreover, $(D, b_D)$ is a fixed B-subpair. For any subgroup P of D, $b_P$ is the block of C(P) such that $(P, b_P) \subset (D, b_D)$; and $N(b_P)$ is its stabilizer in $N_G(P)$.

<u>V.5.6</u> PROPOSITION [$B_3$] *Suppose* $D = \langle x, y : x^{2^{n-1}} = 1, y^2 = 1, x^y = x^{-1} \rangle$ *is a dihedral 2-group of order* $\geq 8$. *A complete set of representatives for the proper subsections of B in G is given by* $\{ (u, b_u): u \in C \}$ *where C is one of the following:*

*(i)* $C = \{ x^i : 1 \leq i \leq 2^{n-2} \}$,

*(ii)* $C = \{ x^i : 1 \leq i \leq 2^{n-2}, y \}$ *[and* $n \geq 3$*]*

*(iii)* $C = \{ x^i : 1 \leq i \leq 2^{n-2}, y, xy \}$.

*Let* $J = x^{2^{n-2}}$. *In particular, an involution* $\neq J$ *of D belongs to the subsection of J if and only if it lies in a Klein 4-group* $V \leq D$ *such that* $N(b_V) - C(V)$ *contains an element of order 3.*

(The possibilities are cases (aa), (ab), (bb) in [$B_3$]).

Proof: We follow the strategy outlined at the end of V.5.2.

(1) <u>Subsections $(u, b_u)$ where u has order > 2:</u> For u we may take $x^i$ for some i with $1 \leq i < 2^{n-2}$. We have to show

(*) If $1 \leq i, j < 2^{n-2}$ and $x^i$, $x^j$ give rise to the same subsection then $i = j$.

Suppose $g \in G$ and $(x^i, b_{x^i})^g = (x^j, b_{x^j})$; then $\langle x^i \rangle^g = \langle x^j \rangle$. Now, $\langle x^i \rangle$ and $\langle x^j \rangle$ are cyclic subgroups of D of the same order, therefore they are equal (III.17), and g belongs to $N_G(\langle x^i \rangle) = C_G(\langle x^i \rangle)\langle y \rangle$. Consequently, $x^i$ and $x^j$ are D-conjugate and then $i = j$.

(2) <u>Subsections $(u, b_u)$ where u has order 2:</u> The D-conjugacy classes of involutions are represented by y, xy and J where $J = x^{2^{n-2}}$ (III.17). Consider the Klein 4-groups $V = \langle y, J \rangle$ and $V' = \langle xy, J \rangle$.

(2.a) <u>y and J define the same subsection if and only if there is an element of order 3 in $N(b_V) - C(V)$:</u> Suppose first that such element g of order 3 exists; then, say, $y^g = J$ and moreover $(y, b_y)^g = (J, b_y{}^g) \subsetneq (V, b_V)^g = (v, b_V) \subsetneq (D, b_D)$. By [3.4; AB] we deduce $b_y{}^g = b_J$. Conversely, suppose that y and J define the same subsection. Then $(y, b_y)$ and $(J, b_J)$ are locally conjugate, by [4.13; AB], see V.5.2. Hence there exists a subpair $(P, b_P)$ such that $(y, b_y) \subset (P, b_P) \subsetneq (D, b_D)$ and there is some $z \in N(b_P)$ with $(y, b_y) \neq (y, b_y)^z$ and moreover, $y^z$ is not D-conjugate to y. Now, P is some dihedral group, by the group structure of D.

We claim that P must be a Klein 4-group. Suppose not; then Aut P is a 2-group and so is N(P)/PC(P). One shows, using V.5.3.1 and 2 that $N(b_P) = N_D(P)C(P)$; hence for $z \in N(b_P)$ the elements $y^z$ and y are already D-conjugate, contrary to the hypothesis above.

Then P = V, say, and Aut V ≅ $S_3$. Suppose there is no element of order 3 in $N(b_V) - C(V)$. Then similarly, using V.5.3.1 and 2 one shows that y and $y^z$ are already D-conjugate; a contradiction.

The same holds with xy instead of y.

V.5.6.1 **COROLLARY** *With the hypothesis of V.5.6, we have*

*$\ell(B) = 3$ if C is as in (i),*

*$\ell(B) = 2$ if C is as in (ii), and*

*$\ell(B) = 1$ if C is as in (iii).*

Proof: By the classification of algebras of dihedral type, $\ell(B) - 1$ is equal to the number of 3-tubes of the block. By V.4.1.1 or V.2.14, this number is the same as the number of G-conjugacy classes $(V, b_V) \subsetneq (D, b_D)$ where V is a Klein 4-group and $N(b_V) - C(V)$ contains an element of order 3.

V.2.6.2 [$B_2$] *Suppose $D = \langle x, y \rangle$ is a Klein 4-group. A complete set of representatives for the proper subsections of B in G is given by $\{ (u, b_u) : u \in C \}$ where C is one of the following:*

*(i) $C = \{ x \}$, or* *(ii) $C = \{ x, y, xy \}$.*

*Moreover in (i) we have $\ell(B) = 3$ and otherwise $\ell(B) = 1$.*

Proof: The arguments in part (2.a) of V.5.6 apply here as well; therefore x and y define the same subsection if and only if there is an element g of order 3 in $N_G(b_D) - C(D)$. Trivially, this happens if and only if x and y define the same subsection; and hence there are two possibilities for *C*. The statements on $\ell(B)$ can be seen similarly as in V.5.6.1.

V.5.7 PROPOSITION [0] *Suppose* $D = \langle x, y: x^{2^{n-1}} = y^2 = 1, x^y = Jx, J = x^{2^{n-2}} \rangle$ *is semidihedral. A complete set of representatives for the proper subsections of B in G is given by* $\{ (u, b_u) : u \in C \}$ *where C is one of the following:*
*(i)* $C = \{ x^i : 1 \leq i \leq 2^{n-2} \}$ *(ii)* $C = \{ x^i : 1 \leq i \leq 2^{n-2}, xy \}$
*(iii)* $C = \{ x^i : 1 \leq i \leq 2^{n-2}, y \}$ *(iv)* $C = \{ x^i : 1 \leq i \leq 2^{n-2}, y, xy \}$.
*In particular, an involution* $\neq J$ *of D belongs to the subsection of J if and only if it for a Klein 4-group* $V < D$, $N(b_V) - C(V)$ *dontains an element of order 3.*

[The possibilities are the cases (aa), (ab), (ba) and (bb) in [0].)

Proof: We follow again the strategy outlined in V.5.2.

(1) Subsections $(u, b_u)$ where $u \in D$ has order > 4 are parametrized by the set $\{ x^i : 1 \leq i < 2^{n-3} \}$. This is clear by the arguments in V.5.6 (1).

(2) Subsections $(u, b_u)$ where u has order 2: The D-conjugacy classes of involutions are represented by y and J. Let V be the Klein 4-group $V = \langle y, J \rangle$ and let $(V, b_V) \subset (D, b_D)$ be the corresponding subpair. Then, as in V.5.6 we have that y and J define the same subsection if and only if there is an element of order 3 in $N(b_V) - C(V)$.

(3) Subsections $(u, b_u)$ where u has order 4: The D-conjugacy classes of elemens of order 4 are represented by xy and w where $w = x^{2^{n-3}}$. Consider the quaternion group Q of order 8 where $Q = \langle xy, w \rangle$, and let $(Q, b_Q) \subset (D, b_D)$ be the corresponding subpair. Then

(3.a) w and xy define the same subsection if and only if there is an element of order 3 in $N(b_Q) - C(Q)$. This is proved by arguments analogue to those of V.5.6,(2.a).

V.5.7.1 COROLLARY *With the hypothesis as in V.5.7, the following holds: If C is as in (i) or (ii) then* $\ell(B) \geq 2$; *and if C is as in (iii) or (iv) then* $\ell(B) \leq 2$.

Proof: By the classification of algebras of semidihedral type, the number of 3-tubes is $\leq 1$. If $\ell(B) = 3$ then it is 1, and for $\ell(B) = 1$ it is 0. This number is 1 if $N(b_V) - C(V)$ contains an element of order 3, and zero otherwise (V.4.2.1).

V.5.8 PROPOSITION [0] *Suppose* $D = \langle x, y : x^{2^{n-1}} = 1, x^{2^{n-2}} = y^2, x^y = x^{-1} \rangle$ *is*

*quaternion. A complete set of representatives for the proper subsections of B in G is given by* $\{ (u, b_u) : u \in C \}$ *where C is one of the following:*
*(i)* $C = \{ x^i : 1 \leq i \leq 2^{n-2} \}$, *(ii)* $C = \{ x^i : 1 \leq i \leq 2^{n-2}, y \}$,
*(ii)* $C = \{ x^i : 1 \leq i \leq 2^{n-2}, y, xy \}$.

(The possibilities are the cases (aa), (ab), (bb) in [O]).This is proved by the arguments in V.5.6, V.5.7. We omit details.

Now we shall determine some numbers $\ell(b)$ for some local blocks.

V.5.9.1 *Let b be a 2-block with cyclic defect groups. Then* $\ell(b) = 1$.
This follows from the general fact that for a cyclic p-block b,the number $\ell(b)$ divides p-1 (see for example [$D_2$] or [F]).

V.5.9.2 *Let b be a block with dihedral defect groups of a group C which has a central involution. Then* $\ell(b) = 1$.

Proof: The central involution, J say, lies in every defect group of C (V.1.2), hence it lies in the centre of a defect group D of b. Let V be a Klein 4-group, $V \leq D$; then $J \in V$ (III.17). Since J is not conjugate to any other element of V it follows. that there is no element $g \in N(b_V) - C(V)$ of order 3. Hence the subsection for b is as in (iii) of V.5.6. Now the statement follows from V.5.6.1.

V.5.9.3 *Let b be a semidihedral block of a group C which has a central involution. Let D be a defect group of b and* $Q < D$ *a quaternion group of order 8 with* $(Q, b_Q) \subset (D, b_D)$. *Then* $\ell(b) = 2$ *if* $N(b_Q) - C(Q)$ *contains an element of order 3, and* $\ell(b) = 1$ *otherwise.*

Proof: Let J be the central involution of D; then $\langle J \rangle \subseteq Z(C)$. The simple b-modules are also modules for $\bar{C} := C/(J)$ and belong as such to a block $\bar{b}$ with $\ell(\bar{b}) = \ell(b)$. Moreover $\bar{b}$ has dihedral defect groups, so we may determine $\ell(\bar{b})$ by applying V.5.6.

The D-conjugacy classes of involutions are represented by $\bar{y}$, $\overline{xy}$ and $\bar{w}$ where $w = x^{2^{n-3}}$. Clearly $\bar{y}$ is not conjugate in $\bar{C}$ to $\overline{xy}$ or $\bar{w}$, since $\bar{y} = \{y, J\}$ does not contain an element of order 4. If $\overline{xy}$ and $\bar{w}$ belong to the same subsecton of $\bar{b}$ then $\ell(\bar{b}) = 2$ by V.5.6 and V.5.6.1 (applied to b); and this occurs if and only if $N(b_Q) - C(Q)$ contains an element of order 3. Otherwise, $\ell(\bar{b}) = 1$.

V.5.9.4 *Let b be a block with quaternion defect groups of a group C which has a central involution. Let D be a defect group of b. If $|D| > 8$ then take $Q_i \subset D$ to be quaternion groups of order 8 such that $(Q_1, b_{Q_1})$ and $(Q_2, b_{Q_2})$ are not G-conjugate. For $|D| = 8$ set $D = Q_1 = Q_2$. Then*

$\ell(b) = 3$ *if $N(b_{Q_i}) - C(Q_i)$ contain elements of order 3 for $i = 1,2$*

$= 2$ *if $N(b_{Q_i}) - C(Q_i)$ contains an element of order 3 for one value of i*

$= 1$ *otherwise.*

Proof: As in V.5.9.3, let $\bar{b}$ be the block of $\bar{C} := C/\langle J\rangle$ containing the simple b-modules; then $\ell(\bar{b}) = \ell(b)$. Since $\bar{b}$ has dihedral defect groups, we may calculate $\ell(\bar{b})$ by applying V.5.6 and V.5.6.1. The three possibilities for the subsections of $\bar{b}$ lead to the three stated cases.

V.5.10 COROLLARY $[B_{2,3}]$ *Let B be a block with dihedral defect groups of order $2^n \geq 4$. Then $k(B) = 2^{n-2} + 3$.*

Proof: Let $(u, b_u) \subset (D, b_D)$ where $(D, b_D)$ is a Sylow B-subpair of G. Then $b_u$ is a block of C(u), and $\langle u \rangle \leq \delta(b_u) \leq_G D$ (see V.5.3); in particular, $b_u$ is a cyclic or a block with dihedral defect groups. Moreover, the group C(u) has a central involution; hence it follows from V.5.9.1 and 2 that $\ell(b_u) = 1$. We deduce from V.5.1 and V.5.6, V.5.6.1 and 2 that $k(B) = \ell(B) + |\mathcal{C}| = 2^{n-2} + 3$.

V.5.11 COROLLARY [0] *Let B be a block with semidihedral defect group D of order*

$2^n$. *Then*

$$k(B) = \begin{cases} \ell(B) + 2^{n-2} + 1 & \text{if } y \notin C \text{ and } \ell(B) \geq 2 \\ \ell(B) + 2^{n-2} + 2 & \text{if } y \in C \text{ and } \ell(B) \leq 2 \end{cases}.$$

Proof: Let $(u, b_u) \subset (D, b_D)$ where $u \in C$ and $C$ is as in V.5.7. Then $b_u$ is a block of $C_G(u)$, and $\langle u \rangle \leq \delta(b_u) \leq_G D$. Moreover, $C(u)$ has a central involution.

(1) <u>If u has order > 4 then $\ell(b_u) = 1$:</u> The defect group of $b_u$ is a subgroup of a semidihedral group and contains $\langle u \rangle$ as a central subgroup. Therefore it is cyclic, and (1) follows from V.5.9.1.

(2) <u>$\ell(b_J) = 2$ if y does not belong to $C$ and = 1 otherwise:</u> We have by V.5.3.3 that the block $b_J$ has a semidihedral defect group. Now the statement follows from V.5.9.3.

(3) <u>If $y \in C$ then $\ell(b_y) = 1$:</u> By V.5.3.3, we deduce $\delta(b_y) \lneq_G D$. Then it follows from V.5.3.1 that $b_y$ is a $V_4$-block, and it has a central involution. By V.5.9.2 we have that $\ell(b_y) = 1$. Consequently setting $\ell(b_z) = 0$ if $z \notin C$ we have

$$\ell(b_y) + \ell(b_J) + \ell(b_{xy}) = \begin{cases} 2 & \text{in (i), (i) of V.5.7,} \\ 3 & \text{in (iii) and (iv) of V.5.7,} \end{cases}$$

and we obtain $k(B) = \begin{cases} \ell(B) + 2^{n-2} + 1 & \text{if } y \text{ does not belong to } C, \\ \ell(B) + 2^{n-2} + 2 & \text{otherwise.} \end{cases}$

If the first case arises then $\ell(B) \geq 2$, by V.5.7.1; and $\ell(B) \leq 2$ otherwise.

<u>V.5.12</u> COROLLARY [O] *Let B be a block with quaternion defect group oforder $2^n$. Then $k(B) = \ell(B) + 2^{n-2} + 2$.*

Proof: Let $(u, b_u) \subset (D, b_D)$ where $u \in C$ and $C$ is as in V.5.8. Then $b_u$ is a block of $C(u)$ and $\langle u \rangle \leq \delta(b_u) \leq_G D$. Moreover, $C(u)$ has a central involution.

(1) <u>If u has order > 4 then $\ell(b_u) = 1$:</u> This is the same as V.5.11 part (1).

Consider now $u = J$. By V.5.9.4 we have that $\ell(b_J) = 1$ if y and xy both belong to $C$; and $\ell(b_J) = 2$ if $y \in C$ but $xy \notin C$. Otherwise, $\ell(b_J) = 3$. Hence in all cases we have that $\ell(b_J) + \ell(b_w)$ [ + $\ell(b_y) + \ell(b_{xy})$ ] = 4. Consequently $k(B) = \ell(B) + 2^{n-2} + 2$. Moreover $\ell(B) \leq 3$, and the statement follows.

<u>V.5.13</u> LEMMA *Suppose B is a block with dihedral defect groups, and $(D, b_D)$ is a Sylow B-subpair. Let V and V' be Klein 4-subgroups of D which are not D-conjugate.*

*Then the subpairs $(V, b_V)$ and $(V', b_{V'})$ are not G-conjugate.*

Proof: We apply results from [AB; chapter 4] and show that $(V, b_V)$ cannot be locally conjugate (that is, F-conjugate) to $(V', b_{V'})$. This is similar to the method we used in classifying subsections.

Suppose $(V, b_V) \subset (R, b_R) \subsetneq (D, b_D)$, and let $g \in N(b_R)$. It suffices to show that $(V, b_V)^g$ is D-conjugate to $(V, b_V)$. This will follow directly if we prove

(*) $\underline{N_G(b_R) = N_D(R)C_G(R)}$: By [AB; 3.4] there is a subpair $(N_D(R), b') \subsetneq (D, b_D)$, and then $(R, b_R) \vartriangle (N_D(R), b')$. In particular $b_R$ is stable in $N_D(R)$; and one inclusion of (*) holds.

If $R \subset D$ then by group theory, $R \subset N_D(R)$. Therefore equality in (*) will follow if we show

$$(**)\quad |N_G(b_R) : C_G(R)R| = \begin{cases} 2 & \text{if } R \subset D \\ 1 & \text{if } R = D \end{cases}.$$

We note first that

(1) $\underline{N_G(b_R)/ C_G(R)R}$ is a 2-group.

This is clear since R is a dihedral group of order $\geq 8$ and $N_G(R)/ C_G(R)R$ is isomorphic to a subgroup of Aut R.

To prove (**), we apply V.5.3.1. Let $D_1$ be a defect group of B satisfying the conditions in V.5.3.1 with R = P. We claim now that

(2) $\underline{N(b_R) = N_{D_1}(R)C_G(R)}$: By V.5.3.2 and since $N_{D_1}(R)$ isa a defect group of $b_R$ in $N(b_R)$, we have that the index of $N_{D_1}(R)C_G(R)$ in $N(b_R)$ is odd. On the other hand it is a 2-power by (1), and (2) follows.

By elementary group theory one sees that $|N_{D_1}(R)C_G(R) : C_G(R)R| = |N_{D_1}(R) : C_{D_1}(R)R|$. Hence, by (2), the index in (**) is $|N_{D_1}(R) : C_{D_1}(R)R|$, which equals 2 if $R \subset D$ and 1 otherwise; since $D_1 \geq R$ and both are dihedral groups of order $\geq 8$.

# VI *Algebras of dihedral type*

## VI.1 *A class of symmetric algebras where hearts are decomposable*

First we shall study basic algebras which are symmetric and indecomposable and have the following properties:

VI.1.0 (1) For each vertex f of the quiver $Q$ of $\Lambda$, $|s^{-1}(f)| = |e^{-1}(f)| = 1$ or 2.

(2) If f is a vertex of $Q$ with $|s^{-1}(f)| = 2$ then rad $f\Lambda$/soc $f\Lambda = U \oplus V$ where U and V are non-zero.

Most quivers in the list IV.2.4 satisfy (1). To motivate the second condition, observe that this means, the standard AR-sequence (I.7.7) of $f\Lambda$ has two indecomposable middle terms. Note also that if $\Lambda$ satisfies VI.1.0 then so does $\Lambda^{op}$. We will investigate the following:

- VI.1.1 A direct-sum decomposition of rad $\Lambda$/soc $\Lambda$.
- VI.1.2 A permutation $\pi$ describing $\Omega(\alpha\Lambda)$ for arrows $\alpha$ of $Q$.
- VI.1.3 A permutation $\pi^*$ describing composition factors of $\Lambda$.
- VI.1.4 Relations for the algebra $\Lambda$.
- VI.1.5 Non-singularity of the Cartan matrix of $\Lambda$.

Assume $\Lambda$ is an algebra satisfying the above hypotheses. We fix arrows of the quiver such that whenenver two arrows , $\gamma$ and $\delta$ say, start at vertex f then $\gamma\Lambda \cap \delta\Lambda =$ soc $f\Lambda$.

### VI.1.1 *Another decomposition of (rad $\Lambda$)/(soc $\Lambda$).*

Take a vertex f in $Q$, and let S be the corresponding simple module. Assume first that f is the endpoint of a unique arrow $\alpha$ in $Q$. Then by I.6.5, we have $R_\alpha \simeq S$, hence $\alpha\Lambda \simeq f\Lambda/S$.

LEMMA *Suppose f is a vertex of Q such that two arrows $\gamma$ and $\delta$ end at f. Then* $R_\gamma \cap R_\delta = S$ *and* $R_\gamma + R_\delta = rad(f\Lambda)$. *Hence* $rad(f\Lambda)/S \simeq R_\gamma/S \oplus R_\delta/S$ *where* $S = S_f$.

Proof: As we have seen in I.6.5, the intersection $R_\gamma \cap R_\delta$ is equal to soc $f\Lambda$. Clearly, $R_\gamma + R_\delta$ is contained in rad($f\Lambda$). We shall prove that both modules have the same length. Consider the right ideal $R = \sum R_\alpha$ where the sum is taken over all arrows in $Q$. Thus $R = \sum_f \oplus \left( \sum_{\alpha=\alpha f} R_\alpha \right)$. For a fixed vertex f, we have that $\underline{\dim} \sum_{\alpha f=\alpha} R_\alpha$ = $\underline{\dim}\, S_f$ if only one arrow ends at f; and $= \underline{\dim}\, R_\alpha + \underline{\dim}\, R_\beta - \underline{\dim}\, S_f$ if two arrows $\alpha$, $\beta$ end at f.

Let $I \subseteq Q_0$ be the set of all vertices f with $|s^{-1}(f)| = 2$. Then

$$\underline{\dim}\, R = \sum_\alpha \underline{\dim}\, R_\alpha - \sum_I \underline{\dim}\, S_f = \sum [\underline{\dim}\, e(\alpha)\Lambda - \underline{\dim}\, \alpha\Lambda] - \sum_I \underline{\dim}\, S_f$$

(using $\alpha\Lambda \simeq e(\alpha)\Lambda/R_\alpha$)

$$= \sum \underline{\dim}\, e(\alpha)\Lambda - \sum \underline{\dim}\, \alpha\Lambda - \sum_I \underline{\dim}\, S_f.$$

Now, the number of arrows starting at a fixed vertex is the same as that ending at it; therefore $\sum \underline{\dim}\, e(\alpha)\Lambda = \sum \underline{\dim}\, s(\alpha)\Lambda$ . Thus it follows that

$$\underline{\dim}\, R = \sum_\alpha (\underline{\dim}\, s(\alpha)\Lambda - \underline{\dim}\, \alpha\Lambda) - \sum_f \underline{\dim}\, S_f$$

$$= \sum_f \Big[ \sum_{\alpha=f\alpha} \underline{\dim}\, f\Lambda - \underline{\dim}\, \alpha\Lambda \Big] - \sum_I \underline{\dim}\, S_f.$$

If $|s^{-1}(f)| = 1$ then $\underline{\dim}\, f\Lambda - \underline{\dim}\, \alpha\Lambda = \underline{\dim}\, S_f$. Suppose two arrows $\alpha$ and $\beta$ start at f, then the corresponding summand is $2\underline{\dim}\, f\Lambda - \underline{\dim}\, \alpha\Lambda - \underline{\dim}\, \beta\Lambda - \underline{\dim}\, S_f$ which is equal to $\underline{\dim}$ rad $f\Lambda$. It follows that $\underline{\dim}$ rad $f\Lambda = \underline{\dim}\, [R_\gamma + R_\delta]$.

<u>VI.1.2</u> *A permutation associated to* $\Lambda$:

We define a map $\pi$ on the set $Q_0 \cup$ { vertices of $Q$ with one arrow } by

$$\pi(\alpha) = \begin{cases} \gamma & \text{if } R_\alpha = \gamma'\Lambda \text{ where } \gamma-\gamma' \in J^2 \\ f & \text{if } R_\alpha = S_f \end{cases}$$

and

$$\pi(f) = \gamma \quad \text{where } \gamma \text{ starts at f, when } |s^{-1}(f)| = 1.$$

By VI.1.1, this is a map. Moreover, $\pi$ is surjective, hence a permutation: Let $f \in Q_0$ with $|e^{-1}(f)| = 1$. Then $\pi(\alpha) = f$ where $\alpha$ ends at f. Now let $\gamma \in Q_1$, and suppose $\gamma$ starts at vertex f. If $|s^{-1}(f)| = 1$ then $\gamma\Lambda = \Omega^{-1}(S_f)$ and $\pi(f) = \gamma$. Otherwise, let $\alpha$ and $\beta$ end at f. By VI.1.1, say, $\gamma\Lambda/\text{soc } f\Lambda \cong R_\alpha/\text{soc } f\Lambda$ and then $\pi(\alpha) = \gamma$.

VI.1.3 *The permutation* $\pi^*$ We define this permutation on the set of arrows of $Q$. Let $\alpha$ be an arrow ending at vertex f. Define

$$\pi^*(\alpha) = \begin{cases} \text{the arrow starting at f if } |s^{-1}(f)| = 1; \\ \text{the arrow } \delta \text{ if } \beta, \delta \text{ start at f and } \pi(\alpha) = \beta. \end{cases}$$

Then $\pi^*$ is onto, hence is a permutation.

The cycles of $\pi$ and also of $\pi^*$ correspond to closed paths in the quiver.

VI.1.4.1 LEMMA *Let $\alpha$ and $\gamma$ be arrows such that $\alpha = \alpha f$ and $f\gamma = \gamma$ for $f \in Q_0$. Then $\pi(\alpha) = \gamma$ implies $\alpha\gamma \in \alpha J^2$. Suppose in addition that $\alpha J \not\subseteq soc\ \Lambda$. Then $\pi(\alpha) = \gamma$ if and only if $\alpha\gamma \in J^3$. In particular, the algebra $\Lambda/J^3$ is special biserial.*

Proof: It suffices to consider a vertex f at which two arrows start. Say the quiver contains

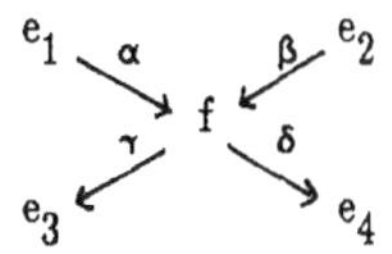


Here $\alpha$ and $\beta$ are independent modulo $J^2$; similarly $\gamma$ and $\delta$; otherwise we allow arrows or vertices to be equal or dependent. Let $S = S_f$, and put $H = rad\ P_f/S$. Then $H \cong \gamma\Lambda/S \oplus \delta\Lambda/S \cong R_\alpha/S \oplus R_\beta/S$. Consequently, say, $\gamma\Lambda/S \cong R_\alpha/S$ and $\delta\Lambda/S \cong R_\beta/S$. Then $R_\alpha = \gamma'\Lambda$ where $\gamma' = \gamma + x$ with $x \in J^2$. Since $\alpha\gamma' = 0$, it follows that $\alpha\gamma \in \alpha J^2$; similarly $\beta\delta$ lies in $\beta J^2$. The dual condition also holds: For some choice $\alpha'$, $\beta'$, $\gamma'$ we know that one of $\alpha'\gamma'$, $\beta'\gamma'$ lies in $J^3$; this is true since general hypotheses also hold for $\Lambda^{op}$. Moreover $\alpha'\gamma' \equiv \alpha\gamma$ modulo $J^3$, and $\beta'\gamma' \equiv \beta\gamma$ modulo $J^3$.

Now suppose that $\alpha\gamma$ and $\alpha\delta$ both lie in $J^3$. If $|s^{-1}(e_1)| = 1$ then it would follow that $\alpha J \subseteq \alpha J^2$ and $\alpha J = 0$, therefore $\alpha \in soc\ \Lambda$, a contradiction to I.3.8. So there must be another arrow, $\rho$ say, starting at $e_1$. Write $\alpha\gamma = \alpha x + \rho y$ where x and $y \in J^2$. Then $\alpha(\gamma - x)$ lies in $\alpha\Lambda \cap \rho\Lambda \subseteq soc\ \Lambda$. Similarly $\alpha(\delta - x') \in soc\ \Lambda$ for some $x' \in J^2$. Now, $fJ = (\gamma - x)\Lambda + (\delta - x')\Lambda$ and $\alpha J = \alpha fJ \subseteq soc\ \Lambda$, as required.

<u>VI.1.4.2</u> LEMMA *Suppose $\alpha$ is an arrow, and let $\sigma = (\alpha_1\ \alpha_2\ \dots\ \alpha_r)$ be the cycle of $\pi^*$ with $\alpha_1 = \alpha$. Then the following holds:*

*(a) The module $\alpha\Lambda$ is uniserial.*

*(b) Let $m$ be the largest integer such that $\zeta := (\alpha_1 \dots \alpha_r)^m \neq 0$. Then $soc(\alpha\Lambda) = \langle \zeta \rangle$.*

*(c) A vector space basis of $\alpha\Lambda$ is given by the set of subwords of $\zeta$ starting with $\alpha$.*

Proof: Let $\alpha$ start at f.

(a) By VI.1.4.1, the module $\alpha\Lambda$ is spanned by all monomials of the form $\omega = w_1 w_2 \dots w_k$ where the $w_i$ are arrows, such that $w_1 = \alpha$ and $w_{i+1} = \sigma(w_i)$. For any k, there is only one such word of length k. This generates $\alpha J^{k-1}$; hence $\alpha\Lambda$ is uniserial.

(b) Suppose (b) is false. Then by (a) [and the choice of m], the socle of $\alpha\Lambda$ is spanned by some element $\zeta' = \zeta(\alpha_1\alpha_2\dots\alpha_k)$ and $k < r$. Since $\Lambda$ is symmetric, we know that $\langle \zeta' \rangle = \mathrm{soc}\ f\Lambda$ and moreover $\zeta' = \zeta' f$; therefore $\alpha_k$ must end at vertex f. In particular the cycle $\sigma$ goes twice through f:

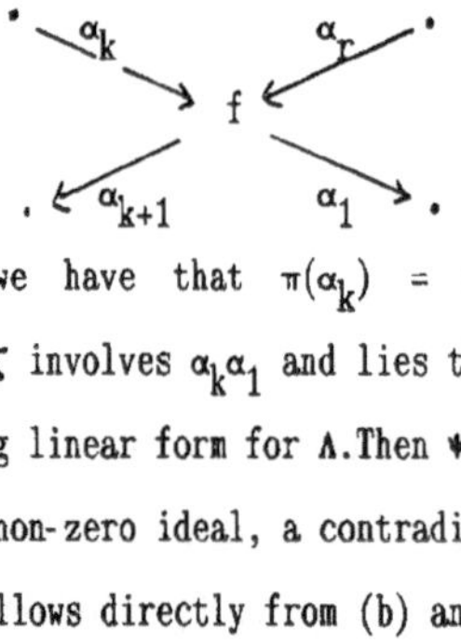


Therefore we have that $\pi(\alpha_k) = \alpha_1$, and $\alpha_k\alpha_1 \in J^3$, by VI.1.4.1. The element $(\alpha_1\alpha_2\dots\alpha_k)\zeta$ involves $\alpha_k\alpha_1$ and lies therefore in $\alpha_1 J^{rm+k} = (\mathrm{soc}\ \alpha\Lambda)J = 0$. Let $\psi$ be a symmetrizing linear form for $\Lambda$.Then $\psi[\zeta'] = \psi[\ (\alpha_1\dots\alpha_k)\zeta] = \psi[0] = 0$. Hence ker $\psi$ contains a non-zero ideal, a contradiction.

Part (c) follows directly from (b) and (a).

<u>VI.1.4.3</u> *Let $\sigma$ and $m$ be as in VI.1.4.2. The number $m$ depends only on $\sigma$. We say that $m$ is the <u>multiplicity</u> of $\sigma$.*

Proof: We proceed by induction on j, $1 < j \leq r$. Put $\upsilon = (\alpha_j \dots \alpha_r\alpha_1 \dots \alpha_{j-1})^m$; we shall prove that $\upsilon J = 0$ but $\upsilon \neq 0$. By induction hypothesis, $\rho \neq 0$ where $\rho = (\alpha_{j-1} \dots \alpha_r\alpha_1 \dots \alpha_j)^m$ but $(\alpha_{j-1} \dots \alpha_r\alpha_1 \dots \alpha_j)^{m+1} = 0$. By VI.1.4.2 we know that $\langle \rho \rangle = \mathrm{soc}\ \alpha_{j-1}\Lambda$. Consequently, by VI.1.4.1, $\upsilon J = \langle \upsilon\alpha_j \rangle \subseteq J(\mathrm{soc}\ \alpha_{j-1}\Lambda) = 0$. Let $\psi$ be a symmetrizing linear form for $\Lambda$; then $\psi[\upsilon] = \psi[\rho] \neq 0$ and therefore $\upsilon \neq 0$.

VI.1.4.4 *Suppose $e$ is a vertex of $Q$ at which two arrows $\alpha$, $\beta$ end and two arrows $\gamma$, $\delta$ start, and $\pi(\alpha) = \gamma$, $\pi(\beta) = \delta$.*

*(a) If none of the arrows is a loop then there are arrows $\gamma'$ and $\delta'$ with $\gamma \equiv \gamma'$ and $\delta \equiv \delta'$ (mod $J^2$) such that $\alpha\gamma' = 0$ and $\beta\delta' = 0$, and moreover $\gamma'\Lambda \cap \delta'\Lambda \subseteq$ soc $\Lambda$..*

*(b) Suppose that $\alpha$ is a loop and $\alpha = \gamma$ but $\beta$ is not a loop. Then there are arrows $\alpha'$ and $\delta'$ with $\alpha \equiv \alpha'$ and $\delta \equiv \delta'$ (mod $J^2$) such that $(\alpha')^2$ lies in soc $\Lambda$ and $\beta\delta' = 0$, and moreover $\alpha'\Lambda \cap \delta'\Lambda \subseteq$ soc $\Lambda$.*

*(c) Suppose $\alpha$ and $\beta$ are loops and $\alpha = \gamma$, $\beta = \delta$. Then there are independent arrows $\alpha'$ and $\beta'$ whose squares lie in soc $\Lambda$; and moreover $\alpha'\Lambda \cap \beta'\Lambda \subseteq$ soc $\Lambda$.*

Proof: By VI.1.1 we have that $\gamma\Lambda$/soc $e\Lambda \cong R_\alpha$/soc $e\Lambda$ and $\delta\Lambda$/soc $e\Lambda \cong R_\beta$/soc $e\Lambda$; and moreover $R_\alpha + R_\beta =$ rad $e\Lambda$ and $R_\alpha \cap R_\beta \subseteq$ soc $\Lambda$.

(a) Choose generators $\gamma'$ and $\delta'$ of $R_\alpha$ and $R_\beta$ respectively, with $\gamma' \equiv \gamma$ and $\delta' \equiv \delta$ (mod $J^2$), and $\gamma' = e\gamma' f$, $\delta' = e\delta' g$ where $f = e(\gamma)$ and $g = e(\delta)$. This gives (a).

(b) Let $\psi$: $R_a \to a\Lambda$/soc $a\Lambda$ be an epimorphism with ker $\psi$ = soc $R_a$. Then there is a generator $a'$ of $R_a$ with $\psi(a') = a +$ soc $a\Lambda$. Since $aa' = 0$ we have that $\psi[(a')^2] = [\psi(a')]a' = 0$ and $(a')^2 \in$ ker $\psi$ = soc $\Lambda$. For $\delta'$ we take a generator of $R_\beta$.

(c) In this case, $\Lambda$ is local. Let $\psi$: $R_\alpha \to \alpha\Lambda$/soc $\Lambda$ and $\zeta$: $R_\beta \to \beta\Lambda$/soc $\Lambda$ be epimorphisms with ker $\psi$ = soc $\Lambda$ and ker $\zeta$ = soc $\Lambda$. Then there are generators $\alpha'$ and $\beta'$ of $R_\alpha$ and $R_\beta$ respectivey with $\psi(\alpha') = \alpha +$ soc $\Lambda$ and $\zeta(\beta') = \beta +$ soc $\Lambda$. Moreover, $\alpha'$ and $\beta'$ are independent since $R_\alpha + R_\beta$ = rad $\Lambda$. The argument in (b) shows that $(\alpha')^2$ and $(\beta')^2$ both lie in soc $\Lambda$.

VI.1.4.5 *Suppose $\pi^*$ has a cycle $(\alpha)$ where $\alpha$ is a loop at some vertex $e$, and let $\beta$, $\gamma$ be arrows starting and ending at $e$ respectively; and suppose $\beta$ and $\gamma$ are not loops. Then $\alpha\beta = 0$ and $\gamma\alpha = 0$.*

Proof: The hypothesis implies that $\alpha\Lambda$ is spanned by $\alpha, \alpha^2, \alpha^3, \ldots$ ; in particular $\alpha\Lambda f = 0$ for any primitive idempotent $f$ with $ef = 0$. Consider now the $\Lambda$-homomorphism $\mu$: $e\Lambda \to \gamma\Lambda$ defined by $\mu(x) = \gamma x$. Since soc $\gamma\Lambda$ = soc $\gamma\Lambda f$ for $f = s(\gamma)$ and $ef = 0$ and moreover $\alpha\Lambda f = 0$, we deduce that $\mu_{|\alpha\Lambda} = 0$ and $\gamma\alpha = 0$.

VI.1.5.1 *The Cartan matrix of* $\Lambda$ We fix a vertex e of $Q$.

(a) Suppose $\alpha$ is an arrow starting at e. Let $\sigma$ be the cycle of $\pi^*$ containing $\alpha$, and let m be the multiplicity of $\sigma$. Then for $f \in Q_0$

$$\dim \alpha\Lambda f = m\ |\{i: \alpha_i \in \sigma \text{ and } \alpha_i f = \alpha_i\}|.$$

(**) Note that this number depends only on the cycle, not on $\alpha$.

(b) Suppose only one arrow starts at e. Then

$$\dim e\Lambda f = \begin{cases} \dim \alpha\Lambda f & \text{if } f \neq e, \\ m+1 & \text{if } f = e. \end{cases}$$

(c) Suppose two arrows start at e, say $\alpha$ and $\beta$. Then for any f,

$$\dim e\Lambda f = \dim \alpha\Lambda f + \dim \beta\Lambda f.$$

This gives information about singularity of the Cartan matrix.

VI.1.5.2 LEMMA *Suppose* $\pi^*$ *has cycles* $\sigma_1$ *and* $\sigma_2$ *which intersect in* $\geq$ *two vertices. Then the Cartan matrix is singular.*

Proof: Suppose e and f are vertices which occur in $\sigma_1$ and $\sigma_2$, with $e \neq f$. Then two arrows start at e and two arrows start at f. By VI.1.5.1(c) we see that $\dim e\Lambda g = \dim f\Lambda g$ for any $g \in Q_0$; consequently $\underline{\dim}\, e\Lambda = \underline{\dim}\, f\Lambda$ and the Cartan matrix has two identical rows.

VI.1.5.3 LEMMA *Assume e and f are distinct vertices of* $Q$ *such that* $|s^{-1}(e)| = |s^{-1}(f)| = 2$. *Suppose that all arrows starting or ending at e or f lie in the same cycle of* $\pi^*$. *Then the Cartan matrix of* $\Lambda$ *is singular.*

The proof is similar to that of VI.1.5.2.

## VI.2 *Algebras of dihedral type*

We will now introduce algebras of dihedral type and then derive the classification, up to Morita equivalence. That is, we determine their basic algebras via quivers and relations.

The first main step is to bound the number of simple modules of these algebras, see VI.3; then we obtain restrictions for the quiver, see VI.4. After these preparations we determine the list of possible algebras, in VI.7 to VI.9. We are able to show that this is a complete classification (see VI.10), using results from the theory of representation of algebras.

VI.2.1 The algebra $\Lambda$ is *of dihedral type* if it satisfies the following conditions:

(1) $\Lambda$ is symmetric and indecomposable.

(2) The Cartan matrix of $\Lambda$ is non-singular.

(3) The stable AR-quiver $\Gamma_s(\Lambda)$ consists of the following components:

  (i) 1-tubes,

  (ii) at most two 3-tubes,

  (iii) non-periodic components of tree class $A_\infty^\infty$ or $\tilde{A}_{1,2}$.

VI.2.1.1 EXAMPLES Suppose B is a 2-block of some group algebra with dihedral defect groups; then B is an algebra of dihedral type: Conditions (1) and (2) hold for arbitrary blocks [F]; see also I.3.7. In V.4.1 we have proved that (3) is true.

To have some explicit examples in mind, the following algebras $\Lambda$ are of dihedral type:

(i) $\Lambda$ = the Kronecker algebra. It has one non-periodic component $\mathbb{Z}\tilde{A}_{1,2}$ and no 3-tube, see II.7.3.

(ii) The algebras in $\Lambda$ in III.1(c) and (c'). The stable AR-quiver has no 3-tubes and infinitely many components $\cong \mathbb{Z}A_\infty^\infty$, see II.7.6. As a special case this includes the group algebra of a dihedral 2-group when char K = 2.

(iii) The algebra $\Lambda$ in II.7.4; a special case is the group algebra of the alternating group $A_4$ over characteristic 2. Then $\Gamma_s(\Lambda)$ has two 3-tubes and one non-periodic component $\cong \mathbb{Z}\tilde{A}_5$, of tree class $A_\infty^\infty$, see II.7.4.

(iv) The algebra $\Lambda$ in II.7.5; this includes as a special case the group algebra of the symmetric group $S_4$ over characteristic 2. Then $\Gamma_s(\Lambda)$ has one 3-tube and infinitely many components $\cong \mathbb{Z}A_\infty^\infty$.

From now, we fix an algebra $\Lambda$ of dihedral type, and we assume that $\Lambda$ is basic.

<u>VI.2.2</u> NOTATION, IMPORTANT FACTS We say that a $\Lambda$-module is *exceptional* if it lies at the end of a 3-tube. There are therefore at most six exceptional $\Lambda$-modules.

Let S be a simple $\Lambda$-module, with projective cover P. By IV.3.1 and the hypothesis on $\Gamma_s(\Lambda)$, we know that (rad P)/S is the direct sum of $\leq$ two indecomposable modules, which are the successors of rad P in $\Gamma_s(\Lambda)$. We write (rad P)/S = U $\oplus$ V, where we always take U $\neq$ 0. Thus V = 0 if and only if rad P lies at the end of some tube. Also, U is indecomposable, and V is indecomposable or zero.

We define modules $\tilde{U}$, $\underset{\sim}{V}$ and $\tilde{V}$, $\underset{\sim}{U}$ as in IV.5.1. These modules will appear frequently in the following.

<u>VI.3</u> THEOREM *Let $\Lambda$ be an algebra of dihedral type. Then $\Lambda$ has a most three simple modules.*

This is proved in $[E_8]$. The reason why there is such a small bound is as follows: Consider the modules $\tilde{U}$, $\tilde{V}$. By IV.5.5, they lie at the end of some components. Since $\Lambda$ is of dihedral type, the only ends of components are those of tubes.

<u>V.3.1</u> *If $\Lambda$ has more than one simple module then for any simple module S at least one of $\tilde{U}$ ( = $\tilde{U}(S)$ ) or $\tilde{V}$ is exceptional*:

Suppose $\Omega^2\tilde{U} \cong \tilde{U}$, then there is an exact sequence $0 \to \tilde{U} \to Q \to P \to \tilde{U} \to 0$ with Q projective, and top Q $\cong$ top $\underset{\sim}{V} \cong$ top V. By the exactness, <u>dim</u> Q = <u>dim</u> P; and then Q $\cong$ P since the Cartan matrix is non-singular; and therefore top V $\cong$ S. Now, if also $\tilde{V} \cong \Omega^2\tilde{V}$ then similarly top U $\cong$ S, and the quiver of $\Lambda$ consists of one vertex with two loops, and S is the only simple $\Lambda$-module.

This shows already that $\Lambda$ cannot have more that six simple modules. A more detailed analysis which exploits the non-singularity of the Cartan matrix improves the bound.

Note that this is best possible, as the example (iii) in VI.2.1.1 shows.

## VI.4 *The quiver of* $\Lambda$.

VI.4.1 LEMMA *Let Q be the quiver of* $\Lambda$. *Then there are no double arrows e f where e, f are distinct vertices.*

Proof: Let e, f be orthogonal primitive idempotents of $\Lambda$, and assume that $\beta, \gamma \in eJf - eJ^2f$. We have to show that $\beta$ and $\gamma$ must be dependent (modulo $J^2$).

For $c \in K$, define $\omega_c = \beta + c\gamma$. The module $\omega_c\Lambda$ (which is indecomposable) must lie at the end of its component in $\Gamma_s(\Lambda)$, by IV.4. We claim that $\omega_c\Lambda$ is exceptional. Otherwise, $\omega_c\Lambda$ lies in a tube of rank 1, and then there is an exact sequence $0 \to \omega_c\Lambda \to e\Lambda \to f\Lambda \to \omega_c\Lambda \to 0$. By exactness, $\underline{\dim}\, e\Lambda = \underline{\dim}\, f\Lambda$, and then $e\Lambda \cong f\Lambda$ since the Cartan matrix is non-singular. But then e and f are not orthogonal, contradiction.

Now, the number exceptional modules is finite but K is infinite. Hence there is some non-zero $c \in K$ such that $\omega_0\Lambda \cong \omega_c\Lambda$. This implies that $\omega_0$ and $\omega_c$ are dependent (modulo $J^2$), hence so are $\beta$ and $\gamma$.

VI.4.2 LEMMA *Assume P is indecomposable projective, such that rad P/soc P = U* $\oplus$ *V with U, V non-zero. Then top U and top V are multiplicity-free.*

Proof: Let $P = e\Lambda$, and assume that $\beta\Lambda + \gamma\Lambda \subseteq \underset{\sim}{U}$ where $\beta$ and $\gamma$ lie in $e\Lambda f$ but not in $J^2$. We have to show that $\beta$ and $\gamma$ are dependent (modulo $J^2$). For $c \in K$, let $\omega_c = \beta + c\gamma$. Then $\omega_c\Lambda$ lies at the end of some tube, and so does $e\Lambda/\omega_c\Lambda$. Note that $\operatorname{rad}(e\Lambda/\omega_c\Lambda) \cong \operatorname{rad}(e\Lambda)/\omega_c\Lambda \cong U/\omega_c\Lambda \oplus V$.

Assume for contradiction that $\beta$ and $\gamma$ are independent (modulo $J^2$). Then $\omega_c\Lambda \subsetneq U$, and $\operatorname{soc}(e\Lambda/\omega_c\Lambda)$ is not simple. It follows that $\omega_c\Lambda$ cannot lie in a 1-tube. Also, if $\beta$ and $\gamma$ are independent then the modules $\omega_c\Lambda$ are non-isomorphic for different $c \in K$. Thus there are infinitely many modules at ends of components other than 1-tubes, and $\Lambda$ is not of dihedral type.

VI.4.3 PROPOSITION *Assume that $S$ is a simple $\Lambda$-module with projective cover $P = e\Lambda$, and let $(\text{rad } P)/S = U \oplus V$. Then the tops and socles of $U$ and $V$ are simple (if non-zero).*

Proof: It suffices to consider the tops, by duality.

Assume first that U and V are both non-zero. Suppose $|\text{top } U| > 1$. Then there are i, j with $i \neq j$ and $\alpha \in e\Lambda e_i$, $\beta \in e\Lambda e_j$ such that $\alpha$ and $\beta$ are independent (modulo $J^2$), and $\alpha\Lambda + \beta\Lambda \subseteq \underset{\sim}{U}$, by VI.2.2. (In particular, $\Lambda$ has more than one simple module.). Then we have the following exceptional modules

(*) $\alpha\Lambda$ and $e\Lambda/\alpha\Lambda$, $\beta\Lambda$ and $e\Lambda/\beta\Lambda$, $\underset{\sim}{U}$.

(1) These modules are pairwise non-isomorphic: For example, $\alpha\Lambda \subsetneq \underset{\sim}{U}$, then $\text{top}(\alpha\Lambda) \neq \text{top}(\beta\Lambda)$. Also $\text{soc}(\alpha\Lambda)$ is simple but the socles of $e\Lambda/\alpha\Lambda$ and $e\Lambda/\beta\Lambda$ are not simple. Hence $\alpha\Lambda$ is not isomorphic to any of the other modules listed in (*).

Since $S_i \neq S_j$, at least one of them is different from S. Say $S_j \neq S$. The modules $U_j$ and $\underset{\sim}{V}_j$ are also exceptional. Since $\text{soc}(\underset{\sim}{V}_j) \cong S_j$ and the modules in (*) whose socles are simple, have socles $\cong S$, we have that $\underset{\sim}{V}_j$ is not in the list (*). Therefore $\underset{\sim}{V}_j$ must be the last exceptional module.

The socles of two of the exceptional modules ( $e\Lambda/\alpha\Lambda$ and $e\Lambda/\beta\Lambda$ ) are not simple, so there must be two whose top is not simple. It follows that $|\text{top } \underset{\sim}{V}_j| > 1$. Hence

(2) There is some $\gamma \in e_j\Lambda e_k$ (for some primitive idempotent $e_k$) such that $\gamma\Lambda \subset \underset{\sim}{V}_j$.

Then $\gamma\Lambda$ is also exceptional since $\text{soc}(e_j\Lambda/\gamma\Lambda)$ is not simple. Only one of the exceptional modules, namely $\underset{\sim}{V}_j$, has socle $\cong S_j$. Thus we must have that $\gamma\Lambda \cong \underset{\sim}{V}_j$. This contradicts (2).

Now we consider the case when $V = 0$. Then P/S and S lie at the end of a tube and S is periodic. Consequently $\Lambda$ has more than one simple module, by IV.3.1, hence there is some arrow $\alpha \in e\Lambda e_i$ where $e \neq e_i$. Assume for contradiction that $|\text{top } U| > 1$. Then $\alpha\Lambda \subsetneq \underset{\sim}{U}$. We claim that

(1) The modules $\alpha\Lambda$, $e\Lambda/\alpha\Lambda$, $\Omega^{-1}S$, S and $\underset{\sim}{U}$ are exceptional and pairwise non-isomorphic:

We have $\mho \cong \Omega^{-1}S$ and $\underset{\sim}{U} \cong \Omega S$, so S, $\mho$ and $\underset{\sim}{U}$ are pairwise non-isomorphic. Now $\alpha\Lambda \subsetneq \underset{\sim}{U}$ and $\alpha\Lambda$ is not simple; so by the lengths, $\alpha\Lambda \neq \mho$, S, $\underset{\sim}{U}$. Consider $e\Lambda/\alpha\Lambda$; this is a proper quotient of $\mho$ which is not simple. So $1 < |e\Lambda/\alpha\Lambda| < |\mho|$ and $|\underset{\sim}{U}|$. Moreover, $\alpha\Lambda$

$\neq e\Lambda/\alpha\Lambda$ since these modules have different tops.

(2) top $U \cong S_i \oplus nS$ $(n > 1)$: Take some arrow $\beta \neq \alpha$ starting at vertex e and ending at $e_j$. Assume for contradiction that $e_j \neq e$; then $\beta\Lambda$ is exceptional. By VI.4.1, $e_j \neq e_i$, and thus $\beta\Lambda$ does not appear in the list (1), consequently $\beta\Lambda$ is the last exceptional module. Now consider the module $e\Lambda/\beta\Lambda$. It is not isomorphic to $\beta\Lambda$ or any of the modules in (1) (by the arguments in the proof of (1)); on the other hand, $e\Lambda/\beta\Lambda$ is exceptional. This is a contradiction.

(3) $soc(\Omega^{-1}S)$ is not simple: By (2), S occurs in $soc(\Omega^{-1}S)$; and soc $\Omega^{-1}S$ must contain a simple module $\neq S$; otherwise $\Lambda$ would be local, see IV.1.4.

(4) $\Omega(\alpha\Lambda)$ is the last exceptional module: Clearly, $\Omega(\alpha\Lambda)$ is exceptional. We have to show that it is not isomorphic to any of the modules in (1). Since $soc(\Omega(\alpha\Lambda)) \cong S_i$, the module could only be isomorphic to $e\Lambda/\alpha\Lambda$, by (3). This is not possible since then it would be fixed by $\Omega^2$. The modules $\underset{\sim}{U}_i$ and $\underset{\sim}{V}_i$ are also exceptional (see IV.5.5). Now, top $\underset{\sim}{U}_i \cong S_i$ and therefore either $\underset{\sim}{U}_i \cong \alpha\Lambda$ or $\underset{\sim}{U}_i \cong \Omega(\alpha\Lambda)$.

(5) $\underset{\sim}{U}_i \cong \alpha\Lambda$ leads to a contradiction: If $\underset{\sim}{U}_i \cong \alpha\Lambda$ then $\Omega(\alpha\Lambda) \cong \underset{\sim}{V}_i$, hence has a simple top. (Here we apply the first part of the Proposition in case $V_i \neq 0$). Thus the top of only one exceptional module (namely $\underset{\sim}{U}$) is not simple, hence $\Omega(\underset{\sim}{U})$ is the only exceptional module whose socle is not simple. It follows that $\Omega(\underset{\sim}{U}) \cong \Omega^{-1}(S)$, hence the module has a simple top. Also, the exceptional modules lie in $\Omega$-orbits of length 3.

Let $\underset{\sim}{U} = \alpha\Lambda + \Sigma\beta_i\Lambda$ where $\beta_i \in e\Lambda e$. Consider the projective cover $\pi$: $P_i \oplus nP \to \underset{\sim}{U}$ given by $\pi(x, y_1, \ldots, y_n) = \alpha x + \Sigma\, \beta_i y_i$. By the hypothesis, $\underset{\sim}{V}_i \cong \{x \in e_i\Lambda\colon \alpha x = 0\}$. Thus if $\underset{\sim}{V}_i = \gamma\Lambda$ then $(\gamma, 0, \ldots, 0)$ lies in $Ker(\pi)$. Clearly, the element does not lie in rad $Ker(\pi)$ (since $\gamma \notin J^2$). As we saw above, $Ker(\pi)$ has a unique maximal submodule. It follows that $\gamma\Lambda \cong Ker(\pi) \cong \Omega^{-1}(S)$, and $soc(\Omega^{-1}S)$ is simple, contradicting (3).

Hence we can only have that $\underset{\sim}{U}_i \cong \Omega(\alpha\Lambda)$. But then $soc(U_i) \cong top(\alpha\Lambda) \cong S_i$. Thus $soc(V_i) \neq S_i$, and $\underset{\sim}{V}_i$ must be exceptional as well. On the other hand, the arguments in (5) show that $\underset{\sim}{V}_i \neq \alpha\Lambda$; and also $\underset{\sim}{V}_i \neq \underset{\sim}{U}_i$. Consequently $\underset{\sim}{V}_i$ is not in the list of exceptional modules. This contradiction shows that top U must be simple.

VI.4.4 COROLLARY *The quiver Q of $\Lambda$ has at most three vertices, and for any vertex f of Q we have that $|s^{-1}(f)| = |e^{-1}(f)| = 2$ or 1.*

<u>VI.5</u> PROPOSITION  *The quiver of* Λ *is one of the following:*

(1)    X  ↺·↻  Y

(2A)    α ↺ $\overset{0}{\cdot} \underset{\gamma}{\overset{\beta}{\rightleftarrows}} \overset{!}{\cdot}$          (2B)    α ↺ $\overset{0}{\cdot} \underset{\gamma}{\overset{\beta}{\rightleftarrows}} \overset{!}{\cdot} ↻$ η

(3A)    $\overset{1}{\cdot} \underset{\gamma}{\overset{\beta}{\rightleftarrows}} \overset{0}{\cdot} \underset{\eta}{\overset{\delta}{\rightleftarrows}} \overset{2}{\cdot}$

(3B)    α ↺ $\overset{!}{\cdot} \underset{\gamma}{\overset{\beta}{\rightleftarrows}} \overset{0}{\cdot} \underset{\eta}{\overset{\delta}{\rightleftarrows}} \overset{2}{\cdot}$

(3D)    α ↺ $\overset{1}{\cdot} \underset{\gamma}{\overset{\beta}{\rightleftarrows}} \overset{0}{\cdot} \underset{\eta}{\overset{\delta}{\rightleftarrows}} \overset{2}{\cdot} ↻$ ξ

(3K)     (3L)     (3Q)     (3Z)

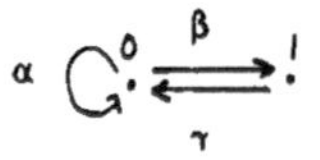

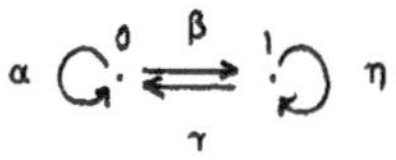

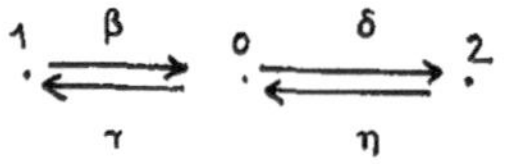

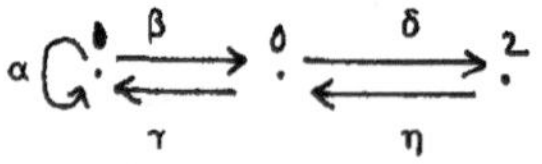

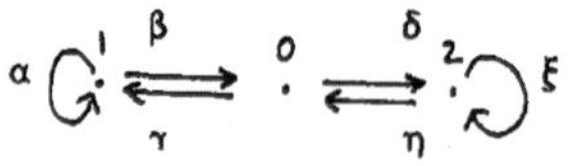

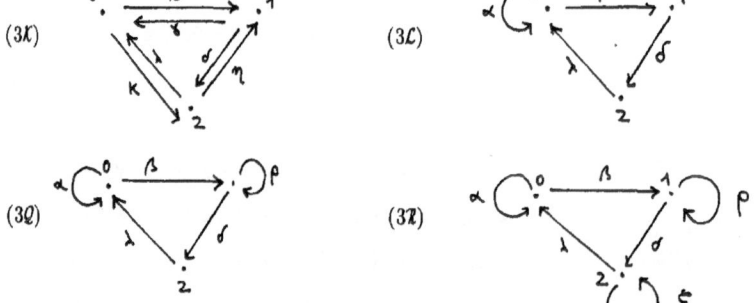

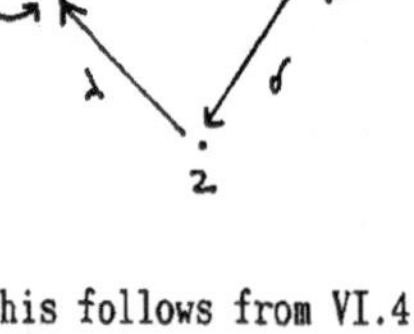

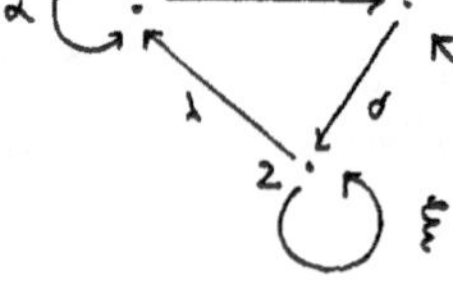

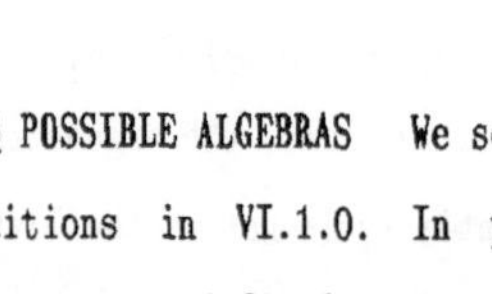

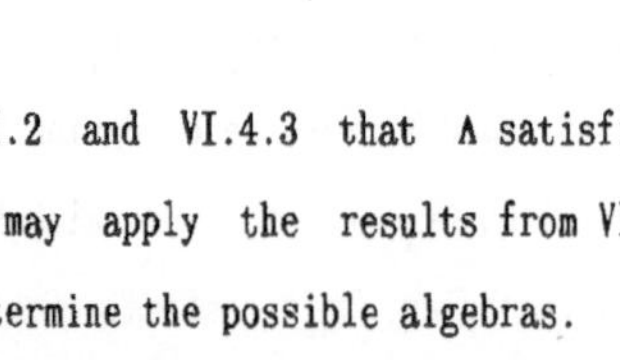

Proof: This follows from VI.4.4 and IV.2.4.

<u>VI.6</u> POSSIBLE ALGEBRAS   We see from VI.5, VI.2 and VI.4.3 that Λ satisfies the conditions in VI.1.0. In particular, we may apply the results from VI.1. The permutation π defined in VI.1.2 allows to determine the possible algebras.

By definition, π describes the Ω-action on modules of the form αΛ where α is an

arrow. On the other hand, by IV.4 and the structure of $\Gamma_s(\Lambda)$, the module $\alpha\Lambda$ lies at the end of some tube of rank 1 or 3; and since $\tau \cong \Omega^2$, we deduce that $\alpha\Lambda$ is $\Omega$-periodic of period dividing 6. Moreover, if $\alpha\Lambda$ has period 2 then $\alpha$ must be a loop (IV.1.3). Therefore, $\pi$ consists of either one 6-cycle, or $\leq$ two 3-cycles and otherwise only 1-cycles. Moreover, each cycle corresponds to a closed path in the quiver. This means that the possibilities for $\pi$ are limited.

For each of the quivers in the list VI.5 we will now write down all possible permutations ( up to notational symmetry). We also list the permutation $\pi^*$ defined in VI.1.3 which describes the composition series of the modules $\alpha\Lambda$, for any arrow $\alpha$. (This module is uniserial, see VI.1.4.2.)

We exclude some cases at once which give rise to a singular Cartan matrix, by applying VI.1.5. In chapter III we have already studied the local algebras. The remaining algebras in the list below will be considered afterwards.

| Quiver | permutation $\pi$ | permutation $\pi^*$ | classified |
|---|---|---|---|
| (1) | $(\alpha)(\beta)$ | $(\alpha\ \beta)$ | III |
| | $(\alpha\ \beta)$ | $(\alpha)(\beta)$ | III |
| $(2\mathcal{A})$ | $(\alpha)(\beta\ e_1\gamma)$, | $(\alpha\ \beta\ \gamma)$ | VI.8.1 |
| $(2\mathcal{B})$ | $(\alpha)(\beta\ \eta\ \gamma)$ | $(\alpha\ \beta\ \gamma)(\eta)$ | VI.8.2 |
| $(3\mathcal{A})$ | $(\gamma\ e_1\beta)(\delta\ e_2\eta)$ | $(\gamma\ \beta\ \delta\ \eta)$ | VI.9.1 |
| | $(\gamma\ e_1\beta\ \delta\ e_2\eta)$ | $(\gamma\ \beta)(\delta\ \eta)$ | -"- |
| $(3\mathcal{B})$ | $(\alpha\ \beta\ \gamma)(\delta\ e_2\eta)$ | $(\alpha)(\beta\ \delta\ \eta\ \gamma)$ | VI.9.2 |
| | $(\alpha\ \beta\ \delta\ e_2\ \eta\ \gamma)$ | $(\alpha)(\beta\ \gamma)(\delta\ \eta)$ | -"- |
| $(3\mathcal{D})$ | $(\alpha\ \beta\ \gamma)(\delta\ \xi\ \eta)$ | $(\alpha)(\beta\ \delta\ \eta\ \gamma)(\xi)$ | VI.9.3 |
| | $(\alpha\ \beta\ \delta\ \xi\ \eta\ \gamma)$ | $(\alpha)(\beta\ \gamma)(\delta\ \eta)(\xi)$ | -"- |
| $(3\mathcal{K})$ | $(\beta\ \delta\ \lambda\ \kappa\ \eta\ \gamma)$ | $(\beta\ \gamma\ \kappa\ \lambda)(\delta\ \eta)$ | C singular |
| | $(\beta\ \delta\ \lambda)(\gamma\ \kappa\ \eta)$ | $(\beta\ \gamma)(\delta\ \eta)(\kappa\ \lambda)$ | VI.9.4 |
| $(3\mathcal{L})$ | $(\alpha\ \beta\ e_1\delta\ e_2\ \lambda)$ | $(\alpha)(\beta\ \delta\ \lambda)$ | VI.9.5 |
| $(3\mathcal{Q})$ | $(\alpha\ \beta\ \rho\ \delta\ e_2\ \lambda)$ | $(\alpha)(\beta\ \delta\ \lambda)(\rho)$ | VI.9.6 |
| $(3\mathcal{R})$ | $(\alpha)(\rho)(\xi)(\beta\ \delta\ \lambda)$ | $(\alpha\ \beta\ \rho\ \delta\ \xi\ \lambda)$ | C singular |
| | $(\alpha\ \beta\ \rho\ \delta\ \xi\ \lambda)$ | $(\alpha)(\beta\ \delta\ \lambda)(\rho)(\xi)$ | VI.9.7 |

## VI.7 *Algebras of dihedral type with one simple module*

We will prove the following result:

THEOREM *Suppose that* Λ *is a local algebra of dihedral type. Then* Λ *is isomorphic to one of the algebras in III.1(a) to (c').*

This implies directly, by III.1.2:

COROLLARY *Suppose* Λ *is a local algebra of dihedral type. Then* Λ/soc Λ *is special biserial, and* Λ *is tame.*

Proof of the Theorem: It follows from VI.1.4.1 that dim $J^2/J^3 \leq 2$. We may assume that this dimension is 2; otherwise we apply III.4(ii) and obtain directly that Λ is one of the algebras in III.1(a) to (b').

Let $\pi$ be the permutation associated to Λ, according to VI.1.2. The possibilities for $\pi$ are $(\alpha\ \beta)$ and $= (\alpha)(\beta)$.

Assume first that $\pi = (\alpha\ \beta)$. From the definition of $\pi$ we have that, without loss of generality, $\Omega(\alpha\Lambda) = \beta\Lambda$; thus $\alpha\beta = 0$; and moreover, since $\pi(\beta) = \alpha$, we have by VI.1.4 that $\beta\alpha \in J^3$. That is, $\Lambda/J^3$ satisfies III.6(i); therefore we obtain from III.9 that Λ is isomorphic to one of the algebras in III.1(a).

Now assume that $\pi = (\alpha)(\beta)$. Then by VI.1.4.4

(*) <u>we may assume that $\alpha^2$ and $\beta^2$ lie in soc Λ</u>.

Now we see that $\Lambda/J^3$ satisfies III.6(iv). Therefore we may apply III.10; moreover (*) shows that we are in Case 1 in III.10; and we deduce that Λ is isomorphic to an algebra in III.1(c) or (c').

## VI.8 *Algebras of dihedral type with two simple modules*

Let Λ be a basic algebra of dihedral type, with two simple modules. There are two possibilities for the quiver of Λ, which are given in VI.5.

VI.8.1 THEOREM *Assume that $\Lambda$ is a basic algebra of dihedral type, with quiver (2A). Then $\Lambda \cong KQ/I$ where $I$ is one of the following ideals:*

*(i) $I = [\gamma\beta,\ \alpha^2,\ (\alpha\beta\gamma)^k-(\beta\gamma\alpha)^k\ ]$ where $k \geq 1$.*

*(ii) char $K = 2$ and $I = [\gamma\beta,\ \alpha^2-(\alpha\beta\gamma)^k,\ (\alpha\beta\gamma)^k-(\beta\gamma\alpha)^k\ ]$, where $k \geq 1$.*

Proof: Let $\pi$ be the permutation associated to $\Lambda$, as defined in VI.1.2. Then $\pi$ must be of the form $\pi = (\alpha)(\beta\ e_1\ \gamma)$, by VI.6. As we have seen in VI.1.4.4, we may assume that $\gamma\beta = 0$ and that $\alpha^2$ lies in soc $\Lambda$. Moreover, let $\pi^* = (\beta\ \gamma\ \alpha)$ have multiplicity $k$; then we have by VI.1.4.2 that $\langle (\gamma\alpha\beta)^k \rangle = \text{soc}\ e_1\Lambda$ and $\text{soc}\ e_0\Lambda = \langle (\beta\gamma\alpha)^k \rangle = \langle (\alpha\beta\gamma)^k \rangle$. Since $\Lambda$ is symmetric we deduce that $(\alpha\beta\gamma)^k = (\beta\alpha\gamma)^k$; and it remains to study the relation for $\alpha^2$. We know that

(*) <u>there is some $c \in K$ such that $\alpha^2 = c(\alpha\beta\gamma)^k$.</u>

Assume first that char $K \neq 2$. Set $\alpha' = [\alpha - c/2(\beta\gamma\alpha)^{k-1}\beta\gamma]$; then $(\alpha')^2 = 0$; and replacing $\alpha$ by $\alpha'$ does not change the other relations. Thus in this case, $\Lambda$ is an algebra as in (i), and clearly also when char $K = 2$ and $c = 0$.

Now assume that char $K = 2$ and that $c \neq 0$. By replacing $\beta$ by an appropriate scalar multiple of itself, one obtains a relation (*) with $c = 1$, and $\Lambda$ is an algebra as in (ii).

REMARK *The algebras in (i) and (ii) are not isomorphic:*

Let $\Lambda$ be as in (i). Assume for contradiction that $\Lambda$ is isomorphic to an algebra in (ii). Then char $K = 2$, and there is some $\omega \in e_0Je_0 - J^2$ whose square is zero. We may write $\omega = \alpha + \alpha x_1\alpha + \alpha x_2\gamma + \beta x_3\alpha + \beta x_4\gamma$ for $x_i \in \Lambda$. Then $\omega^2 = \alpha^2 + \alpha\beta[x_3\alpha + x_4\gamma] + [\alpha x_2 + \beta x_4]\gamma\alpha = 0$. Since $\alpha^2 = (\alpha\beta\gamma)^k$, it follows that each of the summands must lie in soc $\Lambda$. Hence $\alpha\beta x_3\alpha = 0 = \alpha\beta x_3\alpha = 0 = \alpha x_2\gamma\alpha$, and $\beta x_4\gamma \in \text{soc}_2\Lambda$. But then $\beta x_4\gamma$ lies in $Z(\Lambda)$ and we conclude that $\alpha\beta x_4\gamma + \beta x_4\gamma\alpha = 2\alpha\beta x_4\gamma = 0$ since char $K = 2$. Thus $\omega^2 = \alpha^2$ which is non-zero, a contradiction.

VI.8.2 THEOREM *Assume that $\Lambda$ is basic of dihedral type, with quiver (2B). Then $\Lambda \cong KQ/I$ where $I$ is one of the following ideals:*

*(i) $I = [\ \beta\eta,\ \eta\gamma,\ \gamma\beta,\ \alpha^2,\ (\alpha\beta\gamma)^k-(\beta\gamma\alpha)^k,\ (\gamma\alpha\beta)^k-\eta^s]$.*

*(ii) char $K = 2$ and $I = [\beta\eta,\ \eta\gamma,\ \gamma\beta,\ \alpha^2 - (\alpha\beta\gamma)^k,\ (\alpha\beta\gamma)^k - (\beta\gamma\alpha)^k,\ (\gamma\alpha\beta)^k - \eta^s]$, where $k \geq 1$ and $s \geq 2$.*

Proof: Let $\pi$ be the permutation associated to $\Lambda$ defined in VI.1.2; we fix the notation such that $\pi = (\alpha)(\beta\ \eta\ \gamma)$, see VI.6. By VI.1.4.4, we may assume that $\gamma\beta = 0$ and that $\alpha^2 \in \operatorname{soc} \Lambda$. Moreover, we obtain from VI.1.4.5 that $\eta\gamma = 0 = \beta\eta$.

Let k and s be the multiplicities of the cycles $(\beta\ \gamma\ \alpha)$ and $(\eta)$ of $\pi^*$ respectively. By VI.1.4.2 we know that $\operatorname{soc} e_1\Lambda = \langle (\gamma\alpha\beta)^k \rangle = \langle \eta^s \rangle$ and $\operatorname{soc} e_0\Lambda = \langle (\alpha\beta\gamma)^k \rangle = \langle (\beta\gamma\alpha)^k \rangle$. Since $\Lambda$ is symmetric, we have $(\alpha\beta\gamma)^k = (\beta\gamma\alpha)^k$. Moreover $(\gamma\alpha\beta)^k = c\eta^s$ for some $c \in K^*$. We may assume that $c = 1$; otherwise we replace $\eta$ by an appropriate scalar multiple of itself. Note that $s \geq 2$, by I.3.8.

It remains to study the relation for $\alpha^2$. This is done in the same way as in VI.8.1. Similarly, one shows that the algebras in (i) and (ii) are non-isomorphic.

## VI.9 *Algebras of dihedral type with three simple modules*

Our first result is as follows:

<u>VI.9.1</u> THEOREM *Assume that $\Lambda$ is basic of dihedral type, with quiver of the form (3A). Then $\Lambda \cong KQ/I$ where I is one of the following ideals:*

*(i) $I = [\beta\gamma,\ \eta\delta,\ (\gamma\beta\delta\eta)^k - (\delta\eta\gamma\beta)^k]$ where $k \geq 1$.*

*(ii) $I = [\beta\delta,\ \eta\gamma,\ (\gamma\beta)^k - (\delta\eta)^l]$ where $k \geq l \geq 2$.*

Proof: Let $\pi$ be the permutation associated to $\Lambda$, according to VI.1.2. By VI.6 there are two possibilities for $\pi$.

Assume first that $\pi = (\gamma\ e_1\ \beta)(\delta\ e_2\ \eta)$. By VI.1.4.4, we may assume that $\beta\gamma = 0$ and that $\eta\delta = 0$. Let k be the multiplicity of $\pi^* = (\beta\ \delta\ \eta\ \gamma)$. By VI.1.4.2 we have that $\operatorname{soc} e_1\Lambda = \langle (\beta\delta\eta\gamma)^k \rangle$ and $\operatorname{soc} e_2\Lambda = \langle (\eta\gamma\beta\delta)^k \rangle$, and moreover $\operatorname{soc} e_0\Lambda = \langle (\gamma\beta\delta\eta)^k \rangle = \langle (\delta\eta\gamma\beta)^k \rangle$. Since $\Lambda$ is symmetric, we deduce $(\gamma\beta\delta\eta)^k = (\delta\eta\gamma\beta)^k$. We see now that $\Lambda$ is isomorphic to the algebra in (i).

Now consider the case when $\pi = (\gamma\ e_1\ \beta\ \delta\ e_2\ \eta)$. By VI.1.4.4 we may assme that $\beta\delta = 0$ and $\eta\gamma = 0$. Let k and l be the multiplicities of the cycles $(\gamma\ \beta)$ and $(\delta\ \eta)$ of

$\pi^*$ respectively. Then by VI.1.4.2 we have that soc $e_1\Lambda$ = $<(\beta\gamma)^k>$, soc $e_2\Lambda$ = $<(\eta\delta)^l>$ and soc $e_0\Lambda$ = $<(\gamma\beta)^k> = <(\delta\eta)^l>$. Hence $(\gamma\beta)^k = c(\delta\eta)^l$ for some $c \in K^*$. We may assume that c = 1; otherwise we replace $\delta$ by an appropriate scalar multiple of itself. We see now that $\Lambda$ is isomorphic to the algebra in (ii). We must have that k, $l \geq 2$; otherwise $\Lambda$ would be of finite type, by II.8.4.

VI.9.2 THEOREM *Assume that $\Lambda$ is a basic algebra of dihedral type, with quiver of the form (3B). Then $\Lambda \cong KQ/I$ where I is one of the following ideals:*

*(i) $I = [\alpha\beta, \beta\gamma, \gamma\alpha. \eta\delta, (\gamma\beta\delta\eta)^k - (\delta\eta\gamma\beta)^k, (\beta\delta\eta\gamma)^k - \alpha^s]$ where $k \geq 1$ and $s \geq 2$.*

*(ii) $I = [\gamma\alpha, \alpha\beta, \beta\delta, \eta\gamma, (\gamma\beta)^k - (\delta\eta)^l, (\beta\gamma)^k - \alpha^s]$ where $k, l \geq 1$ and $s \geq 2$, and $(k, l) \neq (1, 1)$.*

Proof: Let $\pi$ be the permutation associated to $\Lambda$, according to VI.1.2. By VI.6, there are two possibilities for $\pi$.

Assume first that $\pi = (\alpha\ \beta\ \gamma)(\delta\ e_2\ \eta)$. By VI.1.4.4, we may assume that $\beta\gamma = 0$ and $\eta\delta = 0$; and moreover $\alpha\beta = 0$ and $\gamma\alpha = 0$ (VI.1.4.5). Let k and s be the multiplicities of the cycles $(\beta\ \delta\ \eta\ \gamma)$ and $(\alpha)$ of $\pi^*$ respectively. Then by VI.1.4.2 we know the socle of $\Lambda$, namely soc $e_2\Lambda = <(\eta\gamma\beta\delta)^k>$, soc $e_0\Lambda = <(\gamma\beta\delta\eta)^k> = <(\delta\eta\gamma\beta)^k>$, and moreover soc $e_1\Lambda = <(\beta\delta\eta\gamma)^k> = <\alpha^s>$. Since $\Lambda$ is symmetric, we have $(\gamma\beta\delta\eta)^k = (\delta\eta\gamma\beta)^k$; and as before, we may assume that $(\beta\delta\eta\gamma)^k = \alpha^s$. We see now that $\Lambda$ is isomorphic to the algebra in (i).

Now assume that $\pi = (\alpha\ \beta\ \delta\ e_2\ \eta\ \gamma)$. By VI.1.4.4 we may assume that $\beta\delta = 0$ and $\eta\gamma = 0$. Note that $\pi^*(\alpha) = \alpha$; therefore, by VI.1.4.5 we know that $\alpha\beta = 0$ and $\gamma\alpha = 0$. Let k, l and s be the multiplicities of the cycles $(\beta\ \gamma)$, $(\delta\ \eta)$ and $(\alpha)$ of $\pi^*$ respectively. Then, using VI.1.4.2 we obtain, as before, that $(\gamma\beta)^k = c(\delta\eta)^l$ and $(\beta\gamma)^k = d\alpha^s$ for $0 \neq c, d \in K$. We may assume that c = d = 1; if not we replace $\delta$ and $\alpha$ by appropriate scalar multiples of themselves. Hence $\Lambda$ is isomorphic to the algebra defined in (ii). Moreover, $(k, s) \neq (1, 2)$ since otherwise $\Lambda$ would be of finite type, by II.8.5.

VI.9.3 THEOREM *Assume that $\Lambda$ is a basic algebra of dihedral type with quiver of the form ($\mathcal{SD}$). Then $\Lambda \cong KQ/I$ where I is one of the following ideals:*

*(i)* $I = [\alpha\beta, \beta\gamma, \gamma\alpha, \delta\xi, \xi\eta, \eta\delta, \alpha^s - (\beta\delta\eta\gamma)^k, \xi^t - (\eta\gamma\beta\delta)^k, (\delta\eta\gamma\beta)^k - (\gamma\beta\delta\eta)^k]$ *where* $k \geq 1$ *and* $s, t \geq 2$.

*(ii)* $I = [\alpha\beta, \beta\delta, \delta\xi, \xi\eta, \eta\gamma, \gamma\alpha, (\gamma\beta)^k - (\delta\eta)^l, (\beta\gamma)^k - \alpha^s, (\eta\delta)^l - \xi^t]$ *where* $k, l \geq 1$ *and* $s, t \geq 2$.

Proof: Let $\pi$ be the permutation associated to $\Lambda$, according to VI.1.2. By VI.6, there are two possibilities for $\pi$.

Assume first that $\pi = (\alpha\ \beta\ \gamma)(\delta\ \xi\ \eta)$. By VI.1.4.4 without loss of generality $\beta\gamma = 0$ and $\eta\delta = 0$. Since $\pi^*(\alpha) = \alpha$ and $\pi^*(\xi) = \xi$ we obtain from VI.1.4.5 that $\alpha\beta = 0$ and $\gamma\alpha = 0$, and $\xi\eta = 0 = \delta\xi$. Let k, s, t be the multiplicities of the cycles $(\beta\ \delta\ \eta\ \gamma)$, $(\alpha)$ and $(\xi)$ of $\pi^*$ respectively. Then by VI.1.4.2 we have that $\mathrm{soc}\ e_1\Lambda = \langle (\beta\delta\eta\gamma)^k\rangle = \langle \alpha^s\rangle$, $\mathrm{soc}\ e_2\Lambda = \langle (\delta\eta\gamma\beta)^k\rangle = \langle \xi^t\rangle$ and $\mathrm{soc}\ e_0\Lambda = \langle(\delta\eta\gamma\beta)^k\rangle = \langle (\gamma\beta\delta\eta)^k\rangle$. Since $\Lambda$ is symmetric, we deduce that $(\delta\eta\gamma\beta)^k = (\gamma\beta\delta\eta)^k$. Hence there are $c, d \in K^*$ such that $(\beta\delta\eta\gamma)^k = c(\alpha^s)$ and $(\eta\gamma\beta\delta)^k = d(\xi^t)$. We may assume that $c = d = 1$, otherwise we re-scale $\alpha$ and $\xi$. This proves that $\Lambda$ is isomorphic to the algebra in (i).

Now assume that $\pi = (\alpha\ \beta\ \delta\ \xi\ \eta\ \gamma)$. In this case, without loss of generality $\beta\delta = 0$ and $\eta\gamma = 0$. Moreover, by VI.1.4.5 we see again that $\alpha\beta = 0 = \gamma\alpha$ and $\xi\eta = 0 = \delta\xi$. Let k, l and s, t be the multiplicities of the cycles $(\beta\ \gamma)$, $(\eta\ \delta)$, $(\alpha)$ and $(\xi)$ of $\pi^*$; then $\mathrm{soc}\ e_1\Lambda = \langle (\beta\gamma)^k\rangle = \langle \alpha^s\rangle$, $\mathrm{soc}\ e_2\Lambda = \langle (\eta\delta)^l\rangle = \langle \xi^t\rangle$, and $\mathrm{soc}\ e_0\Lambda = \langle (\beta\gamma)^k\rangle = \langle (\delta\eta)^l\rangle$. Hence there are $b, c, d \in K^*$ such that $(\gamma\beta)^k = b(\delta\eta)^l$ and $(\beta\gamma)^k = c(\alpha^s)$, $(\eta\delta)^l = d\xi^t$. We may assume that $b = 1$ and $c = d = 1$; otherwise we replace first $\delta$, then $\alpha$ and $\xi$ by appropriate scalar multiples of themselves. Thus $\Lambda$ is isomorphic to the algebra in (ii).

VI.9.4 THEOREM *Assume that $\Lambda$ is basic of dihedral type with quiver ($\mathcal{SK}$). Then $\Lambda \cong KQ/I$ where I is the ideal defined by the relations*

$$\beta\delta = \delta\lambda = \lambda\beta = 0 \quad \text{and} \quad \gamma\kappa = \kappa\eta = \eta\gamma = 0$$

$$(\beta\gamma)^a = (\kappa\lambda)^b, \quad (\lambda\kappa)^b = (\eta\delta)^c \quad \text{and} \quad (\delta\eta)^c = (\gamma\beta)^a$$

*where* $a, b, c \geq 1$.

Proof: Let $\pi$ be the permutation associated to $\Lambda$, as in VI.1.2. By VI.6 we must have that $\pi = (\beta\ \delta\ \lambda)(\gamma\ \kappa\ \eta)$. By VI.1.4.4 we may assume that $\beta\delta = 0$ and $\eta\gamma = 0$. Consider the element $\delta\lambda$. Since the cycle of $\pi^*$ containing $\delta$ is $(\delta\ \eta)$, we know that $\delta\Lambda$ is spanned by $\delta$, $\delta\eta$, $\delta\eta\delta$, ... . In particular, $\delta\Lambda e_0 = 0$ and hence $\delta\lambda = 0$. The same argument shows that $\lambda\beta = 0$ and $\gamma\kappa = 0 = \kappa\eta$.

Let a, b and c be the multiplicities of the cycles $(\beta\ \gamma)$, $(\kappa\ \lambda)$ and $(\eta\ \delta)$ of $\pi^*$ respectively. Then by VI.1.4.2 we have soc $e_0\Lambda = \langle (\beta\gamma)^a \rangle = \langle (\kappa\lambda)^b \rangle$, soc $e_1\Lambda = \langle (\gamma\beta)^b \rangle = \langle (\delta\eta)^c \rangle$ and soc $e_2\Lambda = \langle (\eta\delta)^c \rangle = \langle (\lambda\kappa)^b \rangle$. Hence there are r,s $\in$ K such that $(\beta\gamma)^a = r(\kappa\lambda)^b$ and $(\gamma\beta)^a = s(\delta\eta)^c$. We may assume that $r = s = 1$; otherwise we re-scale $\kappa$, $\delta$. Let $\psi$ be a symmetrizing form of $\Lambda$. Then $\psi[(\eta\delta)^c - (\lambda\kappa)^b] = \psi[(\delta\eta)^c - (\kappa\lambda)^b] = 0$. Hence $(\eta\delta)^c = (\lambda\kappa)^b$ since $\langle (\eta\delta)^c - (\kappa\lambda)^b \rangle$ is an ideal which is contained in ker $\psi$. This completes the proof of this theorem.

<u>VI.9.5</u> THEOREM *Assume that $\Lambda$ is basic of dihedral type, with quiver (3C). Then $\Lambda \simeq KQ/I$ where I is the ideal defined by*

$$\alpha\beta = 0 = \lambda\alpha, \qquad (\beta\delta\lambda)^k = \alpha^s, \qquad (\delta\lambda\beta)^k\delta = 0$$

*where $k \geq 2$ and $s \geq 2$.*

Proof: Let $\pi$ be the permutation associated to $\Lambda$, as defined in VI.1.2. Then $\pi = (\alpha\ \beta\ e_1\ \delta\ e_2\ \lambda)$, by VI.6. It follows from VI.1.4.5 that $\lambda\alpha = 0$ and $\alpha\beta = 0$. Let k and s be the multiplicities of the cycles $(\beta\ \delta\ \lambda)$ and $(\alpha)$ of $\pi^*$ respectively. Then by VI.1.4.2 we have that soc $e_0\Lambda = \langle (\beta\delta\lambda)^k \rangle = \langle \alpha^s \rangle$; moreover soc $e_1\Lambda = \langle (\delta\lambda\beta)^k \rangle$ and soc $e_2\Lambda = \langle (\lambda\beta\delta)^k \rangle$. There is some $0 \neq c \in K$ such that $(\beta\delta\lambda)^k = c\alpha^s$; and we may assume that $c = 1$, otherwise we replace $\alpha$ by a scalar multiple of itself. We must have that $k \geq 2$; otherwise $\Lambda$ would be of finite type (II.8.6). This shows that $\Lambda$ is as stated.

<u>VI.9.6</u> THEOREM *Assume that $\Lambda$ is basic of dihedral type, with quiver (3Q). Then $\Lambda \cong KQ/I$ where I is the ideal defined by*

$$0 = \lambda\alpha = \alpha\beta = \beta\rho = \rho\delta, \quad (\beta\delta\lambda)^k = \alpha^s \text{ and } (\delta\lambda\beta)^k = \rho^t,$$

*where $k \geq 1$ and $s, t \geq 2$.*

Proof: Let $\pi$ be the permutation associated with $\Lambda$, as defined in VI.1.2. Then by VI.6 we have $\pi = (\delta\ e_2\ \lambda\ \alpha\ \beta\ \rho)$; and moreover $\pi^* = (\beta\ \delta\ \lambda)(\alpha)(\rho)$. Therefore by VI.1.4.5 we have that $\lambda\alpha = 0 = \alpha\beta$ and $\beta\rho = 0 = \rho\delta$. Let k and s, t be the multiplicities of the cycles $(\beta\ \delta\ \lambda)$, $(\alpha)$ and $(\rho)$ of $\pi^*$ respectively. Then VI.1.4.2 shows that soc $e_0\Lambda$ = $\langle (\beta\delta\lambda)^k \rangle = \langle \alpha^s \rangle$, soc $e_1\Lambda = \langle (\delta\lambda\beta)^k \rangle = \langle \rho^t \rangle$ and soc $e_2\Lambda = \langle (\lambda\beta\delta)^k \rangle$. There are c, $d \in K^*$ such that $(\beta\delta\lambda)^k = c\alpha^s$ and $(\delta\lambda\beta)^k = d\rho^t$; and we may assume that c = d = 1, otherwise we replace $\alpha$ and $\rho$ by appropriate scalar multiples of themselves. This shows that $\Lambda$ is isomorphic to the algebra given.

VI.9.7 THEOREM *Assume that $\Lambda$ is basic of dihedral type, with quiver of the form ($\mathcal{SR}$). Then $\Lambda \cong KQ/I$ where I is the ideal defined by*

$$\alpha\beta = \beta\rho = \rho\delta = \delta\xi = \xi\lambda = \lambda\alpha = 0 \text{ and}$$
$$\alpha^s = (\beta\delta\lambda)^k, \quad \rho^t = (\delta\lambda\beta)^k, \quad \xi^u = (\lambda\beta\delta)^k$$

*where s, t, u $\geq$ 2 and k $\geq$ 1.*

Proof: Let $\pi$ be the permutation associated to $\Lambda$, as defined in V.1.2. Then by VI.6 we must have that $\pi = (\alpha\ \beta\ \rho\ \delta\ \xi\ \lambda)$; and $\pi^* = (\beta\ \delta\ \lambda)(\alpha)(\rho)(\xi)$. Therefore we obtain from VI.1.4.5 that $\alpha\beta = \beta\rho = \rho\delta = \delta\xi = \xi\lambda = \lambda\alpha = 0$. Let k and s, t, u be the multiplicities of the cycles $(\beta\ \delta\ \lambda)$, $(\alpha)$, $(\rho)$ and $(\xi)$ respectively. Then by VI.4.2 we have that soc $e_0\Lambda = \langle (\beta\delta\lambda)^k \rangle = \langle \alpha^s \rangle$, soc $e_1\Lambda = \langle (\delta\lambda\beta)^k \rangle = \langle \rho^t \rangle$ and soc $e_2\Lambda = \langle (\lambda\beta\delta)^k \rangle = \langle \xi^u \rangle$. Hence there are a, b and $c \in K^*$ such that $\alpha^s = a(\beta\delta\lambda)^k$, $\rho^t = b(\delta\lambda\beta)^k$ and $\xi^u = c(\lambda\beta\delta)^k$. We may assume that a, b and c are 1; otherwise we replace $\alpha$, $\rho$, $\xi$. We see now that $\Lambda$ is isomorphic to the algebra given.

We have now found all algebras which could be of dihedral type.

VI.10 *The converse* We have now an important consequence:

VI.10.1 THEOREM *Suppose $\Lambda$ is an algebra of dihedral type. Then $\Lambda$/soc $\Lambda$ is special biserial; in particular, $\Lambda$ is tame.*

Proof: Without loss of generality, $\Lambda$ is basic, then $\Lambda$ is isomorphic to one of the listed algebras. By inspection, we see that $\Lambda/\text{soc}\ \Lambda$ is special biserial. Then it follows from II.3.1 that $\Lambda$ is either tame or of finite type; and on the way we have excluded algebras of finite type.

VI.10.2 **THEOREM** *Let $\Lambda$ be an algebra belonging to the list. Then $\Lambda$ is of dihedral type.*

Proof: Clearly, all algebras in the list have connected quivers, hence are indecomposable. In each case it is easy to define a symmetrizing linear form explicitly, and hence the algebras are symmetric. We have listed the Cartan matrices in the tables and see that they are non-singular (actually, in chapter IX, we will verify that the determinants are $\neq 0$). It remains to study the stable AR-quiver of $\Lambda$. Recall that $\Gamma_s(\Lambda) = \Gamma(\Lambda/\text{soc}\ \Lambda)$, see I.8.11. By VI.10.1, we may apply the results of chapter II and obtain the structure of the AR-components from II.6.4. In particular, every non-periodic module has two predecessors in $\Gamma_s(\Lambda)$. Every band lies in a 1-tube; and there are at most finitely many tubes of rank $> 1$, namely the components of maximal uniserial modules. These are just modules generated by arrows for the basic algebra case. Their $\Omega$-periods are given by the cycle lengths of $\pi$ where $\pi$ is the permutation defined in VI.1.2. Here we have $\Omega^2 \cong \tau$, therefore by VI.6 there are at most two 3-tubes, and no other tubes of rank $> 1$. Therefore $\Gamma_s(\Lambda)$ has the required properties.

VI.10.3 In fact, we know more about the graph structure of $\Gamma_s(\Lambda)$ for algebras of dihedral type. The following was proved in [ES], as a consequence of more general results:

COROLLARY *Let $\Lambda$ be a basic algebra of dihedral type. Then $\Lambda$ is of polynomial growth if and only if $\Lambda$ is one of the following:*

*(i) The 4-dimensial local algebras in III.1(b), (b').*

*(ii) $D(2\mathcal{A})$ with $k = 1$.*

*(iii) $D(3\mathcal{A})_1$ with $k = 1$; $D(3\mathcal{A})_2$ with $k = l = 2$;*
*$D(3\mathcal{B})_2$ with $s = 2$ and $\{k, l\} = \{1, 2\}$; $D(3\mathcal{D})_2$ with $k = l = 1$ and $s = t = 2$,*
*$D(3\mathcal{K})$ with $a = b = c = 1$; $D(3\mathcal{L})$ with $k = s = 2$;*

*$D(3Q)$ with $k = 1$ and $s = t = 2$.*

Otherwise $\Lambda$ does not have a Euclidean component.

# VII. *Algebras of quaternion type*

In this chapter we will study algebras of quaternion type, and the aim is to determine their Morita equivalence classes. We will obtain a list of algebras defined explicitly by quivers and relations, and we will show:

Given any algebra of quaternion type, its basic algebra belongs to this list.

The first main step is to bound the number of simple modules of these algebras, see VII.3, and to obtain the possible quivers, see VII.4; there are only few of them. Then we will determine the list of possible algebras , in VII.6 to VII.8. It does not seem to be possible at present to prove the converse in general; the available theory is only able to deal with special cases, see [$E_6$].

VII.1 DEFINITION We say that the algebra $\Lambda$ is of *quaternion type* if it satisfies the following conditions:

(1) $\Lambda$ is symmetric, indecomposable and tame.

(2) The Cartan matrix of $\Lambda$ is non-singular.

(3) The stable Auslander-Reiten quiver $\Gamma_s(\Lambda)$ consists only of tubes of rank $\leq 2$.

VII.1.1 EXAMPLES Suppose B is a 2-block of some group algebra with quaternion defect groups, then B is an algebra of quaternion type: It is tame, by [BD]; and the other conditions in (1) and (2) hold for arbitrary blocks [F]; see also V.1. In V.4.5 we proved that (3) holds.

To give explicit examples, the following are of quaternion type:

(i) The group algebra KD where D is quaternion of order $2^n$, and char K = 2. More generally, the algebras $\Lambda$ in III.1(d) or (d') are of quaternion type when char K = 2 and $4k = 2^n \geq 8$, see I.8.11.

(ii) The algebra in V.2.7 is of quaternion type when char K = 2. This is the basic algebra for the special linear group $SL_2(3)$ (see V.2.7.1).

VII.1.2 One expects that all algebras in III.1(d), (d') are of quaternion type and also the algebras in V.2.7 over fields of arbitrary characteristic. In fact, it can be proved directly that the simple modules for all these algebras belong to 2-tubes, and this implies that all $\Lambda$-modules have complexity $\leq 1$ (see I.1). However, it remains open how to establish condition (3) for $\Gamma_s(\Lambda)$.

From now on, we fix an algebra $\Lambda$ of quaternion type, and we assume that $\Lambda$ is basic.

VII.2 SUMMARY OF KNOWN FACTS, CONVENTIONS.

(i) All simple $\Lambda$-modules lie in 2-tubes, by IV.1.4.

(ii) Let $Q$ be the quiver of $\Lambda$. Then for any two vertices e, f of $Q$, the number of arrows $e \to f$ is the same as the number of arrows $f \to e$ (IV.1.5.1).

(iii) Suppose e and f are vertices of $Q$. If there is no arrow $e \to f$ then there is no relation of $\Lambda$ involving a path between e and f of length $\leq 2$ (IV.1.6).

(iv) Suppose the quiver $Q$ satisfies $|s^{-1}(f)| = |e^{-1}(f)| = 2$ for each verte f. Given an arrow $\omega$ of $Q$ then there is at most one path $\omega\alpha$ of length 2 which is not involved in a relation (I.10.9).

(v) Suppose $Q$ contains $\overset{f}{\cdot} \underset{\gamma}{\overset{\beta}{\rightleftarrows}} \cdot \vdots$ where no other arrow starts or ends at f. Then we identify $\Omega(S_f)$ with $\beta\Lambda$ and $\Omega^{-1}(S_f)$ with $\gamma\Lambda$ (I.6.5(a)); and $\underline{\dim}\, \beta\Lambda = \underline{\dim}\, \gamma\Lambda$. Moreover, we may assume $\Omega^2(\beta\Lambda) = \gamma\Lambda$.

(vi) The modules $e_iJ/\mathrm{soc}\, e_i\Lambda$ are indecomposable and pairwise non-isomorphic, and all simple modules lie at the end of tubes in $\Gamma_s(\Lambda)$ (IV.1.10).

VII.2.1 NOTATION Sppose $\omega$, $\alpha$ are arrows of $\Lambda$. We say that the element (or the path) $\omega\alpha$ is *independent* if $\omega\alpha$ does not occur in any relation, and *dependent* otherwise.

VII.3 THEOREM *Let $\Lambda$ be an algebra of quaternion type. Then $\Lambda$ has at most three simple modules.*

This is proved in $[E_7]$. The ingredients for the proof are VII.2.(iii) and the fact that quivers satisfying (ii) and (iii) with too many vertices give rise to wild algebras only. For example, consider a quiver

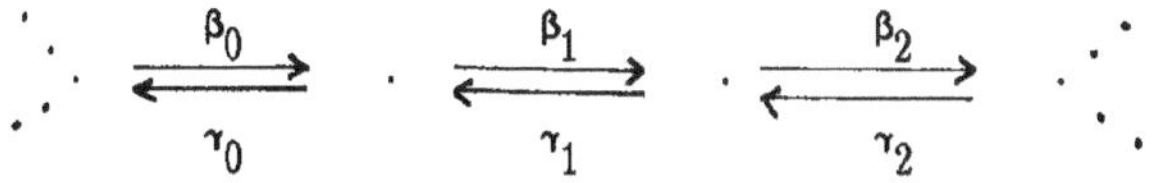

Applying VII.2 (i) and (ii) shows that here all of length 2 are independent. Therefore $\Lambda/J^3$ is generated by zero relations only, and the universal cover of $\Lambda/J^3$ contains a tree without relation of the form $\tilde{E}_7$, namely

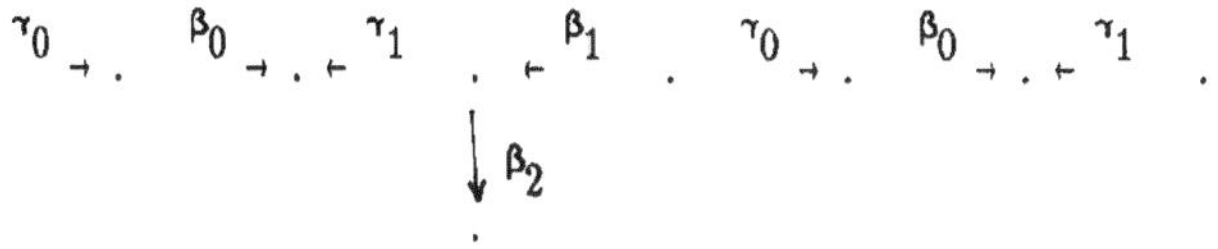

This is the typical argument; for more details see [$E_7$].

<u>VII.4</u> THE POSSIBLE QUIVERS *Let* $\Lambda$ *be of quaternion type. Then the quiver of* $\Lambda$ *is one of the following:*

(1) (2*A*) (2*B*) (3*A*) (3*B*) (3*C*) (3*D*) (3*K*) Proof: First we show

(*) The quiver of $\Lambda$ does not contain double arrows:

Suppose there is a double arrow $e \rightrightarrows f$ and $e \neq f$. Then, by VII.2(ii) the quiver must contain a subquiver of the form $Q'$ $e \;\substack{\longrightarrow\\ \longrightarrow\\ \longleftarrow}\; f$. If there is a loop at vertex e or f then $\Lambda$ is wild, by I.10.8. Otherwise, we apply VII.2(iii) and see that all paths of $Q'$ of length 2 are independent as elements of $\Lambda$. Therefore the universal

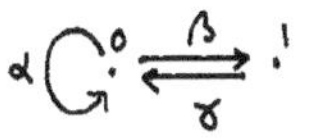

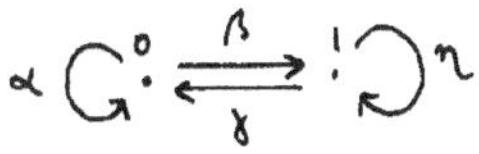

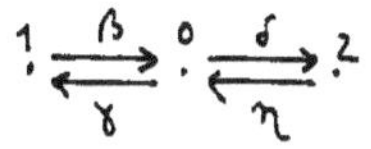

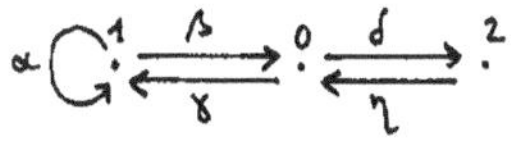

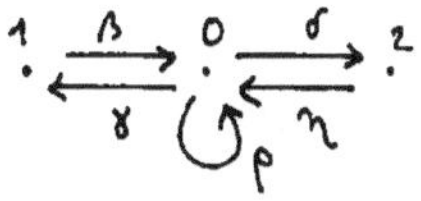

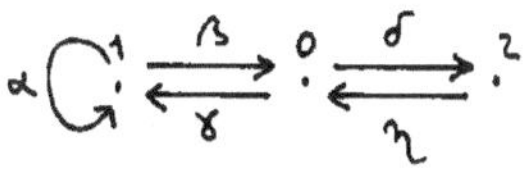

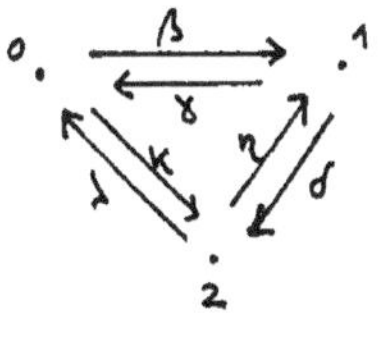

cover of $\Lambda/J^3$ contains $\tilde{E}_7$, namely

$$\begin{array}{ccccccccccccc} & & & & & & e & & & & & & \\ & & & & & & \downarrow & & & & & & \\ f & \to & e & \leftarrow & f & \to & e \leftarrow & f & \to & e & \leftarrow & f & \to e \end{array}$$

and $\Lambda$ is wild, by I.9.5.

Now consider the lists of tame symmetric quivers with at most three vertices in IV.2.4. The connected quivers in these lists which satisfy (*) and also VII.2(ii) are the ones given in VII.4.

<u>VII.5</u> GENERAL PRINCIPLES Our strategy is as follows: We fix one of the possible quivers; and first we study all paths of length 2 (or 3) and determine whether they are independent or not.

Suppose $e_i$ is a primitive idempotent of $\Lambda$; let $\Lambda_i$ be the local algebra $e_i\Lambda e_i$. By exploiting the structure of the algebras $\Lambda_i$, using III.15 and 16, we determine the socle of $e_i\Lambda$ and a vector space basis. Then we determine the remaining relations.

We will make use of a few more general observations.

<u>VII.5.1</u> LEMMA *Suppose $\Lambda$ is of quaternion type, which is not local. Let $\beta\gamma$ and $\alpha\epsilon$ be two paths of length 2 with the same starting points and end points. Suppose each of $\beta\gamma$ and $\alpha\epsilon$ occurs in some relation. Then there is a relation of the form*

$$\beta\gamma \equiv c(\alpha\epsilon) \mod J^3$$

*where c is a non-zero element of K.*

Proof: Suppose $\beta\gamma$ and $\alpha\epsilon$ start at vertex e and terminate at f. Then $\alpha$, $\beta$ must be all arrows starting at e, and $\epsilon$, $\gamma$ are all arrows ending at f, by VII.4. Moreover, since there is a relation in which $\beta\gamma$ occurs there must be an arrow e $\to$ f. Going through the list of quivers, we deduce that either

(i) $\mathcal{Q}$ contains 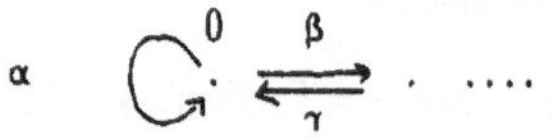

, and $\alpha = \epsilon$, or

(ii) $\mathcal{Q}$ is 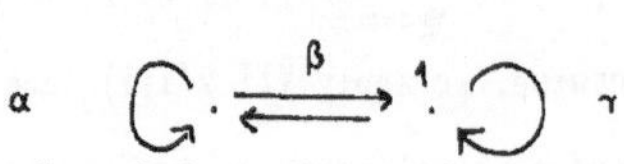

, and $\epsilon = \beta$, or

(iii) $\mathcal{Q}$ is of the form (3$\mathcal{C}$), and the relations start and end at the middle vertex.

Assume first that (i) holds. Consider the module $\Omega^2 S_0 = \omega_0\Lambda + \omega_1\Lambda$, with the usual identification. We have $\alpha^2 = \alpha z + \beta w$ for some $z \in J^2$ and $w \in J$. Therefore we may take $\omega_0 = (\alpha - z, w) \in e_0\Lambda \oplus e_1\Lambda$. Moreover, let $\beta\gamma = \alpha u + \beta v$ with $v \in J^2$; then $\omega_0' = (u, \gamma - v)$ is also a generator of $\Omega^2 S_0$ with $\omega_0' = \omega_0' e_0$. By IV.1.9 we have that $\omega_0'A = \omega_0$ for some unit $A$ of $e_0\Lambda e_0$. Write $A = ce_0 + a$ for $c \in K^*$ and $a \in J$; then $w = (\gamma - v)(ce_0 + a) \equiv c\gamma \bmod J^2$. Therefore we have that $\alpha^2 \equiv \alpha z + \beta(c\gamma) \equiv c(\beta\gamma) \bmod J^3$.

Now consider the second case. Suppose $\alpha\beta = \alpha x + \beta y$ for $x \in J^2$ and $y \in J$, and also $\beta\gamma = \beta z + \alpha w$ for $z \in J^2$ and $w \in J$. Then we may take as a generator $\omega_1$ for $\Omega^2 S_0$ the element $\omega_1 = (\beta - x, y)$, but also $\omega_1' = (w, \gamma - z)$. By IV.1.9 we have that $\omega_1 = \omega_1'A$ for some unit $A$ of $e_1\Lambda e_1$. Write $A = ce_1 + a$ for $a \in J$ and $c \in K^*$; then $y \equiv c\gamma \bmod J^2$ and therefore $\alpha\beta \equiv c(\beta\gamma) \bmod J^3$, as required.

The proof in case (iii) is similar.

<u>VII.5.2</u> LEMMA *Suppose $\Lambda$ is of quaternion type and the quiver contains*

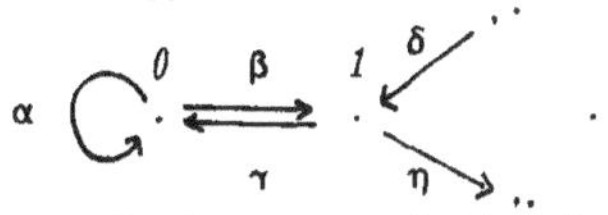


*(a) If $\alpha\beta$ is involved in some relation then so is $\beta\gamma$.*

*(b) If $\alpha\beta$ is independent then so is $\gamma\alpha$. Moreover, $\alpha^2 \in J^3$, and $\beta\gamma$ is independent.*

Proof: (a) Assume $\alpha\beta = \alpha x + \beta y$ for $x \in J^2 e_1$ and $y \in Je_1$. Consider the module $\Omega^2 S_0$, as a generator $\omega_1 = \omega_1 e_1$ we may take $\omega_1 = (\beta - x, y)$. Recall that $\omega_0\alpha' = \omega_1\gamma'$ where $\gamma' \equiv d\gamma \bmod J^2$ with $d \in K^*$; consequently $\beta\gamma$ occurs in some relation.

(b) Suppose that $\gamma\alpha$ is dependent; let $r \in K$ with $\gamma\alpha \equiv r(\eta\gamma) \bmod J^3$, and let $\zeta_0$ and $\zeta_1$ be generators of $\Omega^2 S_1$ with $\zeta_i = \zeta_i e_i$. We may take $\zeta_0 = (\alpha - u, r\gamma - v)$ for $u, v \in J^2$. We also have $\zeta_0\beta' = \zeta_1\eta'$ where $\beta' \equiv \beta \bmod J^2$ and $\eta' \equiv \eta \bmod J^2$; this implies that $\alpha\beta$ is dependent, a contradiction.

Since $\alpha\beta$ is independent, $\alpha^2$ must be involved in some relation (VII.2(iv)). Hence, for some $c \in K$ we have that $\alpha^2 \equiv c(\beta\gamma) \bmod J^3$. Suppose $\omega_0$ and $\omega_1$ are generators of $\Omega^2 S_0$ with $\omega_i = \omega_i e_i$; then we may take $\omega_0 = (\alpha - u, c\gamma - v)$ for $u, v \in J^2$. Since $\alpha\beta$ is independent, $\omega_1$ is of the form $\omega_1 = (z, z')$ where $z \in J^2$. Now we have that $\omega_0\alpha' = \omega_1\gamma'$ and hence $\alpha^2 \in J^3$ (alternatively $c(\gamma\alpha)$ occurs in a relation, hence $c = 0$).

Suppose that $\beta\gamma$ is not independent; then $\beta\gamma$ must lie in $J^3$. We have generators of

$\Omega^2 S_0$ of the form $\omega_0 = (\alpha - u, m')$ and also $\omega_0' = (m, \gamma - v)$ where m and m'and u, v $\epsilon J^2$. By IV.1.9, there is a unit $u = \lambda e_0 +$ a of $e_0 \Lambda e_0$ with $\omega_0 = \omega_0' u$; and therefore $\alpha \in J^2$, a contradiction.

VII.5.2.1 LEMMA *Suppose the quiver of $\Lambda$ is of the form $(\mathcal{K})$. If $\rho\gamma$ is dependent then so is $\gamma\beta$.*

This is proved by the method of VII.5.2.

VII.5.3 LEMMA *Suppose $\Lambda$ is of quaternion type whose quiver contains*

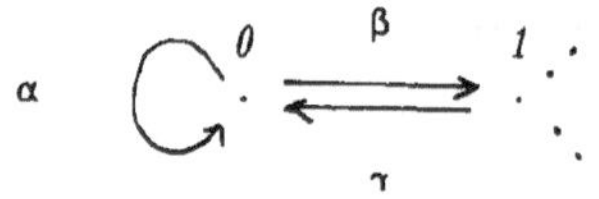

*and assume that the local algebra $\Lambda_0$ is not uniserial. Then $\alpha^2$ does not lie in soc $\Lambda$.*

Proof: Assume that $\alpha^2$ lies in soc $\Lambda$. The hypothesis implies that there is a path $\omega$ such that $\alpha$ and $\beta\omega\gamma$ are independent generators of rad $\Lambda_0$. By III.15 we may assume, after possibly replacing $\alpha$, that soc $\Lambda_0$ is either generated by $\alpha^m$, of by the element $(\alpha\beta\omega\gamma)^k$. Recall that soc $\Lambda_0$ = soc $e_0\Lambda$.

Suppose that the first possibility holds; then m = 2, and it follows that $H(e_0\Lambda)$ is a direct sum, a contradiction to VII.2.(vi).

Hence soc $e_0\Lambda = \langle (\alpha\beta\omega\gamma)^k \rangle$; and since $\Lambda$ is symmetric, we deduce $(\alpha\beta\omega\gamma)^k = (\beta\omega\gamma\alpha)^k$.

Write $\alpha^2 = c(\alpha\beta\omega\gamma)^k$ for some $c \in K$; and set $\alpha' = \alpha - c(\beta\omega\gamma\alpha)^{k-1}\beta\omega\gamma$, and we have $\alpha\alpha' = 0 = \alpha'\alpha$. We identify $\Omega^2 S_0$ with the set $\{(x,y) \in e_0\Lambda \oplus e_1\Lambda: \alpha x + \beta y = 0\}$. Then we have an embedding $\theta: \alpha'\Lambda \to \Omega^2 S_0$ given by $\theta(\alpha' x) = (\alpha' x, 0)$, and im $\theta \not\subseteq$ rad $\Omega^2 S_0$.

On the other hand, $\Omega^{-1} S_0 \cong (\alpha,\gamma)\Lambda$ and $\Omega^{-2} S_0 \cong e_0\Lambda \oplus e_1\Lambda/L$ where $L = (\alpha,\gamma)\Lambda$. Consider the epimorphism $\mu: \Omega^{-2} S_0 \to \alpha'\Lambda$ given by $\mu[(x,y) + L] = \alpha' x$ ( which is well-defined since $\alpha'\alpha = 0$). Then one sees directly that the restriction of $\mu$ to soc $\Omega^{-2} S_0$ is non-zero. Now, $\Omega^{-2} S_0 \cong \Omega^2 S_0$, hence the composition $\mu\theta$ is defined. We have that the restriction of $\mu\theta$ to soc $\alpha'\Lambda$ is non-zero, by the above. On the other hand, soc $\alpha'\Lambda$ is simpe and therefore $\mu\theta$ is 1-1 and then an isomorphism. Consequently $\Omega^{-2} S_0$ is decomposable, a contradiction.

<u>VII.5.4</u> LEMMA *(a) Suppose $\Lambda$ is a tame symmetric basic algebra whose quiver satisfies $|s^{-1}(f)| = |e^{-1}(f)| = 2$ for each vertex $f$. Then $\Omega^{\pm 1}(\beta\Lambda)$ is uniserial, for each arrow $\beta$.*

*(b) Suppose $\Lambda$ is basic and symmetric, assume that $\beta\Lambda$ has $\Omega$-period 4 and that $\Omega(\beta\Lambda)$ is uniserial. Assume also that the quiver contains*

*where $\gamma$ is the only arrow $0 \to 1$. Then $\underline{dim}\,\beta\Lambda = \underline{dim}\,\gamma\Lambda$, and we may assume that $\Omega^2(\beta\Lambda) = \gamma\Lambda$.*

Proof: (a) By duality it suffices to consider the modules $X = f\Lambda/\beta\Lambda$. We show by induction on k that for all such X, the quotient $X/XJ^k$ is uniserial.

The cases k = 1 and 2 are clear. Suppose $X/XJ^{k-1}$ is uniserial, then it is spanned by the cosets of $\{f, \alpha_1, \alpha_1\alpha_2, \ldots, (\alpha_1\alpha_2\ldots\alpha_{k-2})\}$, and $XJ^k$ is generated by $(\alpha_1\alpha_2\ldots\alpha_{k-2}\delta)$ and $(\alpha_1\ldots\alpha_{k-2}\gamma)$ where $\delta$ and $\gamma$ are the arrows starting at $e(\alpha_{k-2})$. By VII.2(iv), at least one of $\alpha_{k-2}\delta$, $\alpha_{k-2}\gamma$ is dependent, $\alpha_{k-2}\gamma$ say, and then $X/XJ^k$ is uniserial.

(b) If $\Omega^4(\beta\Lambda) \cong \beta\Lambda$ then there is an exact sequence

$$0 \to \beta\Lambda \to P_1 \to Q_2 \xrightarrow{\quad \nearrow \Omega^2(\beta\Lambda) \searrow \quad} Q_1 \to P_0 \to \beta\Lambda \to 0$$

where $Q_1$ and $Q_2$ are projective. The sequence is exact and $Q_1$, $Q_2$ are indecomposable by (a), therefore $\underline{dim}\,Q_1 = \underline{dim}\,P_0$ and $\underline{dim}\,Q_2 = \underline{dim}\,P_1$; and since the Cartan matrix is non-singular, we deduce $Q_1 \cong P_0$ and $Q_2 \cong P_1$. Therefore soc $\Omega^2(\beta\Lambda) \cong S_0$ and top $\Omega^2(\beta\Lambda) \cong S_1$, and moreover $\underline{dim}\,\Omega^2(\beta\Lambda) = \underline{dim}\,\beta\Lambda$.

Since $\Omega(\beta\Lambda)$ is uniserial, there is a generator, $\omega$ say, of $Q_1$ which is contained in $\Omega^2(\beta\Lambda)$; and since top $\Omega^2(\beta\Lambda)$ is simple we deduce that $\Omega^2(\beta\Lambda)$ is generated by an arrow. We must have $\omega \equiv \gamma$ and then one replaces $\gamma$ by $\omega$.

## VII.6 *Algebras of quaternion type with one simple module*

The result here is the following:

THEOREM *Suppose that* $\Lambda$ *is a local algebra of quaternion type. Then* $\Lambda$ *is isomorphic to one of the algebras in III.1(e) or (e').*

Proof: Any such algebra is tame, local and symmetric, so it is isomorphic to one of the algebras in the list given in III.1. The algebras there in (a) to (c') are special biserial (modulo socle), so they are excluded here (see II.7.1 and 2). Moreover, the algebras in (d) and (d') are not of quaternion type, see III.1.1 and II.10.2 (and I.8.11).

## VII.7 *Algebras of quaternion type with two simple modules*

There are two possible quivers. For the first quiver, we will now prove the following:

<u>VII.7.1</u> THEOREM *Assume that* $\Lambda$ *is an algebra of quaternion type with quiver of the form (2A). Then* $\Lambda \cong KQ/I$ *where* $I$ *is defined by the relations*

$$\gamma\beta\gamma = (\gamma\alpha\beta)^{k-1}\gamma\alpha, \qquad \beta\gamma\beta = (\alpha\beta\gamma)^{k-1}\alpha\beta$$

$$\alpha^2 = (\beta\gamma\alpha)^{k-1}\beta\gamma + c(\beta\gamma\alpha)^k \quad \text{and} \quad \alpha^2\beta = 0,$$

*where* $k \geq 2$ *and* $c \in K$.

Proof: (1) <u>The elements $\gamma\beta$, $\gamma\alpha$ and $\alpha\beta$ are independent:</u> For $\gamma\beta$, this is true by VII.2(iii). Suppose $\gamma\alpha$ is not independent. Then there is a relation of the form $\gamma\alpha = \gamma x$ with $x \in e_0J^2e_0$. We may assume that $\gamma\alpha = 0$, otherwise we replace $\alpha$ by $\alpha - x$. Then we have by IV.1.5 that $R_\gamma = \alpha\Lambda$. However $\Omega(R_\gamma) \cong \Omega^2(\gamma\Lambda) \cong \Omega^3(S_1) \cong \Omega^{-1}S_1 \cong \beta\Lambda$ (see VII.2(v)), and we may choose $\beta$ such that $\alpha\beta = 0$. Having generators satisfying $\gamma\alpha = 0 = \alpha\beta$ means that $\Lambda$ is special biserial and therefore not of quaternion type [VII.2(vi)], a contradiction. Dually one shows that $\alpha\beta$ is independent.

(2) <u>There is a relation in which $\alpha^2$ occurs:</u> Suppose not, then the universal cover of $\Lambda/J^3$ contains the tree without relations

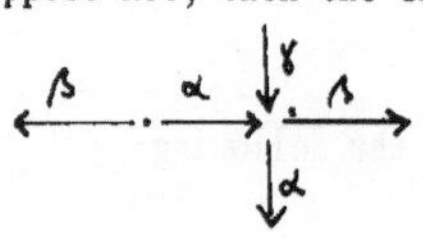

which is wild, by I.9.3.

(3) <u>The element $\alpha^2$ lies in $J^3$</u>: Assume not; then by (2) there is a relation $\alpha^2 = c\beta\gamma + \alpha x + \beta y$ where $0 \neq c \in K$ and $x, y \in J^2$. We may assume that $y = 0$ and $c = 1$, after replacing $\gamma$ if necessary. Then we have that $\beta\gamma \in \alpha\Lambda$, and $H_1 \cong (\beta\gamma)\Lambda \subsetneq \alpha\Lambda \cap \beta\Lambda$. We must have equality, since $\beta\gamma\Lambda$ is the unique maximal submodule of $\beta\Lambda$, and $\alpha\Lambda \neq \beta\Lambda$.

We claim now that $R_\alpha = U(S_1, S_0)$: First we have soc $R_\alpha \cong S_0$. Secondly, identifying $\Omega^2 S_0$ as usual with $\{ (x,y) \in e_0\Lambda \oplus e_1\Lambda\colon \alpha x + \beta y = 0 \}$, there is an exact sequence $0 \to R_\alpha \oplus R_\beta \to \Omega^2 S_0 \to \alpha\Lambda \cap \beta\Lambda \to 0$. Here $R_\beta \cong S_1$ and $\alpha\Lambda \cap \beta\Lambda \cong H_1$; moreover $\underline{\dim}\ \Omega^2 S_0 = \underline{\dim}[P_1 \oplus S_0]$. We deduce $\underline{\dim}\ R_\alpha = \underline{\dim}\ S_0 \oplus S_1$, and the claim follows.

Consequently $R_\alpha \cong e_1\Lambda/\gamma J$, and therefore $\Omega(R_\alpha) \cong \gamma J$. Since $\Omega^4(\alpha\Lambda) \cong \alpha\Lambda$ and $R_\alpha \cong \Omega(\alpha\Lambda)$, there is an exact sequence

$0 \to \alpha\Lambda \to P_0 \to Q \to P_1 \to P_0 \to \alpha\Lambda \to 0$ where $Q$ is projective. By the exactness, $\underline{\dim}\ Q = \underline{\dim}\ P_1$, and $Q \cong P_1$ since the Cartan matrix of $\Lambda$ is non-singular. That is, top$(\gamma J) \cong S_1$, and then $\gamma\alpha \in \gamma J^2$. This is a contradiction to (2). This completes the proof of (3).

(4) <u>$\gamma\beta\gamma$ and $\beta\gamma\beta$ must be involved in some relation:</u> Suppose $\gamma\beta\gamma$ is independent; then in particular, $\gamma\beta$ does not occur in some relation, and $\Lambda$ has a quotient algebra with only zero relations whose universal cover contains the tree $\widetilde{E}_7$, namely

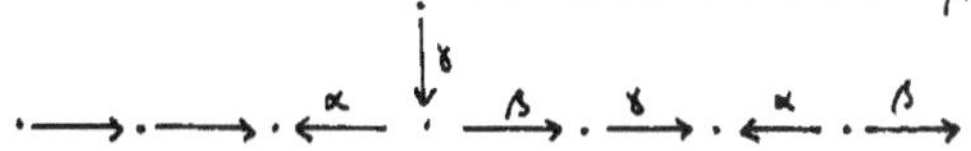

and $\Lambda$ is wild by I.9.5, a contradiction. Dually, one shows the other statement.

Since also $\alpha^2$ lies in $J^3$, we have by (4) that $\gamma\beta\gamma \in J^4$.

(5) <u>The local algebra $\Lambda_1 = e_1\Lambda e_1$ is not uniserial:</u> Suppose it is, then $\Lambda_1 = K[\gamma\beta]$, by (1), and $\gamma\alpha\beta \in (\gamma\beta)^2\Lambda$. But then $e_1J^3 = \langle \gamma\beta\gamma, \gamma\alpha\beta \rangle \subsetneq e_1J^4$ and $e_1J^3 = 0$. Therefore $\gamma J \subsetneq$ soc $\Lambda$ and $\gamma\alpha \in$ soc $\Lambda \cap e_1\Lambda e_0 = 0$, a contradiction to (1).

Hence $\Lambda_1$ has independent generators $\gamma\beta$, $\gamma\alpha\beta$; and by (4) we deduce $\Lambda_1 =$ span $\{ \gamma\beta, (\gamma\beta)^2, \ldots, \gamma\alpha\beta, (\gamma\alpha\beta)^2, \ldots \}$. Consequently $\Lambda_1$ is one of the algebras in III.1(a) to (b'); in particular $\Lambda_1$ is commutative. By III.15 we deduce soc $\Lambda_1 = \langle (\gamma\alpha\beta)^k \rangle =$ soc $e_1\Lambda$ where $k \geq 1$ is the last integer such that $(\gamma\alpha\beta)^k \neq 0$. Note that, by III.15, $k$ is also the last integer such that $(\beta\gamma\alpha)^k \neq 0$.

(6) <u>We may assume that $\gamma\beta\gamma = (\gamma\alpha\beta)^{k-1}\gamma\alpha$:</u> We have that $e_1\Lambda/(\gamma\alpha)J = \mathcal{U}(S_1, S_0, S_1)$, and this must be isomorphic to some submodule of $e_1\Lambda$. Hence $e_1\Lambda$ contains some element

u such that $u = ue_1$ and that $u\Lambda$ has a K-basis $\{u, u\gamma, u\gamma\beta\}$. Then u generates $soc_2(e_1\Lambda)$ and $0 \neq u\gamma \in K(\gamma\alpha\beta)^{k-1}\gamma\alpha$. From the above K-basis for $e_1\Lambda e_1$ we see that $u = c(\gamma\beta) + x$ where $xJ^2 = o$ and $0 \neq c \in K$. Consequently $c(\gamma\beta\gamma) + x\gamma = d(\gamma\alpha\beta)^{k-1}\gamma\alpha$. Since $x\gamma \in soc\ \Lambda \cap e_1\Lambda e_0$ we have $x\gamma = 0$. Now rescale, and (6) follows.

In particular, $\gamma\beta\gamma \in soc_2(e_1\Lambda)$. Note also that $k \geq 2$, by (6) and (1).

(7) $\underline{\beta\gamma\beta = (\alpha\beta\gamma)^{k-1}\alpha\beta}$: Since $\beta\gamma\beta J = \beta\gamma\beta\gamma\Lambda \subseteq soc\ \Lambda$ we deduce that $\beta\gamma\beta \in soc_2\Lambda \cap e_0\Lambda e_1$. Hence $\beta\gamma\beta = c(\alpha\beta\gamma)^{k-1}\alpha\beta$ where $c \in K$; and after multiplying by $\gamma$ we see from (6) that $c = 1$.

Now we study the structure of the local ring $e_0\Lambda e_0$. Its radical is generated by $\alpha$ and $\beta\gamma$. We know that $(\beta\gamma)^2$ lies in its socle. By III.15 we deduce that, for an appropriate choice of $\alpha$, $soc\ e_0\Lambda = \langle (\beta\gamma\alpha)^k \rangle$ and $(\beta\gamma\alpha)^k = (\alpha\beta\gamma)^k$ and moreover $\alpha^2 = d(\beta\gamma\alpha)^{k-1}\beta\gamma + c(\alpha\beta\gamma)^k$. By VII.5.3 we have that $d \neq 0$.

It remains to show that, without loss of generality, $d = 1$. Set $\alpha = A\alpha'$, $\beta = B\beta'$ and $\gamma = \gamma'$ for $A, B \in K^*$. We require $1 = A^kB^{k-2}$ and also $1 = d(A^{k-3}B^k)$, that is, $B^2 = (AB)^k = d^{-1}A^3$. This has solutions, for $k \geq 2$; and replace $\alpha, \beta, \gamma$ by $\alpha', \beta', \gamma'$. This completes the proof.

<u>VII.7.1</u> *Let Λ be one of the algebras defined in VII.7.1. If char K ≠ 2 then we may assume that c = 0. If char K ≠ 2 then for any fixed value of K, there are infinitely many non-isomorphic algebras.*

If char $K \neq 2$ replace $\alpha$ by $\alpha'$ where $\alpha' = \alpha - (c/2)(\beta\gamma\alpha)^{k-1}\beta\gamma$; this gives the same relations except that c is replaced by 0.

Assume char $K = 2$ and $c \neq 0$. Then the scalar transformation in the proof of VII.7 replaces c by $c(AB)^k$. So there are only finitely many values for c which give rise to an algebra isomorphic to a given one, and hence there are infinitely many non-isomorphic algebras.

Suppose that $\Lambda$ is basic of quaternion type, with quiver of the form (2B). To determine the possible algebras, we will use VII.5. If $e_i$ is a primitive idempotent of $\Lambda$ we denote by $\Lambda_i$ the local algebra $e_i\Lambda e_i$.

VII.7.2 THEOREM *Suppose that $\Lambda$ is of quaternion type, with quiver of the form (2B), and assume also that the local algebra $\Lambda_0$ is not uniserial. Then $\Lambda \cong KQ/I$ where I is defined by the following relations:*

$$\gamma\beta = \eta^{s-1}, \quad \beta\eta = (\alpha\beta\gamma)^{k-1}\alpha\beta, \quad \eta\gamma = (\gamma\alpha\beta)^{k-1}\gamma\alpha$$
$$\alpha^2 = (\beta\gamma\alpha)^{k-1}\beta\gamma + c(\beta\gamma\alpha)^k, \quad \alpha^2\beta = 0$$

*where $s \geq 3$, $k \geq 2$ and $c \in K$.*

Proof: (1) <u>One of $\beta\gamma$, $\gamma\beta$ is independent:</u> Assume $\beta\gamma$ and $\gamma\beta$ are not independent. Then by VII.5.2, $\alpha\beta$ is dependent; and by VII.5.2 applied to $\Lambda^{op}$, $\gamma\alpha$ is also dependent; similarly $\beta\eta$ and $\eta\gamma$ are dependent. Therefore, by VII.5.1, $\alpha\beta \equiv c(\beta\eta) \bmod J^3$ and $\gamma\alpha \equiv d(\eta\gamma) \bmod J^3$ for $0 \neq c, d \in K$. Then $\beta\eta\gamma \equiv c(\alpha\beta\gamma)$, and $\Lambda_0$ is generated by $\alpha$ and $\beta\gamma$. However, since $\gamma\beta$ is dependent, there is a relation

$$\beta\gamma = \alpha x_1\alpha + \alpha x_2\gamma + \beta x_3\alpha + \beta x_4\gamma$$

with $x_1 \in \Lambda$ and $x_2$, $x_3$ and $x_4 \in J$. Then necessarily $x_2 = x_2'\beta$, $x_3 = \gamma x_3'$ and $x_4 = \gamma x_4'\beta$, and $\beta\gamma$ lies in $J(\Lambda_0)^2$. We deduce that $\Lambda_0 = K[\alpha]$. This contradicts the hypothesis.

Suppose $\beta\gamma$ is independent. Then by VII.5.2, $\alpha\beta$ is independent, and also $\gamma\alpha$, and $\alpha^2 \in J$. Moreover, we deduce from VII.2 that $\beta\eta$ and $\eta\gamma$ are dependent; and $\gamma\beta$ as well. Consider the local algebra $\Lambda_0$. We claim that $(\beta\gamma)^2$ lies in $J(\Lambda_0)^3$: Write $\gamma\beta = \eta y + \gamma x$ for $y \in J$ and $x \in J^2$; then we see that $\beta\gamma\beta$ lies in $J^4$ (since $\beta\eta \in J^3$). Moreover $(\beta\gamma)^2 \in e_0J^5 = \langle (\alpha\beta\gamma\alpha\beta), (\beta\gamma\alpha\beta\gamma)\rangle$, as required.

Hence either $\dim \Lambda_0 = 4$, or else $\Lambda_0$ is isomorphic to one of the algebras in III.1(c) to (e'). By III.16, we may assume that $\alpha^2$ lies in $\mathrm{soc}_2\Lambda_0$, and $\mathrm{soc}\ \Lambda_0 = \langle (\alpha\beta\gamma)^k\rangle$; this is equal to $\mathrm{soc}\ e_0\Lambda$, and hence $(\alpha\beta\gamma)^k = (\beta\gamma\alpha)^k$. Note that this implies $e_0J^{3k} = \mathrm{soc}\ e_0\Lambda$; and it follows that $(\gamma\alpha\beta)^kJ \subseteq \gamma e_0J^{3k} = 0$. On the other hand, $(\gamma\alpha\beta)^k \neq 0$ since $\Lambda$ is symmetric, and therefore $\mathrm{soc}\ e_1\Lambda = \langle (\gamma\alpha\beta)^k\rangle$. We consider now the local algebra $e_1\Lambda e_1$; it is generated by $\eta$ and $\gamma\alpha\beta$. Moreover, $\eta\gamma\alpha\beta$ and $\gamma\alpha\beta\eta$ lie in $e_1J^5e_1 = \langle \eta^5, (\gamma\alpha\beta)^2\rangle$. We deduce from III.1 that $\Lambda_1$ is commutative. We choose $\eta$ such that $\eta\gamma$ belongs to the highest possible power of J, then soc $\Lambda_1 = \langle \eta^s \rangle$ where s is maximal such that $\eta^s \neq 0$.

(3) <u>We may assume that $\gamma\beta = \eta^{s-1}$</u>: We know a K-basis for $\Lambda$. The element $\gamma\beta$ occurs in some relation which we may write in the form

(*) $\gamma\beta = \Sigma\ a_i(\gamma\alpha\beta)^i + \eta^r v$ where either v = 0, or v is a unit in $K[\eta]$.

Consequently $\gamma\Lambda$ has a basis $\{ \gamma, \gamma\alpha, \gamma\alpha\beta, \ldots, (\gamma\alpha\beta)^k\} \cup S$ where $S \subseteq \{ \eta^r, \eta^{r+1}, \ldots, \eta^{s-1}\}$; and in particular <u>dim</u> $\gamma\Lambda = [2k, k+u]$ where $u = |S|$. Recall that <u>dim</u> $\gamma\Lambda$ = <u>dim</u> $\beta\Lambda$ (VII.5.4). To calculate <u>dim</u> $\beta\Lambda$ we need the relation for $\beta\eta$. We have that $\beta\eta = \Sigma\ c_i(\alpha\beta\gamma)^i\alpha\beta + \beta x$, therefore <u>dim</u> $\beta\Lambda = [2k+2t, k+t+1]$ where t is determined by the first i such that $c_i$ is non-zero. We equate this and obtain t = 0 and then u = 1. It follows that r = s-1 and v ≠ 0; now we write (*) in the form $\gamma z\beta v^{-1} = \eta^{s-1}$ where z is a unit of $e_0\Lambda e_0$, and we replace $\gamma$, $\beta$ by $\gamma z$, $\beta v^{-1}$ respectively. (Since $\eta \notin J^2$, we must have $s \geq 3$). Note that this does not affect the socle conditions.

(4) <u>We may assume that $\eta\gamma = (\gamma\alpha\beta)^{k-1}\gamma\alpha$ and $\beta\eta = (\alpha\beta\gamma)^{k-1}\alpha\beta$:</u> We have that $\eta\gamma \in e_1\Lambda e_0$; hence there are $a_i$ and $b_i \in K$ such that $\eta\gamma = \Sigma\ a_i(\gamma\alpha\beta)^i\gamma + \Sigma\ b_i(\gamma\alpha\beta)^i\gamma\alpha$. Since $\eta\gamma\beta = \eta^s = d(\gamma\alpha\beta)^k$ for some d ≠ 0 and $(\Sigma\ a_i(\gamma\alpha\beta)^i\gamma)\beta \in J^2\gamma\beta \subseteq J^2\mathrm{soc}_2\Lambda = 0$, we deduce $b_i = 0$ for $i < k-1$, and $b_{k-1} = d$. We claim that $a_i = 0$ for all i. If not, we would have $\eta\gamma \notin J^{3k+2}$; but if $\eta' = \eta - \Sigma\ a_i(\gamma\alpha\beta)^i$ then $\eta' \equiv \eta$ and $\eta'\gamma \in J^{3k+2}$, a contradiction to the choice of $\eta$. This gives the first relation; actually we may assume d = 1; otherwise we replace $\alpha$. Now, $\beta\gamma = \Sigma\ c_i(\alpha\beta\gamma)^i\alpha\beta + \Sigma\ d_i(\beta\gamma\alpha)^i\beta\gamma$ for $c_i$ and $d_i \in K$. We obtain two expressions of $\beta\eta\gamma$ in terms of the basis and derive the relation for $\beta\eta$.

We have chosen $\alpha$ such that $\alpha^2$ lies in $\mathrm{soc}_2\Lambda_0$. In fact, we may even assume that $\alpha^2 = a(\beta\gamma\alpha)^{k-1}\beta\gamma + c(\alpha\beta\gamma)^k$ for a, c ∈ K; and by VII.5.3 we must have a ≠ 0. By a scalar transformation it is possible to have a = 1. To complete the proof, observe that k = 1 would imply that $\Lambda_0$ is uniserial.

We note that this shows the following: If $\Lambda_0$ is not uniserial then $\Lambda_1$ is not uniserial as well.

<u>VII.7.3</u> **THEOREM** *Suppose $\Lambda$ is an algebra of quaternion type with quiver (2B), and assume also that $\Lambda_0$ is uniserial. Then $\Lambda \cong KQ/I$ where one of the following holds:*

*(i) I is defined by the relations:*

$$\alpha\beta = \beta\eta \text{ and } \eta\gamma = \gamma\alpha,$$

$$\beta\gamma = \alpha^2 p(\alpha) \text{ and } \gamma\beta = \eta^2 p(\eta) + a\eta^{s-1} + c\eta^s$$

$$\alpha^{s+1} = 0 = \eta^{s+1} \text{ and } \gamma\alpha^{s-1} = 0 = \alpha^{s-1}\beta;$$

*where p is a polynomial with constant term 1 and a, c* $\in K$ *with* $a \neq 0$*. and* $s > 3$*.*

*(ii) I is defined by the relations*

$$\alpha\beta = \beta\eta \text{ and } \eta\gamma = \gamma\alpha$$

$$\beta\gamma = \alpha^2 + c\alpha^3 \text{ and } \gamma\beta = a\eta^{t-1} + d\eta^t$$

$$\alpha^4 = 0 = \eta^{t+1} \text{ and } \gamma\alpha^2 = 0 = \alpha^2\beta;$$

*where* $t \geq 3$ *and* $a, c, d \in K$ *with* $a \neq 0$*. If* $t = 3$ *then* $a \neq 1$*.*

Proof: By the previous theorem, we must have that also $\Lambda_1$ is uniserial, therefore $\Lambda_0 = K[\alpha]$ and $\Lambda_1 = K[\eta]$. We deduce that $\beta\gamma$ and $\gamma\beta$ are both dependent. Consequently, by VII.5.2 (for $\Lambda$ and also for $\Lambda^{op}$) we have that $\alpha\beta$, $\gamma\alpha$ and also $\beta\eta$ and $\eta\gamma$ are dependent. By VII.5.1, there are relations of the form

$$\alpha\beta = u(\beta\eta) \bmod J^3 \text{ and}$$

$$\eta\gamma = v(\gamma\alpha) \bmod J^3 \text{ where } u, v \in K^*.$$

(1) <u>At least one of $\alpha^2$, $\eta^2$ must be dependent:</u>

Suppose that $\alpha^2$ and $\eta^2$ are both independent. Then $\alpha\beta$, $\beta\eta$, $\eta\gamma$ and $\gamma\alpha$ all lie in $J^3$. The elements $\gamma\beta$ and $\beta\gamma$ are dependent but $\alpha^2$ and $\eta^2$ are not, therefore $\gamma\beta$ and $\beta\gamma$ lie in $J^3$. We deduce that $e_0J^2 = \alpha^2\Lambda$ and then $e_0J^k = \alpha^k\Lambda$ for $k \geq 2$, and $e_0J^2e_1 = 0$; similarly $e_1J^2e_0 = 0$. Therefore $\beta\eta = 0 = \eta\gamma$ and $\alpha\beta = 0 = \gamma\alpha$; and $\Lambda$ is special biserial and not of quaternion type (II.7.1 and II.7.2), a contradiction.

Say $\alpha^2$ is dependent; then $\alpha^2 = r(\beta\gamma) \bmod J^3$ for some $r \in K$.

(2) <u>$\alpha^2$ does not lie in $J^3$; in particular $r \neq 0$</u>: Suppose $\alpha^2 \in J^3$; then since $\beta\gamma$ is dependent, it also lies in $J^3$; therefore $e_0J^2 = \langle (\alpha\beta) \rangle$ and $e_0J^3 = \langle (\alpha\beta\gamma), (\alpha^2\beta) \rangle \subseteq e_0J^4$ and $e_0J^3 = 0$ which implies that $e_0J^2 \subseteq \text{soc } e_0\Lambda$; and $H(e_0\Lambda)$ is a direct sum, a contradiction.

We see now that $e_0J = K[\alpha] + \beta K[\eta]$ and $e_1J = \gamma K[\alpha] + K[\eta]$. In particular, $\langle \alpha^s \rangle = \text{soc } e_0\Lambda$ and $\langle \eta^t \rangle = \text{soc } e_1\Lambda$; and $s, t \geq 3$ since $H_0$, $H_1$ are indecomposable. Moreover, since $[\text{soc}_2(e_0\Lambda)/\text{soc } e_0\Lambda]e_0 \neq 0$ we have that $\alpha^{s-1}$ belongs to $\text{soc}_2\Lambda$, and so does $\eta^{t-1}$.

(3) <u>$\gamma\alpha^{s-1} = 0 = \alpha^{s-1}\beta$ and $\gamma\alpha^{s-2} \neq 0 \neq \alpha^{s-1}\beta$:</u> We have $\gamma\alpha^{s-1} \in \text{soc}\,\Lambda \cap e_1\Lambda e_0 = 0$ and $\beta\gamma\alpha^{s-2} \equiv r\alpha^s \neq 0$; similarly one shows the other part.

Since $\beta\gamma \in e_0\Lambda e_0$, we may write $\beta\gamma = \alpha^2 p(\alpha)$ where $p(x)$ is a polynomial $\in K[x]$ with non-zero constant term. In fact, we may assume that $p(0) = 1$; otherwise we replace $\alpha$ by a scalar multiple. Similarly $\gamma\beta = \eta^v q(\eta)$ with $q(x) \in K[x]$ with $q(0) \neq 0$ where $v \geq 2$.

(4) <u>We may assume $\alpha\beta = \beta\eta$ and $\eta\gamma = \gamma\alpha$:</u> We have a relation $\alpha\beta = \beta\eta h$ where $h$ is a unit in $K[\eta]$. Replace $\eta$ by $\eta' = \eta h$; this gives the first statement. Now write $\eta\gamma = \gamma\alpha z(\alpha)$ where z is a polynomial of degree $\leq s-3$ with non-zero constant term. We have $\beta\eta\gamma = \beta\gamma\alpha z(\alpha) = \alpha^2 p(\alpha)z(\alpha)$, but also $\beta\eta\gamma = \alpha\beta\gamma = \alpha^3 p(\alpha)$. Since $p(\alpha)$ is a unit in $K[\alpha]$ we deduce $\alpha^3 = \alpha^3 z(\alpha)$; and $z(\alpha) = 1$ since $\alpha^3, \alpha^4, \ldots, \alpha^s$ are linearly independent.

<u>Case 1</u> $s > 3$: Then we claim first that $v = 2$: We have two expressions for $\beta\gamma\beta$, namely $\beta\gamma\beta = \alpha^2 p(\alpha)\beta \equiv \beta(c\eta)^2 p(c\eta)$, but also $\beta\gamma\beta = \beta\eta^v q(\eta)$. The term of lowest degree is $0 \neq \alpha^2\beta \equiv \beta(c\eta)^2$, by (3); and therefore $v = 2$.

We will now show that $t = s$. If $v = 2$ then $\gamma\beta\eta^{t-2} \equiv \eta^t \neq 0$, so $\beta\eta^{t-2} \neq 0$. On the other hand, $\beta\eta^{t-2} = \alpha^{t-2}\beta$ and by (3) $t \leq s$. Conversely $0 \neq \alpha^{s-2}\beta = \beta\eta^{s-2}$ and hence $s \leq t$.

It remains to derive the details about $\gamma\beta$. Comparing two expressions for $\beta\gamma\beta$ we obtain $\beta\eta^2 p(\eta) = \beta\eta^2 q(\eta)$. Therefore $\eta^2 q(\eta) = \eta^2 p(\eta) + a\eta^{s-1} + b\eta^s$ for $a, b \in K$. If a were 0 then we would evidently have that $H(e_0\Lambda) \cong H(e_1\Lambda)$. This is not possible, by VII.2.(vi).

<u>Case 2</u> $s = 3$. We are only left to study the relation for $\gamma\beta$.

If $s = 3$ then $\beta\gamma = \alpha^2 p(\alpha)$ lies in $\text{soc}_2\Lambda$; hence $\gamma\beta\gamma \in \text{soc}\,\Lambda \cap e_1\Lambda e_0 = 0$; similarly $\beta\gamma\beta = 0$, and $\gamma\beta J = \gamma\beta\eta J$; and $(\gamma\beta)J^2 = \gamma\beta\eta^2\Lambda$. Moreover, we have $\gamma\beta\eta^2 = \gamma\alpha^2\beta \in J\text{soc}_2\Lambda J = 0$. Therefore $(\gamma\beta)J^2 = 0$ and $(\gamma\beta)$ lies in $\text{soc}_2\Lambda$. Therefore $\gamma\beta = c\eta^{t-1} + u\eta^t$. If c were 0 then we would have $0 = \eta\gamma\beta = \gamma\alpha\beta$. On the other hand, $0 \neq \beta\gamma\alpha = \alpha^3 \in \text{soc}\,\Lambda$. Let $\psi$ be a symmetrizing form for $\Lambda$; then $\psi[\gamma\alpha\beta] = \psi[0] = 0 = \psi[\beta\gamma\alpha]$, and ker $\psi$ contains the ideal soc $e_0\Lambda$, a contradiction.

Now suppose that $t = 3$; then we must have $c \neq 1$, since otherwise we would have $H(e_0\Lambda) \cong H(e_1\Lambda)$, a contradiction to VII.2.(vi).

In case $t > 3$ one obtains relations with $a = 1$ by a scalar transformation.

## VII.8 *Algebras of quaternion type, 3 simple modules*

First we will study algebras whose quiver is of the form

(3A) 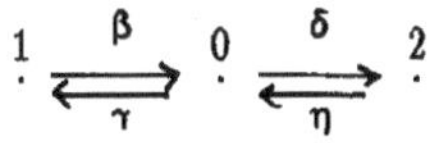

VII.8.1 (1) By VII.2(v) we may assume that $\Omega^2(\beta\Lambda) = \gamma\Lambda$ and that $\Omega^2(\eta\Lambda) = \delta\Lambda$, and we have $\underline{\dim}\ \beta\Lambda = \underline{\dim}\ \gamma\Lambda$ and $\underline{\dim}\ \eta\Lambda = \underline{\dim}\ \delta\Lambda$. Moreover, by VII.2(iii) we know that there are no relations involving a path of length $\leq 2$.

(2) There is a relation involving $\beta\gamma\beta$ or $\beta\delta\eta$ ( or both): The universal cover of $\Lambda/J^3$ contains lines of arbitrary length ( alternating $\gamma\beta$ and $\delta\eta$ ). Therefore, if (2) were not true the universal cover of an appropriate algebra $\Lambda/I$ would contain $\tilde{E}_7$, namely

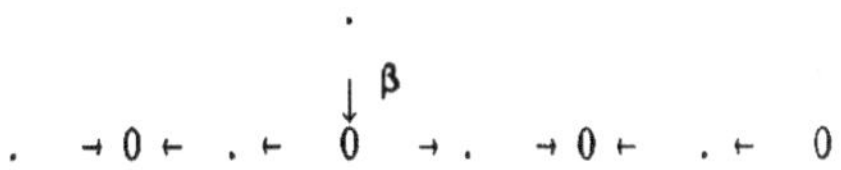

There are two possibilities:

(3*) Either $\beta\gamma\beta \equiv c(\beta\delta\eta)$ modulo $J^4$ for some $0 \neq c \in K$, or one of $\beta\gamma\beta$ and $\beta\delta\eta$ lies in $J^4$.

Similarly

(4*) Either $\eta\delta\eta \equiv d(\eta\gamma\beta)$ modulo $J^4$ for some $0 \neq d \in K$, or one of $\eta\delta\eta$ and $\eta\gamma\beta$ lies in $J^4$.

Moreover, $\eta\delta\eta$ and $\eta\gamma\beta$ are not both in $J^4$: Then we would have that $e_2J^3 = \langle \eta\delta\eta, \eta\gamma\beta \rangle \subseteq e_2J^4$ and $e_2J^3 = 0$. Therefore $\eta\delta J = \langle \eta\delta\eta \rangle = 0$ and $\eta\delta$ belongs to soc $\Lambda$. Since $\eta\delta$ is independent, we deduce $e_2\Lambda = \mathcal{U}(S_2, S_0, S_2)$ and then $S_0 \cong H_2$ and $S_0$ does not lie at the end of a component, a contradiction to VII.2(vi).

VII.8.2 LEMMA *If the local algebra $\Lambda_2 = e_2\Lambda e_2$ is uniserial then $\dim e_2\Lambda e_1 = 1$.*

Proof: Suppose $\Lambda_2$ is uniserial, then its radical is generated by $\eta\delta$ and hence $\eta\gamma\beta\delta$ lies in $(\eta\delta)^2\Lambda$. We claim that $\eta\gamma\beta$ must occur in some relation. Suppose not; then by (4*) we know that $\eta\delta\eta$ belongs to $J^4$ and therefore $e_2J^3 = (\eta\gamma\beta)\Lambda$. We deduce that $e_2J^4 = (\eta\gamma\beta\delta)\Lambda \subseteq (\eta\delta)^2\Lambda \subseteq e_2J^5$ and $e_2J^4 = 0$. This implies $\eta\gamma\beta \in \text{soc}\ \Lambda \cap e_2\Lambda e_0 = 0$; a contradiction since we assumed that $\eta\gamma\beta$ is independent.

Now we have $e_2J^3$ = span { $\eta\delta\eta$, $(\eta\delta)^2$, $(\eta\delta)^2\eta$, ... } and $e_2J^3e_1 = 0$. This shows that $e_2\Lambda e_1$ is spanned by $\eta\gamma$.

The same conclusion holds for an algebra of quaternion type with quiver (3*B*) when $\gamma\alpha$ and $\alpha\beta$ lie in $J^3$.

VII.8.3 THEOREM *Suppose $\Lambda$ is basic of quaternion type with quiver (3A), and assume also that the local ring $\Lambda_1$ is not uniserial. Then $\Lambda \cong KQ/I$ where I is defined by the following relations:*

$$\beta\gamma\beta = (\beta\delta\eta\gamma)^{k-1}\beta\delta\eta \quad and \quad \eta\delta\eta = (\eta\gamma\beta\delta)^{k-1}\eta\gamma\beta,$$

$$\gamma\beta\gamma = (\delta\eta\gamma\beta)^{k-1}\delta\eta\gamma \quad and \quad \delta\eta\delta = (\gamma\beta\delta\eta)^{k-1}\gamma\beta\delta,$$

$$\beta\gamma\beta\delta = 0 = \eta\delta\eta\gamma,$$

*where $k \geq 2$. That is, $\Lambda$ belongs to the family $Q(3A)_2$.*

Proof: (1) The local algebra $\Lambda_2$ is not uniserial: Suppose $\Lambda_2$ is uniserial; then by the above Lemma, dim $e_2\Lambda e_1 = 1$ and since $\Lambda$ is symmetric, dim $e_1\Lambda e_2 = 1$ as well. We deduce that $e_1\Lambda e_2$ is spanned by $\beta\delta$, and $\beta(\delta\eta)^a\delta = 0$ for $a \geq 1$; in particular $J(\Lambda_1)$ = $\langle \beta\gamma, \beta\delta\eta\gamma \rangle$. There is a module $\mathcal{U}(S_2, S_0, S_1)$ which must occur as a submodule of $e_1\Lambda$; generated by some element of $e_1\Lambda e_2$, hence by $\beta\delta$. This implies that $\beta\delta\eta\gamma$ lies in soc $e_1\Lambda$ = soc $\Lambda_1$, and consequently $\Lambda_1$ is uniserial, a contradiction.

Now we apply (3*) and (4*) of VII.8.1 and deduce that the elements $\beta\gamma\beta$, $\gamma\beta\gamma$ and also $\eta\delta\eta$ and $\delta\eta\delta$ lie in $J^4$. Therefore $J(\Lambda_1)$ has independent generators $\beta\gamma$ and $\beta\delta\eta\gamma$; and $J(\Lambda_2)$ is generated by $\eta\gamma$ and $\eta\gamma\beta\delta$.

(2) soc $e_1\Lambda = \langle (\beta\delta\eta\gamma)^k \rangle$, and $k > 2$: We have that $e_1J^3 = (\beta\delta\eta)\Lambda$, $e_1J^4 = (\beta\delta\eta\gamma)\Lambda$ and so on, and therefore $\dim[e_1J^k/e_1J^{k+1}] \leq 1$ for $k \geq 3$. Consequently $(\beta\gamma)^2 \in (\beta\delta\eta\gamma)^2\Lambda$. Now we deduce from III.15 that soc $\Lambda_1 = \langle (\beta\delta\eta\gamma)^k \rangle$ = soc $e_1\Lambda$, where k is maximal such that $(\beta\delta\eta\gamma)^k \neq 0$. Since $\Lambda_2$ is not uniserial, $k \geq 2$.

Then k is also the last integer such that $(\eta\gamma\beta\delta)^k \neq 0$, and therefore soc $\Lambda_2 = \langle (\eta\gamma\beta\delta)^k \rangle$. Moreover, we must have that soc $e_0\Lambda = \langle (\gamma\beta\delta\eta)^k \rangle$ and since $\Lambda$ is symmetric, $(\gamma\beta\delta\eta)^k = (\delta\eta\gamma\beta)^k$.

(3) We may assume that $\beta\gamma\beta = (\beta\delta\eta\gamma)^{k-1}\beta\delta\eta$: There is a module of the form $\mathcal{U}(S_1, S_0,$

$S_1$) which occurs as a submodule of $e_1\Lambda$; let $\omega$ be a generator with $\omega = \omega e_1$. Then $0 \neq \omega\beta \in \mathrm{soc}_2 e_1\Lambda = \langle (\beta\delta\eta\gamma)^{k-1}\beta\delta\eta \rangle$. Write $\omega = c(\beta\gamma) + \sum_{i\geq 1} a_i(\beta\delta\eta\gamma)^i$; this gives $\omega\beta = c\beta\gamma\beta + \sum a_i(\beta\delta\eta\gamma)^i\beta$, and we deduce $a_1 = \ldots = a_{k-1} = 0$ and $c(\beta\gamma\beta) = d(\beta\delta\eta\gamma)^{k-1}\beta\delta\eta$ for $c, d \in K^*$. We may assume that $c = d = 1$; otherwise we replace $\beta$ by an appropriate scalar multiple.

The relation for $\gamma\beta\gamma$ follows from (3) by exploiting a K-basis for $e_0\Lambda$. Similarly one shows that, without loss of generality, $\eta\delta\eta = c(\eta\gamma\beta\delta)^{k-1}\beta\delta\eta$ for some $0 \neq c \in K$; and it follows that $\delta\eta\delta = c(\gamma\beta\delta\eta)^{k-1}\gamma\beta\delta$. It is possible to have $c = 1$, after a scalar transformation (since $k \neq 1$).

<u>VII.8.4</u> **THEOREM** *Suppose $\Lambda$ is basic of quaternion type with quiver (3A). Assume also that the local ring $\Lambda_1$ is uniserial. Then*
*$\Lambda \cong KQ/I$ where $I$ is defined by the following relations:*

$$\beta\delta\eta = (\beta\gamma)^{a-1}\beta \quad \text{and} \quad \eta\gamma\beta = d(\eta\delta)^{b-1}\eta,$$
$$\delta\eta\gamma = (\gamma\beta)^{a-1}\gamma \quad \text{and} \quad \gamma\beta\delta = d(\delta\eta)^{b-1}\delta,$$
$$\beta\delta\eta\delta = 0 = \eta\gamma\beta\gamma,$$

*where $0 \neq d \in K$ and $a, b \geq 2$. If $(a, b) = (2, 2)$ then $d \neq 1$; and otherwise $d = 1$. That is, $\Lambda$ belongs to the family $Q(3A)_1$.*

Proof: By the previous theorem, we know that also $\Lambda_1$ is uniserial, hence $J(\Lambda_1) = \langle (\beta\gamma) \rangle$ and $J(\Lambda_2) = \langle (\eta\delta) \rangle$. Moreover, by VII.8.2, $\dim e_1\Lambda e_2 = 1$ and $\dim e_2\Lambda e_1 = 1$. Therefore $e_1\Lambda e_2$ is spanned by $\beta\delta$, and $e_2\Lambda e_1$ by $\eta\gamma$; consequently $\beta\delta\eta\delta = 0 = \eta\gamma\beta\gamma = \beta\gamma\beta\delta = \eta\delta\eta\gamma$. Moreover $\beta\delta\eta\gamma$ lies in $(\beta\gamma)^2\Lambda$, and $\eta\gamma\beta\delta$ belongs to $(\eta\delta)^2\Lambda$.

Let a, b be maximal such that $(\beta\gamma)^a \neq 0$ and $(\eta\delta)^b \neq 0$. Then $\mathrm{soc}\, e_1\Lambda = \mathrm{soc}\, \Lambda_1 = \langle (\beta\gamma)^a \rangle$, and $\mathrm{soc}\, e_2\Lambda = \langle (\eta\delta)^b \rangle$. We claim that $a, b \geq 2$: Suppose for example that $a = 1$; then $\mathrm{soc}\, e_1\Lambda = \langle (\beta\gamma) \rangle$. Since $0 \neq \beta\delta\Lambda \subsetneq a_1\Lambda$ and $\mathrm{soc}\, e_1\Lambda$ is simple, we deduce $0 \neq \gamma\beta \in \beta\delta\Lambda$, and $\gamma\beta$ is not independent, a contradiction.

(1) <u>$\beta\delta\eta \in (\beta\gamma\beta)\Lambda$ and $\eta\gamma\beta \in \eta\delta\eta\Lambda$:</u> Suppose $\beta\delta\eta \notin \beta\gamma\beta\Lambda$; then by VII.8.1 (3*) we have $\beta\gamma\beta \in e_1J^4$, and then $\beta\delta\eta\gamma \in (\beta\gamma)^2\Lambda \subsetneq e_1J^5$, and $e_1J^4 \subsetneq e_1J^5 = 0$. Consequently $(\beta\gamma)^2 = 0$ which is a contradiction to (1). The other part is proved similarly.

This shows that $e_1\Lambda$ has a K-basis $\{ e_1, \beta, \beta\delta, \beta\gamma, \beta\gamma\beta, \ldots, (\beta\gamma)^a \}$. In particular $\dim e_1J^k/e_1J^{k+1} \leq 1$ for $k = 3, 4, \ldots$ .

There exists a uniserial module $\mathcal{U}(S_2, S_0, S_1)$ which must be isomorphic to a submodule of $e_1\Lambda$. It is generated by an element of $e_1\Lambda e_2$, hence by $\beta\delta$. We deduce that $0 \neq \beta\delta\eta \in \mathrm{soc}_2(e_1\Lambda) = \langle (\beta\gamma)^{a-1}\beta \rangle$. Also, $\beta\delta\eta \in e_1\Lambda e_0$, therefore $\beta\delta\eta = c(\beta\gamma)^{a-1}\beta$ for some $c \in K^*$; and without loss of generality $c = 1$; otherwise replace $\beta$.
Similarly there is some $d \in K^*$ with $\eta\gamma\beta = d(\eta\delta)^{b-1}\eta$. Using now the obvious K-basis for $e_0\Lambda$ we obtain as consequences the relations $\delta\eta\gamma = c(\gamma\beta)^{a-1}\gamma$ and $\gamma\beta\delta = d(\delta\eta)^{b-1}\delta$. We deduce that $(\gamma\beta)^a = \delta\eta\gamma\beta = d(\delta\eta)^b$; and therefore $(\gamma\beta)^aJ = \langle (\gamma\beta)^a\delta, d(\delta\eta)^b\gamma \rangle = 0$.

Assume that $a = b = 2$; if $d$ were $\neq 1$ then we would have that $H_1 \cong H_2$, a contradiction to VII.2(iv). On the other hand, if $(a, b) \neq (2, 2)$ then there is a scalar transformation which gives relations with $d = 1$.

Now we will study algebras of quaternion type with quiver (3B). Here we have only one family:

<u>VII.8.5</u> THEOREM *Assume $\Lambda$ is of quaternion type with quiver (3B). Then $\Lambda \cong KQ/I$ where I is defined by the following relations:*

$$\beta\gamma = \alpha^{s-1}, \qquad \eta\delta\eta = (\eta\gamma\beta\delta)^{k-1}\eta\gamma\beta,$$
$$\alpha\beta = (\beta\delta\eta\gamma)^{k-1}\beta\delta\eta, \qquad \delta\eta\delta = (\gamma\beta\delta\eta)^{k-1}\gamma\beta\delta,$$
$$\gamma\alpha = (\delta\eta\gamma\beta)^{k-1}\delta\eta\gamma \quad \text{and} \quad \alpha^2\beta = 0 = \beta\delta\eta\delta,$$

*where $k \geq 1$ and $s \geq 3$.*

Proof: By VII.2(iii), $\delta\eta$ is independent. Let $e = e_0 + e_1$, then the algebra $e\Lambda e$ has the quiver

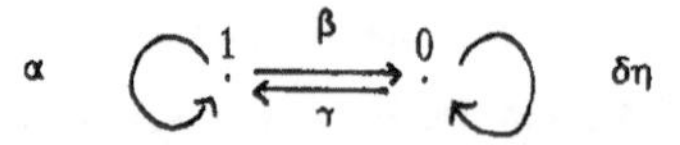


and consequently, by VII.5.4(a) we deduce that the module $\Omega_{e\Lambda e}(\beta\Lambda)$ is uniserial; and then $\Omega(\beta\Lambda)$ must also be uniserial. Therefore. by VII.5.4, <u>dim</u> $\beta\Lambda$ = <u>dim</u> $\gamma\Lambda$ and we may assume $\Omega^2(\beta\Lambda) = \gamma\Lambda$. We choose $\delta$, $\eta$ according to VII.2(v); by VII.2(iii), the paths $\delta\eta$, $\eta\delta$, $\beta\delta$, $\eta\gamma$, $\gamma\beta$ are independent.

(1) <u>The elements $\gamma\alpha$ and $\alpha\beta$ both lie in $J^3$, and $\beta\gamma$ is dependent:</u> The universal cover

of $\Lambda/J^3$ (modulo an appropriate ideal) contains lines

of arbitrary length. Attaching a $\gamma$-arrow at * gives $E_7$, and $\Lambda$ would be wild, by I.9.5 and I.4.7, unless one of $\gamma\beta$, $\gamma\alpha$ is involved in some relation. Hence $\gamma\alpha$ occurs in a relation; and since there are no other paths of length 2 from vertex $e_0$ to vertex $e_1$ it follows that $\gamma\alpha$ lies in $J^3$. Similar arguments show that $\alpha\beta$ lies in $J^3$. The last statement follows from VII.5.2.

(2) <u>There is a relation involving $\eta\delta\eta$ and also one with $\delta\eta\delta$:</u> Using the above line and attaching a $\eta$-arrow at ** shows that one of $\eta\delta\eta$, $\eta\gamma$ occurs in some relation (otherwise $\Lambda$ would be wild, as above), and it must be be $\eta\delta\eta$. The same argument applies to $\delta\eta\delta$.

Therefore $\eta\delta\eta \equiv c(\eta\gamma\beta)$ mod $J^4$ and $(\eta\delta)^2 \in (\eta\gamma\beta\delta)\Lambda$. We deduce that $\dim[e_2J^r/e_2J^{r+1}] \leq 1$ for $r \geq 3$; note also that $\Lambda_2$ is commutative, by III.16. Moreover, soc $\Lambda_2 = \langle (\eta\gamma\beta\delta)^k \rangle$ where k is maximal such that $(\eta\gamma\beta\delta)^k \neq 0$; in particular, this shows that $e_2J^{4k} = \text{soc } e_2\Lambda$.

(3) <u>We may assume that $\eta\delta\eta = (\eta\gamma\beta\delta)^{k-1}\eta\gamma\beta$:</u> There is a module of the form $\mathcal{U}(S_2, S_0, S_2)$ which is isomorphic to a submodule of $e_2\Lambda$ and (3) follows by the argument of (3) in the proof of VII.8.3.

We study now the socle of $e_0\Lambda$. We have that $(\delta\eta\gamma\beta)^kJ \subseteq \delta e_2J^{4k} \subseteq J[\text{soc } e_2\Lambda] = 0$; and $\langle (\delta\eta\gamma\beta)^k \rangle \subseteq \text{soc } e_0\Lambda$. Let $\psi$ be a symmetrizing form for $\Lambda$; then $\psi[(\delta\eta\gamma\beta)^k] = \psi[(\eta\gamma\beta\delta)^k] \neq 0$ since $(\eta\gamma\beta\delta)^k$ spans a non-zero ideal. Similarly soc $e_0\Lambda = \langle (\gamma\beta\delta\eta)^k \rangle$ and then $(\gamma\beta\delta\eta)^k = (\delta\eta\gamma\beta)^k$ since $\Lambda$ is symmetric.

Now we see thta $e_0J$ has a K-basis $\{ \gamma, \gamma\beta, \gamma\beta\delta, \gamma\beta\delta\eta, \gamma\beta\delta\eta\gamma, \ldots, (\gamma\beta\delta\eta)^k \} \cup \{ \delta, \delta\eta, \ldots, (\delta\eta\gamma\beta)^{k-1}\delta\eta\gamma \}$. Using this and the relation for $\eta\delta\eta$ we deduce that $\delta\eta\delta = (\gamma\beta\delta\eta)^{k-1}\gamma\beta\delta$.

Note that $e_0J^{4k} = \text{soc } e_0\Lambda$. Therefore $J(\beta\delta\eta\gamma)^k \subseteq J^{4k}e_0 = 0$ and $\langle (\beta\delta\eta\gamma)^k \rangle \subseteq$ soc $e_1\Lambda$; and equality holds since $\Lambda$ is symmetric and $0 \neq (\delta\eta\gamma\beta)^k \in$ soc $\Lambda$. We see now that $\Lambda_1$ is commutative, by III.16; and we may choose $\alpha$ to have soc $e_1\Lambda = \langle \alpha^s \rangle$ and $\alpha^{s-1} \in \text{soc}_2\Lambda$.

(4) <u>We may assume that $\beta\gamma = \alpha^{s-1}$:</u> We use the argument of VII.7.1, where we exploited that fact that <u>dim</u>$\beta\Lambda$ = <u>dim</u> $\gamma\Lambda$. Here we have that <u>dim</u> $\beta\Lambda$ = [2k, k+u, k]

and $\underline{\dim}\ \gamma\Lambda = [\ 2k+2v,\ k+v+1,\ k+v]$ and $v = 0$ and $u = 1$; we obtain (4).

Let $\alpha^s = c(\beta\delta\eta\gamma)^k$ with $0 \neq c \in K$. Using (4) and a K-basis, the relations for $\alpha\beta$ and $\gamma\alpha$ follow, after a scalar transformation.

Now we will study algebras of quaternion type with quiver (3C). Here our result is as follows:

VII.8.6 **THEOREM** *Suppose $\Lambda$ is a basic algebra of quaternion type with quiver (3C). Then $\Lambda \cong KQ/I$ where I is defined by the relations*

$$\beta\rho = 0 = \rho\gamma \quad \text{and} \quad \eta\rho^2 = 0 = \rho^2\delta$$

$$\delta\eta - \gamma\beta = \rho^{s-1}, \quad \eta\rho = (\eta\delta)^{k-1}\eta, \quad \rho\delta = (\delta\eta)^{k-1}\delta \quad \text{and}$$

$$(\beta\gamma)^{k-1}\beta\delta = 0 = (\eta\delta)^{k-1}\eta\gamma,$$

*where $k \geq 2$, $s \geq 3$.*

Proof: By VII.2(iii), the elements $\beta\delta$, $\beta\gamma$, $\eta\gamma$ and $\eta\delta$ are independent. Moreover,

(1) We may assume that $\beta\rho = 0$ and $\rho\gamma = 0$: Assume for contradiction that $\beta\rho$ is independent. Then $\Lambda/J^2$ has a covering which contains the quiver

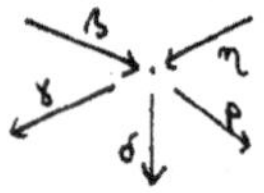


with no relations, except possibly $\eta\rho = 0$. This is a wild category (I.10.7); and $\Lambda$ is wild. Hence $\beta\rho \in e_1J^3e_0 = \beta e_0J^2e_0$ and there is some $x \in e_0J^2e_0$ with $\beta(\rho - x) = 0$. We may replace $\rho$ by $\rho' = \rho - x$. Recall that $\Omega(\beta\Lambda) \cong \Omega^2 S_1$ (VII.2(v)), and this module has a simple top $\cong S_0$; consequently $\beta\rho = 0$ implies that $\rho\Lambda = \Omega(\beta\Lambda)$. Now $\Omega(\rho\Lambda) \cong \Omega^3 S_1 \cong \Omega^{-1}S_1 \cong \gamma\Lambda$; so we may choose $\gamma$ to get $\rho\gamma = 0$.

Similarly, for some $\tilde{\rho}$ with $\tilde{\rho} \equiv \rho \bmod J^3$ and $\tilde{\rho} \in e_0Je_0$ we have $\eta\tilde{\rho} = 0$ and $\tilde{\rho}\delta = 0$; in particular, $\eta\rho$ and $\rho\delta$ are dependent. (However, $\eta\rho \neq 0$ since otherwise $\eta\Lambda \cong \Omega^{-1}(\rho\Lambda) \cong \beta\Lambda$.) By VII.5.2.1 we know that then also $\gamma\beta$ and $\delta\eta$ are dependent.

(2) It is not possible that $e_0J^2 = \rho^2\Lambda$: Suppose this happens; then $\gamma\beta$ and $\delta\eta$ lie in $\rho^2\Lambda$. Since $\rho\delta \in e_0J^3$ and $\rho\gamma = 0$, it follows that $\rho^2\Lambda = \text{span}\ \{\ \rho^2,\ \rho^3,\ \ldots,\ \}$ and $e_0J^2e_1 = 0 = e_0J^2e_2$. Consequently $H(e_0\Lambda)$ is a direct sum; a contradiction to VII.2(vi).

(3) <u>The element $\delta\eta$ does not lie in $J^3$</u>: Suppose $\delta\eta \in J^3$. If also $\gamma\beta \in J^3$ then we would have $e_0J^2 = \rho^2\Lambda$, contrary to (2). Hence $J(e_0\Lambda e_0) = \langle \gamma\beta, \rho \rangle$, with two independent generators. Then $\gamma\beta$ and $\rho$ are also independent in the algebra $e\Lambda e$ modulo its radical, where $e = e_0 + e_2$. Hence the quiver of $e\Lambda e$ is of the form

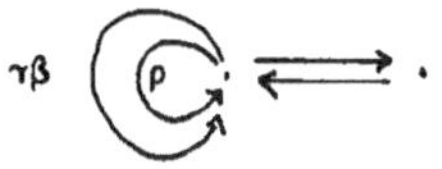


and $e\Lambda e$ is wild; and then $\Lambda$ as well (I.4.7), a contradiction.

Similarly one shows that $\gamma\beta \notin J^3$. We deduce that there is a relation $c_1\gamma\beta + c_2\delta\eta + c_3\rho^2 \equiv 0 \bmod J^3$, with $c_1 \neq 0$ and one of $c_2$, $c_3 \neq 0$. Also there is a relation $d_1\gamma\beta + d_2\delta\eta + d_3\rho^2 \equiv 0 \bmod J^3$ with $d_2 \neq 0$ and one of $d_1$, $d_3 \neq 0$. By VII.5.1, there must be a relation with both $d_1$ and $d_2 \neq 0$, say; and then, without loss of generality,

$$\delta\eta \equiv \gamma\beta - c\rho^2 \bmod J^3 \quad \text{for some } c \in K.$$

We have now that $J(\Lambda_0)$ is spanned by the powers of $\gamma\beta$ and of $\rho$; in particular $\Lambda_0$ is commutative (III.16); and $\operatorname{soc} \Lambda_0 = \langle \rho^s \rangle = \langle (\gamma\beta)^k \rangle$ ; and we may assume that $\rho^s = (\gamma\beta)^k$.

(4) <u>$\rho^2\delta = 0$ and $\eta\rho^2 = 0$:</u> Let $\tilde{\rho}$ be as above with $\tilde{\rho}\delta = 0 = \eta\tilde{\rho}$; writing $\tilde{\rho}$ in terms of the K-basis we have $\tilde{\rho} = \rho u + \Sigma a_i(\gamma\beta)^i$ where u is a unit in $K[\rho]$. Hence $0 = \tilde{\rho}\delta = \rho u\delta + \Sigma a_i(\gamma\beta)^i\delta$ and $0 = \rho\tilde{\rho}\delta = \rho^2u\delta = u\rho^2\delta$. Since u is a unit, $\rho^2\delta = 0$; and similarly $\eta\rho^2 = 0$. We write now

(*) $\delta\eta = \gamma\beta z + \rho^v w$ where z is a unit in $K[\gamma\beta]$, w is a unit in $K[\rho]$ and $v \geq 2$.

(5) <u>$v = s-1$ and hence $s \geq 3$:</u> We have $\delta\eta\rho^2 = \rho^{v+2}w = 0$ and hence $\rho^{v+2} = 0$, so $v \geq s-1$. Assume for contradiction that $v \geq s$. Then $\delta\eta\rho = 0$ and $J\eta\rho = 0$, and consequently $\eta\rho \in \operatorname{soc} \Lambda \cap e_2\Lambda e_0 = 0$. This is not possible, as we noted above.

(6) <u>We may assume that $\delta\eta = \gamma\beta + \rho^{s-1}$</u>: By (5) and since $\rho^s = (\gamma\beta)^k$, we may rewrite (*) in the form $\delta\eta = \gamma\beta y + \rho^{s-1}$ where y is a unit in $K[\gamma\beta]$. Then $\rho y^{-1} = \rho$; and we replace $\eta$ by $\eta' = \eta y^{-1}$. Note that this does not affect the previous relations.

We have $\rho\delta J = \rho\delta\eta\Lambda = \operatorname{soc} e_0\Lambda$, therefore $\rho\delta \in [\operatorname{soc}_2(e_0\Lambda 0]e_2$. Therefore $\rho\delta = c(\gamma\beta)^{k-1}\delta$, and $c = 1$ since $\rho\delta\eta = \rho^s$. In terms of a basis, we have $\eta\rho = \Sigma a_i(\eta\delta)^i\eta$ for $a_i \in K$. Then $\eta\rho\delta = \eta(\gamma\beta)^{k-1}\delta = \eta[\delta\eta + \rho^{s-1}]^{k-1}\delta = (\eta\delta)^k = \Sigma a_i(\eta\delta)^{i+1}$, and the relation for $\eta\delta$ follows.

Finally, $(\beta\gamma)^{k-1}\beta\delta J = (\beta\gamma)^{k-1}\beta\delta\eta\Lambda = (\beta\gamma)^k\beta\Lambda$, by (6), and this is 0. Therefore

$(\beta\gamma)^{k-1}\beta\delta \in \mathrm{soc}\ \Lambda \cap e_1\Lambda e_2 = 0$. Similarly $(\eta\delta)^{k-1}\eta\gamma = 0$.

Now we will determine the possible algebras of quaternion type with quiver (3*D*)

VII.8.7 THEOREM *Suppose* $\Lambda$ *is a basic algebra of quaternion type with quiver (3D) Then* $\Lambda \cong KQ/I$ *where I is defined by the following relations:*

$$\beta\gamma = \alpha^{s-1}, \qquad \eta\delta = \xi^{t-1},$$
$$\gamma\alpha = (\delta\eta\gamma\beta)^{k-1}\delta\eta\gamma, \qquad \delta\xi = (\gamma\beta\delta\eta)^{k-1}\gamma\beta\delta,$$
$$\alpha\beta = (\beta\delta\eta\gamma)^{k-1}\beta\delta\eta, \qquad \xi\eta = (\eta\gamma\beta\delta)^{k-1}\eta\gamma\beta,$$
$$\alpha^2\beta = 0 = \delta\eta\delta$$

*where* $k \geq 1$ *and* $s,\ t \geq 3$.

Proof: We use the conventions for $\beta$, $\gamma$ and $\delta$, $\eta$ of VII.5.4. By VII.2 there are no relations involving $\beta\delta$, $\eta\gamma$, $\gamma\beta$, $\delta\eta$; and therefore, by VII.2.4, each of the elements $\beta\gamma$, $\eta\delta$, $\delta\xi$, $\xi\eta$ and also $\gamma\alpha$ and $\alpha\beta$ are dependent. In particular,the elements $\gamma\alpha$, $\alpha\beta$, $\delta\xi$ and $\xi\eta$ lie in $J^3$ since in each case there is only one path of length 2 between the vertices in question. Moreover, by VII.5.1 we have $\beta\gamma \equiv r\alpha^2 \bmod J^3$ where $r \in K$; and also $\delta\eta \equiv u\xi^2 \bmod J^3$ for $u \in K$.

Consider the local algebra $\Lambda_1$, its radical is generated by $\alpha$ and $(\beta\delta\eta\gamma)$. It is tame symmetric and must be isomorphic to one of the algebras in III.1(a) to (b'), so by III.15 we may assume that $\mathrm{soc}\ e_1\Lambda = \mathrm{soc}\ \Lambda_1 = \langle \alpha^s \rangle = \langle (\beta\delta\eta\gamma)^k \rangle$, for an appropriate choice of $\alpha$. Using the fact that $\Lambda$ is symmetric, we have $(\eta\gamma\beta\delta)^k \neq 0$, and clearly $(\eta\gamma\beta\delta)^{k+1} = 0$. Then we see by a similar argument that $\mathrm{soc}\ e_2\Lambda = \langle \xi^t \rangle = \langle (\eta\gamma\beta\delta)^k \rangle$. Also $(\gamma\beta\delta\eta)^k \neq 0$ but $(\gamma\beta\delta\eta)^{k+1} = 0$, and $\mathrm{soc}(e_0\Lambda) = \langle (\gamma\beta\delta\eta)^k \rangle = \langle (\delta\eta\gamma\beta)^k \rangle$, and moreover $(\gamma\beta\delta\eta)^k = (\delta\eta\gamma\beta)^k$. The argument in VII.7.(2*A*) which uses $\underline{\dim}\ \beta\Lambda = \underline{\dim}\ \gamma\Lambda$ gives that we may assume $\beta\gamma = \alpha^{s-1}$; and similarly, without loss of generality, $\delta\eta = \xi^{t-1}$. The other relations are straighforward, including a scalar transformation.

It remains to investigate algebras of quaternion type with quiver (3*K*).

VII.8.8 **THEOREM** *Suppose Λ is a basic algebra of quaternion type with quiver (3K). Then Λ ≅ KQ/I where I is defined by the following relations:*

$$\beta\delta = (\kappa\lambda)^{a-1}\kappa, \qquad \delta\lambda = (\gamma\beta)^{b-1}\gamma, \qquad \lambda\beta = (\eta\delta)^{c-1}\eta,$$

$$\eta\gamma = (\lambda\kappa)^{a-1}\lambda, \qquad \kappa\eta = (\beta\gamma)^{b-1}\beta, \qquad \gamma\kappa = (\delta\eta)^{c-1}\delta,$$

$$\gamma\beta\delta = 0 = \delta\eta\gamma = \lambda\kappa\eta,$$

*where a, b, c ≥ 2.*

(Note that the relations listed imply that all paths of length 3 which change direction once are 0 and also the nilpotence for the socle.)

Proof: We use the convention of VII.5.4. By VII.2.(iii) we know that all closed paths of length 2 are independent, and then, by VII.2(iv), any other path of length 2 must occur in a relation, hence lies in $J^3$.

Consider the local algebra $\Lambda_0$. Its radical has independent generators $\beta\gamma$ and $\lambda\kappa$; and moreover $(\beta\gamma)(\kappa\lambda) \in e_0J^5e_0 \subseteq \langle (\beta\gamma)^3, (\kappa\lambda)^3 \rangle$. Similarly $(\kappa\lambda)(\beta\gamma) \in J(\Lambda_0)^3$. We deduce from III.15 that soc $e_0\Lambda = \langle (\kappa\lambda)^a \rangle = \langle (\beta\gamma)^b \rangle$; in fact we may assume that $(\kappa\lambda)^a = (\beta\gamma)^b$.

This implies $(\gamma\beta)^b \neq 0$ but $(\gamma\beta)^{b+1} = 0$, and $(\lambda\kappa)^a \neq 0 = (\lambda\kappa)^{a+1}$; using the fact that Λ is symmetric. A similar argument applies to $\Lambda_1$ and $\Lambda_2$, and hence soc $e_1\Lambda = \langle (\gamma\beta)^b \rangle = \langle (\delta\eta)^c \rangle$, and without loss of generality $(\gamma\beta)^b = (\delta\eta)^c$.
Also, soc $e_2\Lambda = \langle (\eta\delta)^c \rangle = \langle (\lambda\kappa)^a \rangle$, and it follows that $(\eta\delta)^c = (\lambda\kappa)^a$; since the element $(\eta\delta)^c - (\lambda\kappa)^a$ spans an ideal contained in the kernel of a symmetrizing linear form for Λ.

Combining the information we see that $e_0J$ has a K-basis
$\{ \beta, \beta\gamma, \beta\gamma\beta, \ldots, (\beta\gamma)^b, \kappa, \kappa\lambda, \kappa\lambda\kappa, \ldots, (\kappa\lambda)^{a-1}\kappa \}$, and there are similar bases for $e_1J$ and $e_2J$.

(1) <u>We have $\beta\delta = u(\kappa\lambda)^{a-1}\kappa$ and $\gamma\kappa = v(\delta\eta)^{c-1}\delta$ for $u, v \in K^*$:</u>

Recall that <u>dim</u> βΛ = <u>dim</u> γΛ. Using the above K-bases we have that <u>dim</u> βΛ = [a,a,0] + [0,r,r+1] and <u>dim</u> γΛ = [a,a,0] + [0,r,r+1] and consequently r = 0, and βδ lies in $soc_2(e_0\Lambda)$. It also belongs to $e_0\Lambda e_2$, consequently $\beta\delta = u(\kappa\lambda)^{a-1}\kappa$ for u ∈ K. If u were 0 then we would have δ ∈ Ω(βΛ) - rad Ω(βΛ), and $S_2$ would occur in top Ω(βΛ). On the other hand, from VII.5.4 we know that top Ω(βΛ) ≅ $S_1$. Similarly the other relation holds.

The same argument shows that $\delta\lambda = u'(\gamma\beta)^{b-1}\gamma$ and $\eta\gamma = v'(\lambda\kappa)^{a-1}\lambda$. Comparing two expressions for $\beta\delta\lambda$ shows that $u = u'$; and similarly $v = v'$. By the same method one derives relations for $\lambda\eta$ and $\kappa\beta$, namely $\kappa\eta = v(\beta\gamma)^{b-1}\beta$ and $\lambda\beta = u(\eta\delta)^{c-1}\eta$.

We have that $\beta\delta$ lies in $soc_2\Lambda$, hence $\gamma\beta\delta \in soc\, \Lambda \cap e_1\Lambda e_2 = 0$. Similarly $\delta\eta\gamma = 0 = \kappa\lambda\eta$. The relations imply that a, b, $c \geq 2$, for otherwise $\lambda$, $\gamma$, $\eta$ would lie in $J^2$. Now it is possible, by a scalar transformation, to obtain the same relation with $u = v = 1$.

VII.9 *On the converse* It is not possible at present to show in general that the algebras we obtained are in fact of quaternion type. There are some cases where the converse is established, namely blocks (see VII.1.1 ) and also there are some algebras for small parameters where covering methods give results, these have been done in detail in [$E_6$].

Of course, one can verify that the simple modules have $\Omega$-period 4; and this implies that all modules have finite complexity and by [MR] exclude that $\Gamma_s(\Lambda)$ has components $\cong \mathbb{Z}A_\infty^\infty$ or $\mathbb{Z}D_\infty$; which gives some evidence.

# VIII *Algebras of semidihedral type*

In this chapter we study algebras of semidihedral type; the aim is to determine the Morita equivalence classes. We will obtain a list of basic algebras defined explicitly by quivers and relations, and we will show:

Given an algebra of semidihedral type, its basic algebra belongs to this list.

In general terms, this is organized as follows: Suppose $\Lambda$ is basic and of semidihedral type. We start with some general preparation (VIII.2). Then we deal with the cases when $\Lambda$ is local (VIII.3) and when $\Lambda$ has two simple modules (VIII.4). In the next part (VIII.5), we are concerned with those algebras where a simple module has three predecessors in $\Gamma_s(\Lambda)$; this leads quickly to only one family of algebras (namely $SD(3\mathcal{C})_1$).

In VIII.6, we study quivers with "few" arrows; here we come close to algebras having Euclidean components $\mathbb{Z}\tilde{D}_n$, or even to algebras of finite type, whose stable AR-quiver is of the form $\mathbb{Z}D_n/\langle \tau^3 \rangle$.

In the next section we study algebras where distinct projectives has isomorphic hearts (VIII.7); and then in VIII.8 we deal with algebras whose quiver has "end vertices". By an end vertex we mean a vertex f with $|s^{-1}(f)| = |e^{-1}(f)| = 1$ and f is only joined to one other vertex.

It then remains to study algebras whose quiver has "enough" arrows; this will be done in VIII.9. As a by-product we obtain

VIII.10 **THEOREM** *Suppose $\Lambda$ is an algebra of semidihedral type. Then $\Lambda$ has at most three simple modules.*

We will include some comments about the converse in VIII.11.

VIII.1 DEFINITION We say that the algebra $\Lambda$ is *of semidihedral type* if it satisfies the following conditions:

(1) $\Lambda$ is symmetric, indecomposable and tame,

(2) The Cartan matrix of $\Lambda$ is non-singular,

(3) The stable AR-quiver $\Gamma_s(\Lambda)$ has the following components:

(i) Tubes of rank $\leq 3$, and at most one 3-tube,

(ii) Non-periodic components isomorphic to $\mathbb{Z}A_\infty^\infty$ and $\mathbb{Z}D_\infty$.

VIII.1.1 EXAMPLES Suppose B is a 2-block of some group algebra with semidihedral defect groups, then B is an algebra of semidihedral type: It is tame, by [BD] and the other conditions in (1) and (2) hold for arbitrary blocks ([F], see also V.1). In V.4.2 we proved that (3) is satisfied.

More explicitly, the algebras $\Lambda_m$ in II.9 are of semidihedral type, by II.10.1 (and $[CB_3]$; and therefore also the algebras III.1(d) and (d'), by I.8.11. This includes the group algebra of the semidihedral 2-group.

Another example is the group algebra of the group $GL_2(3)$ over a field of characteristic 2.

From now we assume that $\Lambda$ is a basic algebra of semidihedral type.

VIII.2.0 There are a number of steps; we will summarize them now, to be sure that we have dealt with all possible quivers. Suppose $\Lambda$ is a basic algebra of semidihedral type, with quiver $\mathcal{Q}$. If 0 is a vertex of $\mathcal{Q}$, and $e_0$ is the corresponding idempotent then we set $H_0 := \text{rad } e_0\Lambda/\text{soc } e_0\Lambda$. We write $0 \xrightarrow{\alpha} 1$ if $\alpha$ is the only arrow starting at vertex 0, and similarly $0 \xrightarrow{\beta} 1$ if $\beta$ is the only arrow ending at vertex 1.

(i) $\mathcal{Q}$ with $< 2$ vertices: VIII.3 and VIII.4

(ii) Assume $\mathcal{Q}$ has $> 3$ vertices.

There is a vertex 0 with $|s^{-1}(0)| = 1 = |e^{-1}(0)|$

0 is an end vertex: VIII.8 (quivers $(3\mathcal{A})$, $(3\mathcal{B})$).

0 is not an end vertex:

If $H_0 \neq H_k$ for $k \neq 0$ then the quiver is $(3\mathcal{F})$: VIII.6.2

If $H_0 \cong H_k$ for $k \neq 0$ then the quiver is $(3\mathcal{A})$ of $(3\mathcal{C})$: VIII.7.2-4

There is a vertex 0 with $|e^{-1}(0)| = 1 < |s^{-1}(0)|$ (or >). Then no double arrow ends at 0(see VIII.8.5)

If $H_0 \neq H_k$ $(k \neq 0)$ then the quiver is $(3\mathcal{F})$ or $(3\mathcal{Y})$: the dual of VIII.6.3

$H_0 \cong H_k$ for $k \neq 0$: Excluded by VIII.7.5

Otherwise, for each vertex f, $|s^{-1}(f)| = |e^{-1}(f)| = 2$ (see beginning of VIII.9). Then $H_0 \neq H_k$, by VIII.7.1 and VIII.7.6-8.

If $\mathcal{Q}$ contains 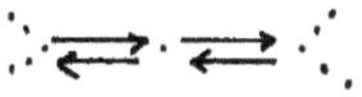 then $\mathcal{Q}$ is $(3\mathcal{D})$ or $(3\mathcal{K})$: VIII.9.3

$\mathcal{Q}$ does not contain 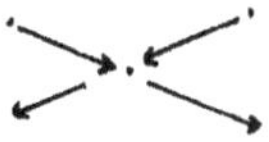 VIII.9.4

$\mathcal{Q}$ does not contain 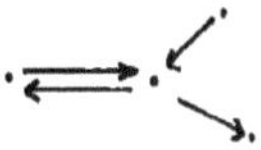 VIII.9.5

$\mathcal{Q}$ does not contain a double arrow VIII.9.6

$\mathcal{Q}$ does not contain  VIII.9.7

This implies that, if $\mathcal{Q}$ contains 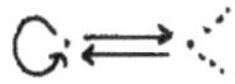 then $\mathcal{Q}$ is of the form $(3\mathcal{D})$. In particular, $\mathcal{Q}$ has at most three vertices.

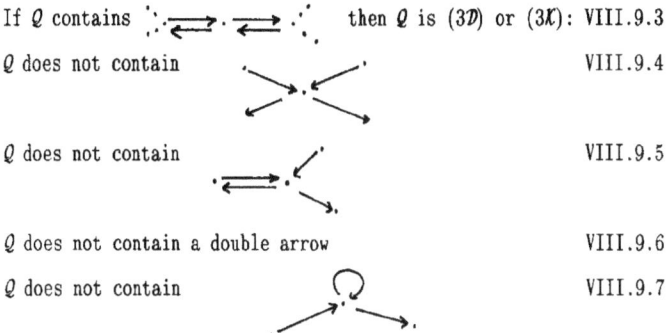

Now we summarize where the algebras are determined:

| Quiver | algebras with $H_1 \neq H_2$ | algebras with $H_1 \cong H_2$ |
|---|---|---|
| $(2\mathcal{A})$ | VIII.4.4, VIII.4.5 | n.a. |
| $(2\mathcal{B})$ | VIII.4.6, VIII.4.7 | VIII.4.8 |
| $(3\mathcal{A})$ | VIII.8.7 | VIII.7.9 |
| $(3\mathcal{B})$ | VIII.8.8.3, VIII.8.8.4 | n.a. |
| $(3\mathcal{C})$ | excluded by VIII.5.2 | VIII.5.1 and VIII.7.10 |
| $(3\mathcal{D})$ | VIII.9.12 | excluded by VIII.7.1 |
| $(3\mathcal{F})$ | VIII.6.4 | n.a. |
| $(3\mathcal{Y})$ | VIII.6.5 | n.a. |
| $(3\mathcal{K})$ | VIII.9.13 | n.a. |

VIII.2.0.1 We will now describe the possibilities for the position in $\Gamma_s(\Lambda)$ of a simple module, and consequences, proved earlier.

Let S be simple and $H = \text{rad } P_S/\text{soc } P_S$; recall that the summands of H are the non-projective middle terms of the standard AR-sequence (see I.7.7). Then one of the following holds:

(1) S lies at the end of a tube, of rank 2 or 3; moreover, H is indecomposable, and top H ≅ soc H (IV.1.4, IV.1.5).

(2) S is *of type a*: $H = U \oplus V$ where U and V are indecomposable and non-zero. The modules $\bar{U}$, $\underset{\sim}{V}$, $\bar{V}$ and $\underset{\sim}{U}$ are defined (IV.5.1), and they lie at ends of components, often at tubes (IV.5.5, and IV.6).

(3) S is *of type d*: $\beta(S) = 1$ and $\beta(H) = 3$; here S lies at the end of a $\mathbb{Z}D_\infty$-component. Then either $H \cong \text{rad } Q/\text{soc } Q$ for Q projective, $\neq P_S$, or else U, V are defined as in IV.3.3(a) with $\tau U \cong V$; and $\bar{U}$, $\underset{\sim}{V}$ are defined (IV.5.1) and $\bar{U}$, $\underset{\sim}{V}$ lie at ends of components, often in tubes (IV.5.5, IV.6).

(4) $\beta(S) = 3$. Then IV.3.2 holds, and $\Lambda$ is determined by IV.3.2, VIII.5.1. Therefore we assume everywhere else that this does not happen.

CONVENTION If U and V are defined in (3) then we very often have that top U and top V are simple. Then we choose arrows $\alpha$, $\beta$ with $V = \alpha\Lambda/\text{soc } \alpha\Lambda$ and $U = \beta\Lambda/\beta\Lambda \cap \alpha\Lambda$, and moreover $\alpha\Lambda \cap \beta\Lambda \subseteq \text{soc}_2\ \alpha\Lambda$, see IV.3.4.1. Note that in case soc U and top V are simple then U and V lie at the end of a tube, by IV.6.2.

VIII.2.1 LEMMA *Let $\Lambda$ be of semidihedral type; then $\Lambda$ must have a simple module which is non-periodic.*

Proof: Suppose that all simple $\Lambda$-modules are periodic; then they have complexity $\leq 1$ (see I.1). If $0 \to X \to Y \to Z \to 0$ is an exact sequence then the complexity of Y is $\leq$ the maximum of the complexities of X, Z. Therefore it follows by induction on the length that all $\Lambda$-modules have complexity $\leq 1$. By the hypothesis, $\Gamma_s(\Lambda)$ has a component $\Theta \cong \mathbb{Z}D_\infty$; and if $M \in \Theta$ then by [MR] the length of the modules in the $\tau$-orbit of M is unbounded, and since $\tau \cong \Omega^2$, it follows that M does not have complexity $\leq 1$, a contradiction.

VIII.2.2 LEMMA *Suppose the quiver of* $\Lambda$ *contains a vertex* $0$ *such that the arrows starting or ending at* $0$ *are as follows:*

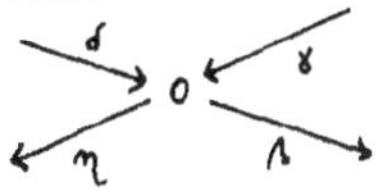

*We assume* $e(\eta) \neq e(\beta)$. *Assume that* $S_0$ *lies at the end of a* $\mathbb{Z}D_\infty$*-component, and that IV.3.8(a) holds. Then*

*(a)* $\underline{dim}\ \beta\Lambda + \underline{dim}\ \eta\Lambda - \underline{dim}\ soc\ V_0 = \underline{dim}\ \gamma\Lambda + \underline{dim}\ \delta\Lambda - \underline{dim}\ top\ U_0$.

*(b)* $R_\delta/S_0 \oplus R_\gamma/S_0 \cong V_0 \oplus rad\ U_0$.

Proof: (a) We have an exact sequence $0 \to D \to \eta\Lambda \oplus \beta\Lambda \to \eta\Lambda + \beta\Lambda \cong \Omega S_0 \to 0$ where $D = \eta\Lambda \cap \beta\Lambda$. The hypothesis implies that $D = soc_2(V_0)$. Dually, there is an exact sequence $0 \to \Omega^{-1}S_0 \cong (\delta, \eta)\Lambda \to \delta\Lambda \oplus \gamma\Lambda \to D' \to 0$, and by the hypothesis, $D' \cong U_0/rad^2U_0$. Now $\underline{dim}\ \Omega S_0 = \underline{dim}\ \Omega^{-1}S_0$, and (a) follows.

(b) We find, using (a), that $\underline{dim}[R_\delta + R_\gamma] = \underline{dim}\ e_0\Lambda - \underline{dim}\ S_0 - \underline{dim}\ top\ U_0$. Therefore $(R_\delta + R_\gamma)/S_0$ is a maximal submodule of $H_0$ which is a direct sum. Since top $H_0$ is multiplicity-free and has length 2, $H_0$ has precisely two maximal submodules. One of them is indecomposable, and the other is $V_0 \oplus rad\ U_0$.

VIII.2.3 LEMMA *Suppose that the quiver of* $\Lambda$ *contains*

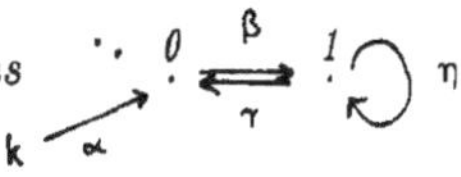

*and assume that* $R_\beta = \underset{\sim}{V}_1$ *(and* $\underset{\sim}{V}_1$ *is defined.) Then* $S_0$ *is non-periodic.*

Proof: Suppose that $S_0$ is periodic. If $\Omega^3S_0 \cong S_0$ then the modules at the end of the 3-tube are $S_0$, $\Omega^{\pm1}S_0$ and none of them can be isomorphic to $\underset{\sim}{V}_1$ (compare socles and tops, and $\underset{\sim}{V}_1$ is not simple.) So in this case $\underset{\sim}{V}_1$ must have period 4. On the other hand, $\underline{Hom}_\Lambda(S_0, \beta\Lambda) \neq 0$ but $\underline{Hom}_\Lambda(S_0, \underset{\sim}{V}_1) = 0$; and since $\underset{\sim}{V}_1 = R_\beta = \Omega(\beta\Lambda)$, this is a contradiction to IV.1.1. Hence $S_0$ can only have period 4. We will now show that

(*) $\beta\eta = 0 = \eta\gamma$.

Using IV.1.1 and the fact that $\underline{\mathrm{Hom}}_\Lambda(S_0, \beta\Lambda) \neq \underline{\mathrm{Hom}}_\Lambda(S_0, \underset{\sim}{V}_1)$ we see that $\underset{\sim}{V}_1$ (and $\beta\Lambda$) cannot have period 3. Hence there is an exact sequence
$0 \to \beta\Lambda \to P_0 \to Q_3 \to Q_2 \to P_1 \to \beta\Lambda \to 0$ where $Q_2$ and $Q_3$ are projective, with top $Q_2 \cong$ top $\underset{\sim}{V}_1$ (which is simple). By the exactness and the non-singularity of the Cartan matrix, $P_1 \oplus Q_3 \cong Q_2 \oplus P_0$ and $Q_2 \cong P_1$, therefore $\underset{\sim}{V}_1 \cong \eta\Lambda$, for some choice of $\eta$. In particular $\beta\eta = 0$, and by IV.1.11, $\eta\gamma = 0$.

We use the identifications of I.6.5 for $\Omega^2 S_0$ and $\Omega^{-2}S_0$. Then by (*) there is an embedding j: $(0, \eta)\Lambda \to \Omega^2 S_0$; and $(0, \eta)$ does not lie in rad $\Omega^2 S_0$. Dually, $\eta\gamma = 0$ implies that there is an epimorphism p: $\Omega^{-2}S_0 \to \eta\Lambda$, and p does not vanish on soc $\Omega^{-2}S_0$. Then p(soc $\Omega^{-2}S_0)e_1 \neq 0$; and since moreover the socle of $\Omega^{-2}S_0$ is multiplicity-free and $\Omega^2 S_0 \cong \Omega^{-2}S_0$, it follows that the composition $p \circ j$ is one-to-one and then an isomorphism. That is, $\Omega^2 S_0 \cong \eta\Lambda$. This is not possible, by IV.1.5.

<u>VIII.3</u> **THEOREM** *Suppose $\Lambda$ is a local algebra of semidihedral type. Then $\Lambda$ is isomorphic to one of the algebras in III.1(d) or (d').*

As we have noted in VIII.1.1, the converse also holds.

Proof: Since $\Lambda$ is local, tame and symmetric, $\Lambda$ must be isomorphic to one of the algebras in III.1. The algebras modulo the socle in (a) to (c') are special biserial, by III.2, so they do not have $\mathbb{Z}D_\infty$-components (II.7.1 and II.7.2) and are therefore excluded here.

Suppose $\Lambda$ is as in III.1(e) or (e'); then by V.4.6 the simple module has $\Omega$-period 4, and by Lemma VIII.2.1, $\Lambda$ is therefore not of semidihedral type.

Hence a local algebra of semidihedral type must be if the form as in III.1(d) or (d').

## VIII.4 *Algebras with two simple modules*

There are four possible quivers for a symmetric tame connected algebra with two simple modules, they are given in IV.2.2. We start by showing that two of them do not occur for algebras of semidihedral type.

<u>VIII.4.1</u> LEMMA *Assume that $\Lambda$ is of semidihedral type. Then the quiver of $\Lambda$ is not of the form (2C).*

Proof: Suppose that $\Lambda$ has such a quiver; we claim that the simple modules are not periodic. Assume this is not so; let S be simple and periodic, then the period of S is $\leq 4$, because $\Gamma_s(\Lambda)$ has only tubes of rank $\leq 3$ (and at most one 3-tube). Then by IV.1.4, $\Omega$-periods 1 or 2 are excluded, and it follows from IV.1.5 that the number of arrows $e \to f$ is the same as the number of arrows $f \to e$, a contradiction.

Let $H = \text{rad } P_0/S_0$, then H is indecomposable since soc H is simple. It follows that $\Omega^{-1}S_0$ and then $S_0$ both have only one predecessor in $\Gamma_s(\Lambda)$, and consequently $S_0$ can only lie at the end of a component of type $\mathbb{Z}D_\infty$. We have that $H \neq \text{rad } P_1/S_1$, for example since the socle of rad $P_1/S_1$ is not simple. Thus, by IV.3.3, there is an AR-sequence $0 \to V \to H \to U \to 0$. It follows from IV.3.4 that U is simple and soc $H \cong$ soc V; and clearly $U \cong S_1$. Now we conclude that top $\Omega U \cong \text{top}(\text{rad } P_1) \cong S_0$. On the other hand, top $\Omega U \cong \text{soc } \Omega^2 U \cong \text{soc } V \cong \text{soc } H \cong S_1$, a contradiction.

<u>VIII.4.2</u> LEMMA *Suppose that $\Lambda$ is of semidihedral type, then the quiver of $\Lambda$ is not of the form (2D).*

Proof: Suppose that the quiver of $\Lambda$ is 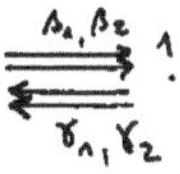

. We claim that the simple modules are not periodic:

Suppose $S_0$ is periodic; then the period of $S_0$ must be $\leq 4$ by the hypothesis on $\Gamma_s(\Lambda)$. The period is not 1 or 2, by IV.1.4, and then it follows from IV.1.6 and IV.1.7 that no path of length 2 starting and ending at vertex 0 is involved in a relation. Consider the algebra $\Lambda/I$ where I is the ideal $J^3 + e_1 J^2$; then $\Lambda/I$ may be generated by zero-relations and has a universal cover, as in I.10.4. This contains the tree without relations of the form $\tilde{E}_7$, namely

and $\Lambda$ is wild (I.9.5, I.10.7, I.4.7). Moreover, rad $P_0/S_0 \neq$ rad $P_1/S_1$, from the quiver. Thus either IV.3.3(a) or IV.3.1 holds; and in any case, the modules $\underset{\sim}{U}_i$ and $\underset{\sim}{V}_i$ are defined. Note that, in case IV.3.3(a), we must have that soc $H_i \cong$ soc $U_i \oplus$ soc $V_i$ and top $H_i \cong$ top $U_i \oplus$ top $V_i$, see IV.3.6. By IV.5.5 the modules $\underset{\sim}{U}_i$ and $\underset{\sim}{V}_i$ must lie at ends of tubes.

(1) $\underline{\Omega^4\underset{\sim}{U}_i \neq \underset{\sim}{U}_i}$: Assume for contradiction that the period of $\underset{\sim}{U}_0$ divides 4. Then there is an exact sequence $0 \to \underset{\sim}{U}_0 \to P_1 \to Q \to P_1 \to P_0 \to \underset{\sim}{U}_0 \to 0$ where $Q$ is projective. Then $\underline{\dim}\ P_0 \oplus Q = \underline{\dim}\ P_1 \oplus P_1$, and since the Cartan matrix is non-singular, it follows that $P_0 \oplus Q \cong P_1 \oplus P_1$; this is not possible.

Hence the modules at the end of the 3-tube are $\underset{\sim}{U}_0$, $\underset{\sim}{V}_0$ and $X = \Omega\underset{\sim}{V}_0 \cong \Omega^{-1}\underset{\sim}{U}_0$. Now, $\underset{\sim}{U}_1$ also lies at the end of the 3-tube. Comparing socles and tops we see that $\underset{\sim}{U}_1 \cong \underset{\sim}{V}_0$. It follows that $\underset{\sim}{V}_1 \cong X$ and $S_0 \cong$ top $V_1 \cong$ top $X \cong$ soc $\underset{\sim}{U}_0 \cong S_1$, a contradiction.

<u>VIII.4.3</u> LEMMA *Let $\Lambda$ be of semidihedral type, with quiver of the form (2A). Then the simple module $S_1$ is periodic, and $S_0$ lies at the end of a $\mathbb{Z}D_\infty$-component, with one predecessor. Moreover, $U_0$ and $V_0$ are defined.*

Proof: We know from IV.3.7 that $S_1$ is periodic. Clearly $H_0 \neq H_1$, and therefore by VIII.2.1, $S_0$ is non-periodic.

Suppose the statement for $S_0$ does not hold; then by VIII.2.0.1 we have that $H_0 \cong U_0 \oplus V_0 = \alpha\Lambda/S_0 \oplus \beta\Lambda/S_0$. Then $\Lambda$ satisfies trivially the hypothesis VI.1.0, and we may apply the results of VI.1. In particular, $H_0 \cong R_\alpha/S_0 \oplus R_\gamma/S_0$, and $\Lambda$ has a permutation $\pi$ as defined in VI.1.2. There are two possibilities for $\pi$, namely $(\alpha)(\beta\ e_1\ \gamma)$ and $(\alpha\ \beta\ e_1\ \gamma)$. In the first case we know from VI.8.1 that $\Lambda/\mathrm{soc}\ \Lambda$ is special biserial, and $\Lambda$ is not of semidihedral type. Consider the other possibility for $\pi$; then $\pi^*$ (see VI.1.3) is of the form $(\alpha)(\beta\ \gamma)$. By V.1.4.5 we have then that $\alpha\beta = 0$ and $\gamma\alpha = 0$ and hence $\Lambda$ is special biserial and not of semidihedral type, a contradiction.

It follows that $S_0$ must lie in a $\mathbb{Z}D_\infty$-component with one predecessor. As we have seen, $H_0 \neq H_1$, and therefore $U_0$ and $V_0$ are defined.

VIII.4.4 PROPOSITION *Let* $\Lambda$ *be of semidihedral type with quiver (2A). Assume that* $S_1$ *has period 3. Then* $\Lambda \cong KQ/I$ *where* $I$ *is the ideal defined by the relations*

$$\gamma\beta = 0, \qquad (\alpha\beta\gamma)^k = (\beta\gamma\alpha)^k$$
$$\alpha^2 = (\beta\gamma\alpha)^{k-1}\beta\gamma + c(\alpha\beta\gamma)^k$$

*where* $k \geq 2$ *and* $c \in K$. *That is,* $\Lambda$ *belongs to the family* $SD(2A)_2$.

Proof: We may take $\gamma\Lambda = \Omega S_1$ and $R_\gamma = \beta\Lambda$, see I.6.4 and thus $\gamma\beta = 0$. By VIII.2.2 we have that $V_0 \oplus \mathrm{rad}\, U_0 \cong R_\alpha/S_0 \oplus R_\gamma/S_0$. Since $R_\gamma = \beta\Lambda \not\subseteq \mathrm{rad}^2(e_0\Lambda)$ we deduce $R_\gamma/S_0 \cong V_0$ and hence

(1) $\underset{\sim}{V}_0 = \beta\Lambda$ and $U_0 \cong \alpha\Lambda/\alpha\Lambda \cap \beta\Lambda$; and $R_\alpha + R_\gamma = \beta\Lambda + \alpha J$.

Moreover, $U_0 = \Omega^{-1}\underset{\sim}{V}_0 \cong \Omega^{-1}(\beta\Lambda) \cong \gamma\Lambda$. In particular, $\underline{\dim}\, U_0 = \underline{\dim}\, V_0$ and soc $U_0 \cong S_1$, soc $V_0 = S_0$. We take the projective cover $\pi\colon P_0 \to U_0$ given by $\pi(x) = \alpha x + \alpha\Lambda \cap \beta\Lambda$ and identify ker $\pi$ with $\Omega U_0$; them $R_\alpha \subsetneqq \Omega U_0$.

(2) $\alpha\beta$ is not involved in any relation: Suppose this is false, then $\alpha\beta = \alpha x + \beta y$ for $x, y \in J^2$, and we deduce $\pi(\beta - x) = 0$ and $(\beta - x)\Lambda \subset \Omega U_0 - \mathrm{rad}\, \Omega U_0$. So $S_1$ occurs in top $\Omega U_0$. This is a contradiction since top $\Omega U_0 \cong$ soc $V_0 \cong S_0$.

(3) There is a relation involving $\alpha^2$: Suppose not; then $\Omega U_0 \subseteq \mathrm{rad}^2(e_0\Lambda)$ and it follows from (2) that $R_\alpha \subseteq \Omega U_0 \subseteq R_\gamma + R_\alpha$. We deduce that $R_\alpha/S_0 \subseteq \Omega U_0/S_0 \subseteq R_\gamma/S_0 \oplus R_\alpha/S_0$. That is, $\Omega U_0/S_0$ is a submodule of a direct sum which contains one of the summands. On the other hand, top $\Omega U_0/S_0$ is simple. So it follows that $\Omega U_0/S_0 = R_\alpha/S_0$ and $\Omega U_0 = R_\alpha$, and $U_0 = \alpha\Lambda$, a contradiction to (1).

[Note that we did not use "tame" here.]

By (3) we may write

(*) $\alpha^2 = c(\beta\gamma) + \alpha x + \beta y$ where $c \in K$ and $x, y \in J^2$. Therefore, since $\gamma\beta = 0$, we have $\gamma\alpha^2 \in J^4$ and $\gamma\alpha^2\beta \in J^5$. We deduce that the local algebra $\Lambda_1 = e_1\Lambda e_1$ is uniserial and its radical is generated by $\gamma\alpha\beta$. Consequently soc $e_1\Lambda$ = soc $\Lambda_1$ = $\langle (\gamma\alpha\beta)^k \rangle$ for some $k \geq 1$ ( $k = 0$ is impossible since dim $e_1\Lambda e_1 \geq 2$). In fact, it follows that $e_1\Lambda$ is a uniserial $\Lambda$-module.

Now we will study the local algebra $\Lambda_0 = e_0\Lambda e_0$; we have that rad $\Lambda_0 = \langle \alpha, \beta\gamma \rangle$.

Case 1 rad $\Lambda_0$ is cyclic. Then it is generated by $\alpha$ and is also commutative. We deduce that $\beta\gamma = \alpha^u v$ where $u \geq 2$ and $v \in K[\alpha]$ (either a unit or 0). Since $S_0$ is of type d, we deduce $\beta\gamma \in \mathrm{soc}_2\Lambda$. On the other hand, $\beta\gamma \notin$ soc $\Lambda$ since otherwise $S_1$ would be a summand of $H_0$; but $H_0$ is indecomposable, by VIII.4.3.

Consequently $\beta\Lambda$ = span { $\beta$, $\beta\gamma$, $\beta\gamma\alpha$ }; and v is a unit in $K[\alpha]$. By (3) we must have u = 2 and we deduce easily the stated relations with k = 1.

Case 2 rad $\Lambda_0$ is not cyclic; then we use III.16. By (3) and since $(\beta\gamma)^2 = 0$, $\Lambda_0$ is isomorphic to an algebra in III.1(d) or (d'). By III.16, we may choose $\alpha$ such that soc $\Lambda_0 = \langle (\alpha\beta\gamma)^k \rangle$ and $(\alpha\beta\gamma)^k = (\beta\gamma\alpha)^k$; and moreover $\alpha^2 = b(\beta\gamma\alpha)^{k-1}\beta\gamma + c(\alpha\beta\gamma)^k$, for b, c $\in$ K. Since $H_0$ is indecomposable, we must have that $b \neq 0$ and then without loss of generality b = 1.

Suppose k = 1; then we have trivially $\Omega V_0 \cong U_0$; but also $V_0 \cong \Omega^2 U_0$ and $U_0$ is periodic. Therefore the component of $U_0$ is $\mathbb{Z}D_n/\langle \tau^3 \rangle$; and by Auslander's theorem (I.8.3), $\Lambda$ is of finite type and is excluded here. This completes the proof of VIII.4.4

We note that in case char K $\neq$ 2 we may assume c = 0; and otherwise c = 0 or 1. This can be seen by the methods in VI.

VIII.4.5 THEOREM *Assume that* $\Lambda$ *is of semidihedral type, with quiver (2A), and that* $S_1$ *has period 4. Then* $\Lambda \cong KQ/I$ *where I is the ideal defined by*

$$\alpha^2 = c(\alpha\beta\gamma)^k, \qquad (\alpha\beta\gamma)^k\alpha = 0,$$
$$\beta\gamma\beta = (\alpha\beta\gamma)^{k-1}\alpha\beta \quad \text{and} \quad \gamma\beta\gamma = (\gamma\alpha\beta)^{k-1}\gamma\alpha,$$

*where* $k \geq 2$ *and* $c \in K$. *That is,* $\Lambda$ *belongs to the family* $SD(2A)_1$.

Proof: We choose arrows according to VIII.2.0.

(1) $R_\alpha/S_0 \oplus R_\gamma/S_0 \cong V_0 \oplus \text{rad } U_0, \text{top } U_0 \cong \text{soc } V_0$ and $\underline{\dim}\, U_0 = \underline{\dim}\, V_0$: The first statement is VIII.2.2(a). Moreover, $\underline{\dim}\, \beta\Lambda = \underline{\dim}\, \gamma\Lambda$ (since $\beta\Lambda \cong \Omega^{-1}S_1$ and $\gamma\Lambda \cong \Omega S_1$); therefore by VIII.2.2(b) we have top $U_0 \cong$ soc $V_0$; then the last part follows from the fact that $\Omega^2 U_0 \cong V_0$.

(2) $U_0 \cong \beta\Lambda/\beta\Lambda \cap \alpha\Lambda$ and $V_0 \cong \alpha\Lambda$: Suppose not, then , without loss of generality, $\beta\Lambda \cong V_0 \cong \Omega^{-1}S_1$; moreover top $U_0 \cong S_0 \cong$ soc $V_0$, and hence soc $U_0 \cong S_1$. Since $V_0 \cong \Omega^{-1}S_1$ it follows that $U_0 \cong \Omega^{-2}S_1$ and soc $U_0 \cong S_0$ (IV.1.5), a contradiction.

(3) $\alpha\Lambda/S_0 \cong R_\alpha/S_0$: If not, then by the first part of (1), $R_\alpha/S_0 \cong$ rad $U_0$ and $R_\gamma/S_0 \cong V_0$; and then $\underline{\dim}\, R_\alpha/S_0 < \underline{\dim}\, R_\gamma/S_0 = \underline{\dim}\, \alpha\Lambda/S_0$. On the other hand, $\underline{\dim}\, R_\alpha = \underline{\dim}\, P_0$

- $\underline{\dim}$ $\alpha\Lambda$ = $\underline{\dim}$ $P_0$ - $\underline{\dim}$ $\underset{\sim}{V}_0$ = $\underline{\dim}$ $U_0$, which is by (1) equal to $\underline{\dim}$ $V_0$ = $\underline{\dim}$ $\alpha\Lambda$; this is a contradiction.

(4) <u>We may assume that $\alpha^2 \in$ soc $\Lambda$:</u> Let $\psi$: $R_\alpha \to \alpha\Lambda/S_0$ be an epimorphism such that ker $\psi$ = soc$(e_0\Lambda)$. There is a generator $\alpha' \in R_\alpha$ satisfying $\psi(\alpha') = \alpha + S_0$. By definition of $R_\alpha$, we have that $\alpha\alpha' = 0$, and consequently $\psi[(\alpha')^2] = [\psi(\alpha')]\alpha' = [\alpha + S_0]\alpha' = \alpha\alpha' + S_0 = 0$ and $(\alpha')^2 \in$ soc $(e_0\Lambda)$. Since $\alpha\Lambda/S_0 \cong \alpha'\Lambda/S_0$, $\alpha'$ must be a generator. So we may replace $\alpha$ by $\alpha'$. (Actually also $\alpha'\Lambda/S_0 \cong V_0$.)

(5) <u>$\alpha\beta$ does not lie in soc$_2\Lambda$:</u> Suppose $\alpha\beta \in$ soc$_2\Lambda$; then by (4), we have that $\underset{\sim}{V}_0 = \alpha\Lambda = \mathcal{U}(S_0, S_1, S_0)$, and by (1), $\underline{\dim}$ $U_0 = [1, 1]$, and conseqently $U_0 = \mathcal{U}(S_1, S_0)$. This implies (by IV.3.4) that $\beta\gamma \in$ soc$_2e_0\Lambda$, and consequently $H_0$ is a direct sum. This is a contradiction to VIII.4.3.

(6) <u>The elements $\gamma\beta$, $\alpha\beta$ and $\gamma\alpha$ are independent, and there are relations involving $\gamma\beta\gamma$ and $\beta\gamma\beta$:</u> By IV.1.6, we know that $\gamma\beta$ is independent. Suppose there is a relation involving $\alpha\beta$; then there are x, y $\in J^2$ with $\alpha\beta = \alpha x + \beta y$ for $x \in e_0J^2e_0$ and $y \in e_1J^2e_0$. By (4) we may assume that $x = \beta x_1$, and then $\alpha\beta(1 - x_1) = \beta y \in$ soc$_2\Lambda$. Since $(1-x_1)$ is a unit, we deduce $\alpha\beta \in$ soc$_2\Lambda$, a contradiction to (5).

Suppose there is a relation involving $\gamma\alpha$, then $\gamma\alpha = \gamma x$ for $x \in e_0J^2e_0$, and $\alpha' = \alpha - x \in R_\gamma$ - rad $R_\gamma$. By IV.1.5 we know that top $R_\gamma$ is simple, consequently $\alpha'\Lambda = R_\gamma$. However, $\Omega(\alpha'\Lambda) = \Omega^3(S_1) = \beta\Lambda$ and $\alpha'\beta = 0$, therefore $\alpha\beta = x\beta$, which contradicts the fact just proved. The remaining part of (6) follows by the argument (4) in the proof of VII.7.1.

As in VII.7.1, part (5), we have that the local algebra $e_1\Lambda e_1$ is not uniseria, but is commutative. Moreover the argument of (6) in VII.7.1 shows that $\gamma\beta\gamma \in$ soc$_2(e_1\Lambda)$ - soc $e_1\Lambda$. We deduce $\langle(\gamma\beta)^2\rangle$ = soc $e_1\Lambda$ and $\gamma\beta\gamma\alpha = 0$. Moreover $e_0\Lambda e_0 = \langle \alpha, \beta\gamma \rangle$ with $\alpha^2$ and $(\beta\gamma)^2$ lying in the socle. Using now III.16 it is easy to determine the relations for $\Lambda$. (Concerning the relation for $\beta\gamma\beta$, note that $H_0$ is indecomposable). Moreover, $k \geq 2$ since otherwise $\alpha\beta$ would lie in soc$_2\Lambda$, which is not possible, by (5).

Now we shall study algebras whose quiver is of the form (2*B*). At least one of the simple $\Lambda$-modules is non-periodic, by VIII.2.1. We consider first algebras where a simple module, $S_1$ say, has two predecessors in $\Gamma_s(\Lambda)$, Then we have the following result:

VIII.4.6 **PROPOSITION** *Let $\Lambda$ be of semidihedral type, with quiver of the form (2B). Assume that rad $P_1/S_1 \cong U_1 \oplus V_1$ where $U_1$ and $V_1$ are indecomposable and non-zero. Then $\Lambda \cong KQ/I$ where I is defined by the relations*

$$\gamma\beta = 0 = \eta\gamma = \beta\eta, \qquad \alpha^2 = (\beta\gamma\alpha)^{k-1}\beta\gamma + c(\beta\gamma\alpha)^k,$$
$$\eta^t = (\gamma\alpha\beta)^k, \qquad (\alpha\beta\gamma)^k = (\beta\gamma\alpha)^k,$$

*where $k \geq 1$, $t \geq 2$ and $c \in K$. That is, $\Lambda$ belongs to the family $SD(2B)_1$.*

Proof: We choose the notation such that $\underset{\sim}{V}_1 = \eta\Lambda$ and $U_1 = \gamma\Lambda$; these lie at ends of tubes, by IV.6.2. Then the module $R_\beta$ is generated by an arrow, and hence $S_0$ is non-periodic, by VIII.2.3.

(1) $H_0$ must be indecomposable: Otherwise $\Lambda$ satisfies VI.1.0 and has permutations $\pi$, $\pi^*$ as defined in VI.1.2, VI.1.3. Since C is non-singular, we know by VI.1.5.3 that $\pi^*$ is not a 4-cycle, hence $\pi^* = (\alpha)(\beta\ \eta\ \gamma)$ or $(\alpha)(\beta\ \gamma)(\eta)$. In the first case $\Lambda$ is the algebra determined in VI.8.2, and $\Lambda$/soc $\Lambda$ is special biserial; and in the second case $\Lambda$ is special biserial, by IV.4.5. In any case, $\Gamma_s(\Lambda)$ does not have $\mathbb{Z}D_\infty$-components (II.6.4), and $\Lambda$ is not of semidihedral type.

It follows now that $S_0$ satisfies IV.3.3(a). Moreover, $H_0 \neq H_1$, so $U_0$ and $V_0$ are defined. Note also that, by VIII.2.2 applied twice, we see that $\underline{\dim}\ \beta\Lambda = \underline{\dim}\ \gamma\Lambda$ and also $\underline{\dim}\ \beta\Lambda$ - $\underline{\dim}$ soc $V_0$ = $\underline{\dim}\ \gamma\Lambda$ - $\underline{\dim}$ top $U_0$. Therefore top $U_0 \cong$ soc $V_0$.

(2) $R_\eta \cong \gamma\Lambda$ and then $R_\beta \cong \eta\Lambda$ and $\eta\gamma = 0 = \beta\eta$: Suppose not, then $R_\beta \cong \gamma\Lambda$; and this module lies in a tube (IV.6.2). If its $\Omega$-period is 4 then there is an exact sequence $0 \to \gamma\Lambda \to P_1 \to P_0 \to Q \to P_0 \to \gamma\Lambda \to 0$ where Q is projective, and the Cartan matrix is singular. So the period can only be 3, then top $\Omega(\gamma\Lambda) \cong$ soc $\Omega(\gamma\Lambda) \cong S_0$. By VII.5.4(a), $\Omega(\gamma\Lambda)$ is uniserial; taking both facts together,it follows that $\Omega(\gamma\Lambda) = \alpha'\Lambda$ where $\alpha' \equiv \alpha \bmod J^2$. Then $\gamma\alpha' = 0$, and $H_0$ is decomposable, by IV.1.11; a contradiction to (1).

From (2) we also see that $\eta\Lambda$ is spanned by $\eta$, $\eta^2$, ..., , consequently $\underset{\sim}{V}_1 \cong \tilde{V}_1$.

(3) $\underline{V_0 \cong \beta\Lambda}$: Suppose not, then $\beta\Lambda = \langle U_0 \rangle$. It follows that $\eta\Lambda = \Omega(\beta\Lambda) \subsetneq \Omega U_0 = \{x: \beta x \in \alpha\Lambda\}$. Since $\eta$ is an arrow, $\eta\Lambda \not\subseteq \mathrm{rad}\ \Omega U_0$, and then top $\Omega U_0$ is not simple, a contradiction.

(4) $\underline{\beta\Lambda/S_0 \cong R_\gamma/S_0}$: We deduce from (3) that $U_0 \cong \alpha\Lambda/\alpha\Lambda \cap \beta\Lambda$. Therefore top $U_0 = S_0$ and then soc $V_0 \cong S_0 =$ top $\Omega U_0$. It follows that $\alpha\beta$ is not involved in any relation, and $\beta\Lambda/S_0 \neq R_\alpha/S_0$. Hence (4) follows.

(5) Without loss of generality, $\underline{\beta\gamma = 0}$: Given $\gamma$, we may choose $\beta$ such that $\beta$ is an arrow $0 \to 1$ and $\beta\gamma = 0$.

Then it is still true that $\beta\eta = 0$, theis follows from the fact that $e_0(\eta\Lambda) = 0$. Now it is straightforward to determine the relations. We omit details.

Now we assume that no simple module of $\Lambda$ has two predecessors in $\Gamma_s(\Lambda)$. By VIII.2.2, $S_0$ (say) is non-periodic; and by VIII.2.0.1 there are two possibilities for $S_0$ to be considered.

VIII.4.7 PROPOSITION *Let $\Lambda$ be of semidihedral type with quiver of the form (2B). Assume that $S_0$ satisfies IV.3.3(a) and that $H_1$ is indecomposable. Then $\Lambda \cong KQ/I$ where I is defined by the relations*

$$\gamma\beta = \eta^{t-1}, \quad \beta\eta = (\alpha\beta\gamma)^{k-1}\alpha\beta, \quad \eta\gamma = (\gamma\alpha\beta)^{k-1}\gamma\alpha, \quad \alpha^2 = c(\alpha\beta\gamma)^k,$$
$$\beta\eta^2 = 0 = \eta^2\gamma,$$

*where $k \geq 1$, $t \geq 3$ and $c \in K$. That is, $\Lambda$ belongs to family $SD(2B)_2$.*

Proof: The hypothesis implies that there is an AR-sequence $0 \to V_0 \to H_0 \to U_0 \to 0$. Moreover, by VIII.2.2, $V_0 \oplus \mathrm{rad}\ U_0 \cong R_\alpha/S_0 \oplus R_\gamma/S_0$. The argument in VIII.4.5 shows that $V_0 \cong R_\alpha/S_0$ and $V_0 \cong \alpha\Lambda$. Thus without loss of generality, $\alpha^2$ lies in soc $\Lambda$. Moreover $U_0 \cong \alpha\Lambda$ and soc $U_0 \cong S_0$. Then soc $V_0 \cong S_1$, and hence top $\Omega U_0 \cong S_1$. This implies now that $\beta\gamma$ does not occur in a relation.

(1) There is an epimorphism $\underline{\pi: \gamma\Lambda \to \mathrm{rad}\ U_0}$: Since $\Lambda$ is tame, we deduce from the above that $\beta\eta$ is involved in some relation, hence rad $U_0 \cong \beta\gamma\Lambda/\beta\Lambda \cap \alpha\Lambda$, and (1) follows.

(2) $\underline{S_1}$ does not satisfy IV.3.2: Suppose it does, then the above arguments also apply to $S_1$; in particular $U_1 \cong \gamma\Lambda/\gamma\Lambda \cap \eta\Lambda$, and there is an epimorphism $\pi_1: \beta\Lambda/S_0 \to$ rad

$U_1$. Note that $\gamma\Lambda = \langle U_1 \rangle$ and $\beta\Lambda = \langle U_0 \rangle$. Hence we have for the composition factors $\underline{\dim}\, U_0 = \underline{\dim}\, S_1 + \underline{\dim}$ rad $U_0 \geq \underline{\dim}\, S_1 + \underline{\dim}\, \gamma\Lambda/S_1 = \underline{\dim}\, S_1 + \underline{\dim}\, [U_1 + \text{soc } V_1]$, and similarly $\underline{\dim}\, U_1 \geq \underline{\dim}\, S_0 + \underline{\dim}\, [U_0 + \text{soc } V_0]$. This implies $|U_0| > |U_0|$, a contradiction.

Hence $S_1$ must be periodic.

(3) $\underline{\beta\eta \in \text{soc}_2(\alpha\Lambda) \text{ and } \langle \beta\eta\gamma \rangle = \text{soc}(e_0\Lambda), \text{ also } \alpha\beta\eta = 0}$: Using (1), we may assume that $\Omega U_0 = \eta\Lambda$, in particular $\beta\eta \in \text{soc}_2(\alpha\Lambda) - \text{soc}(\alpha\Lambda)$. It follows that $\beta\eta^2 = 0$ and then also $\eta^2\gamma = 0$; moreover $\langle \beta\eta\gamma \rangle = \text{soc}(e_0\Lambda)$. Then $\alpha\beta\eta = 0$ as well.

(4) $\underline{\eta\gamma \in \text{soc}_2(\Lambda) \text{ and } \langle \eta\gamma\beta \rangle = \text{soc}(e_1\Lambda)}$: By (3) we have that $J(\eta\gamma) = \text{soc } e_1\Lambda$ and therefore $\eta\gamma \in \text{soc}_2\Lambda$. [ Note: this shows also that $S_1$ must have period 4.]
Then $\eta\gamma J = \text{soc}(e_1\Lambda)$, and (4) follows.

(5) $\underline{\langle \gamma\beta\eta \rangle = \text{soc}(e_1\Lambda)}$: We have that $(\gamma\beta\eta)J = \langle \gamma\beta\eta^2, \gamma\beta\eta\gamma \rangle = 0$. Hence $\gamma\beta\eta$ lies in soc $\Lambda$. Let $\psi$ be a symmetrizing linear form. Then $\psi[\gamma\beta\eta] = \psi[\beta\eta\gamma] \neq 0$ and therefore $\gamma\beta\eta \neq 0$.

(6) $\underline{\gamma\beta \text{ is involved in some relation}}$: Suppose not, then since $\Lambda$ is tame we must have $\gamma\alpha = \gamma x + \eta y$ where $x \in J^2$ and $y \in J$. We know that $\eta\Lambda e_0 = \langle \eta\gamma \rangle$ and $\eta\gamma = \gamma z$ as in (4). So if $\alpha' = \alpha - x - z$ then $\alpha \equiv \alpha' \pmod{J^2}$, and $\alpha'\gamma = 0$. Thus $H_0$ is a direct sum, by IV.1.11, a contradiction.

(7) $\underline{\gamma\beta \text{ lies in } \gamma J^2 + \text{soc}_2\Lambda}$: By (6) we have that $\gamma\beta = \gamma x + \eta y$ where $x \in J^2$ and $y \in J$. Clearly $\beta\eta y \in \text{soc } \Lambda$, so we are done if we can show that $\eta(\eta y) \in \text{soc } \Lambda$. We may write $\gamma x = \gamma\alpha x_1 + \gamma\beta x_2$ for $x_i \in J$. Then $\eta^2 y = \eta\gamma\beta - \eta\gamma\alpha x_1 - \eta\gamma\beta x_2$ and $\eta\gamma\alpha = 0$, $\eta\gamma\beta \in \text{soc } \Lambda$.

It is now straightforward to determine the relations.

<u>VIII.4.8</u> **PROPOSITION** *Suppose that $\Lambda$ is of semidihedral type, with quiver (2B). Assume that $H_0 \cong H_1$. Then $\Lambda \cong KQ/I$ where $I$ is generated by the following relations:*

$$\beta\eta^{s-1}\gamma = q(0)\alpha^{s+1} \quad \text{and} \quad \gamma\alpha^{s-1}\beta = p(0)\eta^{s+1}$$

$$\alpha^s\beta = 0 = \gamma\alpha^s = \alpha^{s+2} = \eta^{s+2} \quad \text{and} \quad \eta^s\gamma = 0 = \beta\eta^s$$

*and, modulo socle,*

$$\beta\gamma = \alpha^2 p(\alpha)$$

$$\eta\gamma = \gamma\alpha p(\alpha)$$

$$\alpha\beta = \beta\eta q(\eta)$$

$$\gamma\beta = \eta^2 q(\eta)$$

*where p and q are polynomials over K with non-zero constant term, and $s \geq 2$.*

Proof: This situation can only occur when IV.3.3(b) holds. In particular, $H_0$ is then indecomposable. Assume that $\varphi$: $H_0 \to H_1$ is an isomorphism. Choose arrows $\alpha$, $\beta$ and $\gamma$, $\eta$ such that $\varphi(\alpha) = \gamma$ and $\varphi(\beta) = \eta$ (modulo socle).

(1) <u>$\beta\gamma = c\alpha^2 + x$ for some $c \in K$ and $x \in e_0J^2$:</u> Assume this is not so. Then there is also no relation of the form $\eta\gamma = c(\gamma\alpha) + x$. Then the universal cover of a factor algebra of $\Lambda/J^3$ contains the tree

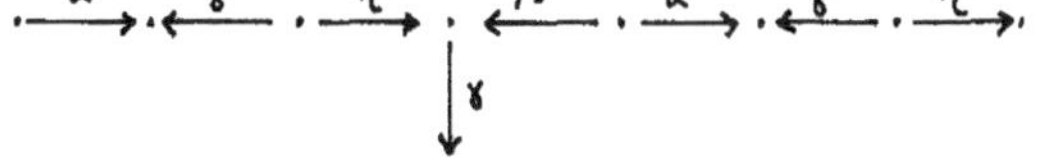


and $\Lambda$ is wild (I.9.5, I.4.7). Hence we may write

(*) $\beta\gamma = c\alpha^2 + \alpha x_1 + \beta x_2$ ($c \in K$ and $x_1, x_2 \in J$)

$\eta\gamma = c\gamma\alpha + \gamma x_1 + \eta x_2 + w$ ($w \in \operatorname{soc} e_1\Lambda$).

Similarly one obtains relations

(**) $\gamma\beta = d\eta^2 + \gamma y_1 + \eta y_2$ ($d \in K$ and $y_1, y_2 \in J$)

$\alpha\beta = d\beta\eta + \alpha y_1 + \beta y_2 + w'$ ($w' \in \operatorname{soc} e_0\Lambda$).

We see now that $e_0J^2$ is generated by $\alpha^2$ and $\beta\eta$, and that $e_1J^2$ is generated by $\gamma\alpha$ and $\eta^2$.

(2) <u>$\eta^2$ does not lie in $J^3$:</u> Suppose not; then $\beta\eta$ lies in $J^3$ as well, and $e_0J^2 = \alpha^2\Lambda$. Let k be maximal such that $\alpha^k \neq 0$. Then $e_0\Lambda$ has a K-basis $\{e_0, \alpha, \alpha^2, \ldots, \alpha^k, \beta\}$. In particular, $e_0\Lambda e_1$ is only 1-dimensional. By the shape of the quiver, $S_1$ occurs in $(\operatorname{soc}_2 e_0\Lambda)/S_0$ but also in $(\operatorname{rad} e_0\Lambda)/(\operatorname{rad}^2 e_0\Lambda)$. This is only possible if $H_0$ has a direct summand $\cong S_1$. This is not so, by IV.3.3(b).

Similarly one proves that $\alpha^2$ does not lie in $J^3$ and then $\gamma\alpha \notin J^3$ as well.

The rest follows using $\varphi$.

## VIII.5 *Simple modules with 3 predecessors and vertices with 3 arrows*

First we will determine algebras of semidihedral type where a simple module has three predecessors in $\Gamma_s(\Lambda)$. We will see that there is one family of algebras, and the quiver is of the form $(3\mathcal{C})$. (After that we assume always that simple modules have at most two predecessors.)

Then we will prove, in VIII.5.2, that given any algebra of semidihedral type with quiver $(3\mathcal{C})$, the projectives for end vertices have isomorphic hearts.

<u>VIII.5.1</u> **PROPOSITION** *Let $\Lambda$ be basic of semidihedral type; and assume that there is a simple $\Lambda$-module which has three predecessors in $\Gamma_s(\Lambda)$. Then $\Lambda \cong KQ/I$ where $Q$ is of the form $(3\mathcal{C})$, and $I$ is defined by the following relations:*

$$\beta\delta = 0 = \beta\rho = \rho\gamma \quad \textit{and} \quad \eta\gamma = 0 = \eta\rho = \rho\delta,$$
$$\rho^s = \gamma\beta = \delta\eta$$

*where $s \geq 3$.*

We note that we do not use that $\Lambda$ is tame.

Proof: Suppose $S_0$ is simple with three predecessors in $\Gamma_s(\Lambda)$. By IV.3.2 there are two simple modules $S_1$ and $S_2$ such that $P_i = \mathcal{U}(S_i, S_0, S_i)$ for $i = 1, 2$. Moreover $\text{rad}\, P_0/S_0 \cong S_1 \oplus S_2 \oplus X$ where $X$ is non-zero and $X$ has two predecessors in $\Gamma_s(\Lambda)$. Hence there is a module $\underset{\sim}{X}$ which is an extension of $S_0$ by $X$, with socle $\cong S_0$. By IV.5.5, $\underset{\sim}{X}$ must lie at the end of some component.

(1) <u>$\underset{\sim}{X}$ is not of type d:</u> We have that $\tau^{-1}\underset{\sim}{X}$ has Loewy length 2, with socle $\cong S_0$ and top $\cong S_1 \oplus S_2$. Suppose that $\underset{\sim}{X}$ is of type d, then the stable AR-quiver contains

$$\begin{array}{ccccc} \underset{\sim}{X} & & & & \tau^{-1}\underset{\sim}{X} \\ & \searrow & & \overset{\lambda}{\nearrow} & \\ \tau R & \longrightarrow & M & \longrightarrow & R. \end{array}$$

(1.a) <u>There is no projective attached to M:</u> By the hypothesis, $\underset{\sim}{X} \neq \text{rad}\, Q/\text{soc}\, Q$ for $Q$ projective. Moreover, $\underset{\sim}{X}$ is tall (IV.5.4), and (1.a) follows from IV.5.3.

(1.b) <u>Suppose $\underset{\sim}{X} \cap \tau R \neq 0$.</u> Then $\text{soc}\, \underset{\sim}{X} \subseteq \tau R$, and $M/S \cong X \oplus \tau^{-1}\underset{\sim}{X} \cong \tau R/S \oplus R$. By the hypothesis, $X$ is indecomposable and therefore $M/S$ has two summands. It follows now that $X \cong R$, which is a contradiction since $X$ has two predecessors in $\Gamma_s(\Lambda)$.

Hence we must have that $\underset{\sim}{X} \cap \tau R = 0$. Then $\lambda$ induces an inclusion of $\tau R$ into $\tau^{-1}\underset{\sim}{X}$. But $\tau^{-1}\underset{\sim}{X}$ has only three proper non-zero submodules, which are $S$, $\Omega^{-1}S_1$ and $\Omega^{-1}S_2$. Since $S$ has three predecessors but $\tau R$ has one, $\tau R \neq S$. Suppose $\tau R \cong \Omega^{-1}S_1$ (say), then the other module at the end must be $\Omega^{-1}S_2$, and $\underset{\sim}{X} \cong \Omega^{-1}S_2$, a contradiction.

It follows that $\underset{\sim}{X}$ must lie at the end of a tube.

(2) $\underset{\sim}{X}$ <u>does not have period dividing 4:</u> Otherwise there is an exact sequence $0 \to \tau^{-1}\underset{\sim}{X} \to P_0 \to Q \to P_0 \to P_1 \oplus P_2 \to \tau^{-1}\underset{\sim}{X} \to 0$ where $Q$ is projective. By exactness, $\underline{\dim}\, P_1 + \underline{\dim}\, P_2 + \underline{\dim}\, Q = 2\underline{\dim}\, P_0$, and the Cartan matrix is singular.

Hence $\underset{\sim}{X}$ must lie at the end of the 3-tube, and consequently $\Omega\underset{\sim}{X} \cong \Omega^{-2}\underset{\sim}{X}$, and top $\underset{\sim}{X}$ = soc $\Omega^{-2}\underset{\sim}{X} = S_0$.

Dually, $\tilde{X}$ lies at the end of the 3-tube, and then necessarily $\underset{\sim}{X} \cong \tilde{X}$.

Now we know that for $i = 0, 1, 2$ the composition factors of rad $P_i/\mathrm{rad}^2P_i$ belong to $\{S_0, S_1, S_2\}$; therefore these are all simple modules (I.5.6). Moreover, the quiver is of the form (3C). Let $\beta\Lambda$ = rad $e_1\Lambda$ and $\eta\Lambda$ = rad $e_2\Lambda$; and rad $e_0\Lambda = \langle \rho, \gamma, \delta \rangle$. Since $P_1$ only has length 3 we deduce $\beta\Lambda$ = span $\{ \beta, \beta\gamma \}$ and $\beta\delta = 0 = \beta\rho$. Similarly $\eta\gamma = \eta\rho = 0$. We may assume that $\rho\Lambda = \underset{\sim}{X}$; by the above we know that rad $\underset{\sim}{X}/\mathrm{rad}^2\underset{\sim}{X} = \mathcal{U}(S_0, S_0)$. Therefore $\rho\gamma$ and $\rho\delta$ lie in $\rho J^2$ (using the direct sum decomposition of $H_0$), and then $\rho\Lambda$ is spanned by the powers of $\rho$ which implies that $\rho\Lambda e_1 = 0 = \rho\Lambda e_2$, and consequently $\rho\gamma = 0 = \rho\delta$.

Let $s$ be the integer such that $\rho^s \neq 0$ but $\rho^{s+1} = 0$. Then $\rho^sJ = 0$ and $\langle \rho^s \rangle$ = soc $e_0\Lambda$. Hence there non-zero elements $c, d \in K$ satisfying $\rho^s = c(\gamma\beta)$ and $\rho^s = d(\delta\eta)$. After re-scaling $\gamma$ and $\delta$ if necessary we obtain such relations with $c = d = 1$.

If $s = 2$ then the component of $S_0$ is $\cong \mathbb{Z}\tilde{D}_5$, and $\Lambda$ is not of semidihedral type: Consider the AR-quiver around $S_0$, it is of the form

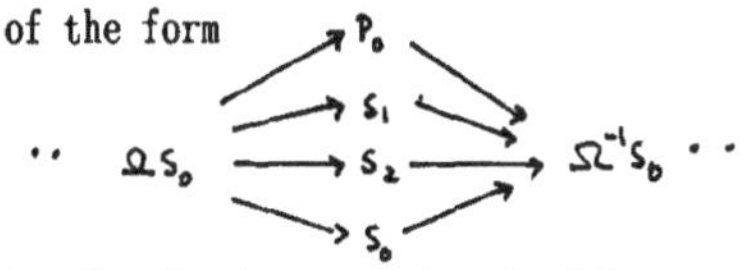


Moreover, since $s = 1$ we have $S_0 \cong H_1 \cong H_2$. So there are irreducible maps $\Omega S_1$ ( $= \mathcal{U}(S_0, S_1)$ ) $\to S_0$ and $\Omega S_2$ ( $= \mathcal{U}(S_0, S_2)$ ) $\to S_0$:. Since there is also an irreducible map $\Omega S_0 \to S_0$ we see that $S_0$ has three predecessors in $\Gamma_s(\Lambda)$, and the components is $\mathbb{Z}\tilde{D}_5$ (see I.8.4).

From now we will always assume that a simple does not have three predecessors in $\Gamma_S(\Lambda)$.

VIII.5.2 PROPOSITION *Assume that* $\Lambda$ *is an algebra of semidihedral type with quiver of the form* (*SC*). *Then* $rad\ P_1/S_1 \cong rad\ P_2/S_2$.

Proof: Set $H_i := \mathrm{rad}\ P_i/S_i$ for $i = 0, 1, 2$; we assume for contradiction that $H_1$ is not isomorphic to $H_2$.

(1) $S_1$ and $S_2$ must be periodic: Suppose, for example, that $S_1$ is not periodic. Since $H_1$ is indecomposable, it follows that $S_1$ can only lie at the end of a $\mathbb{Z}D_\infty$-component, with one predecessor. Then one of IV.3.3(a) or (b) holds.

We assume that $H_1 \neq H_2$; moreover, by comparing tops we see that $H_1 \neq H_0$. This excludes IV.3.3(b), so there is an AR-sequence $0 \to V_1 \to H_1 \to U_1 \to 0$. Since soc $H_1$ and top $H_1$ are both simple, we deduce from IV.3.4 that $U_1$ and $V_1$ must be simple, and then $U_1 \cong \mathrm{top}\ H_1 \cong S_0 \cong \mathrm{soc}\ H_1 \cong V_1$. On the other hand, $V_1 \cong \tau U_1$ and we deduce that $U_1$ is periodic. This is a contradiction since we started with $U_1$ lying in a $\mathbb{Z}D_\infty$-component.

(2) If $i \overset{\omega}{\to} j$ is an arrow in the quiver, $\omega \neq \rho$, then $e_i\Lambda/\omega\Lambda$ has a simple socle: This is clear for $\omega = \beta$ or $\eta$ since then $e_i\Lambda/\omega\Lambda$ is simple. Consider $\omega = \gamma$; then $e_0\Lambda/\omega\Lambda \cong \Omega^{-2}S_2$. Now, soc $\Omega^{-2}S_2 \cong \mathrm{top}\ \Omega^{-3}S_2$ and $\Omega^{-3}S_2 \cong S_2$ or $\Omega S_2$, depending on the period of $S_2$. In both cases, top $\Omega^{-3}S_2$ is simple, as required.

(3) $S_0$ must lie at the end of a $\mathbb{Z}D_\infty$-component, with one predecessor: By (1) and VIII.2.1 we know that $S_0$ is not periodic. Suppose (3) is false; then $H_0 \cong U_0 \oplus V_0$ as in IV.3.1. We may choose the notation such that top $V_0$ has length 2; then at least one of $S_1$, $S_2$ occurs in top $V_0$, say $S_1$. Then we may assume that $\gamma\Lambda \subseteq V_0$; this implies that $(\mathrm{rad}\ P_0)/\gamma\Lambda \cong \underset{\sim}{V}_0/\gamma\Lambda \oplus U_0$. On the other hand, $(\mathrm{rad}\ P_0)/\gamma\Lambda$ is the unique maximal submodule of $P_0/\gamma\Lambda$ which has a simple socle, by (2). Hence $\underset{\sim}{V}_0/\gamma\Lambda \oplus U_0$ has a simple socle, and consequently we must have that $\underset{\sim}{V}_0/\gamma\Lambda = 0$ and $\underset{\sim}{V}_0 = \gamma\Lambda$. Thus top $V_0$ has length 1, a contradiction.

Therefore IV.3.3(a) must hold, and there is an AR-sequence

(*) $0 \to V_0 \to H_0 \overset{\kappa}{\to} U_0 \to 0$. Now, neither $U_0$ nor $V_0$ is isomorphic to $S_0$ since

$S_0$ lies in a different AR-component; and then $U_0$ and $V_0$ are not simple. Hence by IV.3.4 we know that top $H_0 \cong$ top $U_0 \oplus$ top $V_0$ and that soc $H_0 \cong$ soc $U_0 \oplus$ soc $V_0$. We may use the argument in the proof of (3) again and see that top $V_0$ and soc $U_0$ must be simple.

Consequently $|\text{top } U_0| = 2$ and $|\text{soc } V_0| = 2$. Let $\bar{\gamma} = \gamma + \text{soc } \Lambda$ and $\bar{\delta} = \delta + \text{soc } \Lambda \in H_0$, and let $\kappa$ be the map in (*). Then one of $\kappa(\bar{\gamma})$, $\kappa(\bar{\delta})$ is a generator for $U_0$, say $\kappa(\bar{\gamma})$. Define $X := V_0 + \bar{\gamma}\Lambda$. Comparing tops one sees that $\kappa(X)$ is a proper submodule of $U_0$. By the AR-property it follows that $X \cong V_0 \oplus \kappa(\bar{\gamma})\Lambda$. Hence there is a generator $\gamma_1 \in \text{rad } P_0 - \text{rad}^2 P_0$ with $\gamma_1 \equiv \gamma$ (modulo $J^2$) such that $\bar{\gamma}_1\Lambda$ is a complement of $V_0$ in X. Consequently, soc $V_0 \subseteq \text{soc}(P_0/\gamma_1\Lambda)$, and the socle of $P_0/\gamma_1\Lambda$ is not simple, a contradiction to (2).
This completes the proof of VIII.5.2.

We continue the study of algebras with quiver (3C) in VIII.5.7 where we determine all algebras of semidihedral type where projectives have isomorphic hearts.

## VIII.6 *Quivers with few arrows*

We assume throughout this chapter that $\Lambda$ is an algebra of semidihedral type having at least three simple modules.

<u>VIII.6.1</u> Algebras of semidihedral type whose ordinary quiver has few arrows seem to come close to algebras of finite type, such that the stable AR-quiver has tree class $D_n$. This might explain why one needs some extra work, bearing in mind the difficulties arising in $[Ri_{2,3}]$.

Whenever only one arrow starts or ends at some vertex 0 in the quiver $Q$ and if there is an AR-sequence $0 \to V_0 \to H_0 \to U_0 \to 0$ then $U_0$ or $V_0$ is again simple. This occurs always twice, by IV.3.4. By using IV.6.5 we know then three modules at the end of the 3-tube which allows to determine the algebra.
(For finite type, the "end of the 3-tube" is rather the other end of the component. We interfer therefore with the situation where the component is $\mathbb{Z}D_n/\langle \tau^3 \rangle$).
Also, there are some algebras with a component $\mathbb{Z}\mathcal{D}_n$ having tubes of rank 2 and 3; these are similar to algebras of semidihedral type as long as one only uses local properties of $\Gamma_s(\Lambda)$.

<u>VIII.6.2</u> LEMMA *Assume that the quiver of* $\Lambda$ *contains* $\cdots 2 \xrightarrow{\alpha} 0 \xrightarrow{\beta} 1 \cdots$ *where* $H_0 \neq \operatorname{rad} Q/\operatorname{soc} Q$ *for* $Q$ *indecomposable projective,* $Q \neq P_0$. *Then the quiver of* $\Lambda$ *is of the form* ($\mathcal{TF}$).

Proof: (1) <u>$S_0$ is of type d, and $U_0$ and $V_0$ are defined:</u> We have that top $H_0 \neq$ soc $H_0$, therefore $S_0$ is non-periodic. Moreover, $H_0$ is indecomposable and hence $S_0$ must be of type d. The rest is the hypothesis.

Since top $H_0$ and soc $H_0$ are simple we have by IV.3.4 that $V_0$ and $U_0$ are simple, and $V_0 \cong U_2$ and $U_0 \cong V_1$. Now IV.3.5 shows that $S_0 \cong U_2$ and also $S_0 \cong V_1$. Moreover, for $i \leq 2$, the modules $\mathfrak{U}_i$, $\mathfrak{V}_i$ lie at the end of the 3-tube (IV.5.6 and IV.6.4), and they are $\mathcal{U}(S_0, S_1) = \mathfrak{U}_0 \cong \underset{\sim}{V}_1$ and $\underset{\sim}{V}_0 = \mathfrak{U}_2 \cong \mathcal{U}(S_2, S_0)$ ;and moreover $\mathfrak{U}_1 = \underset{\sim}{V}_2$. In particular,

(2) <u>top $V_2 \cong S_1$ and soc $U_1 \cong S_2$.</u>

(3) <u>$V_2$ and $U_1$ are not simple:</u> Suppose $V_2$ is simple, then $V_2 \cong S_1$ and $U_1 \cong S_2$ (IV.3.5). Then we have that the modules $P_i$ for $i \leq 2$ are all uniserial, and $\Lambda$ is of finite type (II.8.2), a contradiction. The same argument shows that $U_1$ is not simple.

Consequently, by IV.3.4 we have that top $H_2 \cong$ top $V_2 \oplus$ top $U_2 \cong S_1 \oplus S_0$ and soc $H_1 \cong$ soc $V_1 \oplus$ soc $U_1 \cong S_0 \oplus S_2$. Moreover, since $U_2 \cong S_0$ we have $\Omega H_0 \cong H_2$ and therefore soc $H_2 \cong$ top $H_0 \cong S_1$. Similarly, since $V_1 \cong S_0$ we have that $\Omega^{-1} H_0 \cong H_1$ and top $H_1 \cong$ soc $H_0 \cong S_2$. This shows tht the quiver is as stated.

We record from the proof:

<u>VIII.6.2*</u> With the hypothesis of VIII.6.2 we have that $P_0 = \mathcal{U}(S_0, S_1, S_2, S_0)$.

<u>VIII.6.3</u> LEMMA *Assume that the quiver of* $\Lambda$ *contains* $\cdots 0 \longrightarrow 1 \cdots$ *and* $H_0 \neq \operatorname{rad} Q/\operatorname{soc} Q$ *for* $Q$ *indecomposable projective with* $Q \neq P_0$. *Then the quiver of* $\Lambda$ *is of the form either* ($\mathcal{TF}$) *or* ($\mathcal{TN}$).

Proof: We have, as in VIII.6.2, part (1), that $S_0$ is of type d, and $U_0$ and $V_0$ are defined. Since top $H_0$ is simple, we deduce that $V_0$ is simple, $\cong S_2$ say (see IV.3.4). Consequently $U_2$ is defined and $\cong S_0$; and the module $\underset{\sim}{V}_0 \cong \mathcal{U}(S_2, S_0)$ which is also $\cong \mathfrak{U}_2$, lies at the end of the 3-tube (IV.6.4).

Assume first that $V_2$ is simple, $\cong S_i$ say. Then $P_2 = \mathcal{U}(S_2, S_0, S_i, S_2)$. Therefore there is an arrow $0 \to i$ and consequently $i = 1$, and the quiver contains $2 \to 0 \to 1$, with $H_0 \neq \mathrm{rad}\, Q/\mathrm{soc}\, Q$ for $Q$ projective. This situation has been dealt with in VIII.6.2, and from there we deduce that the quiver is of the form $(3\mathcal{F})$. Similarly, if $U_0$ is simple the quiver is of the form $(3\mathcal{F})$.

Now assume that $V_2$ and $U_0$ are both not simple. Then we obtain from IV.3.4 that top $H_2 \cong S_0 \oplus$ top $V_2$ and also soc $H_0 \cong$ soc $U_0 \oplus$ soc $V_0 \cong$ soc $U_0 \oplus S_2$.

Since $\Omega \underset{\sim}{V}_2 \cong \tilde{U}_0$, we have now that top $V_2 \cong$ soc $U_0$. Recall from IV.3.5 that $\Omega H_0 \cong H_2$. Since here top $H_0 \cong S_1$ it follows that soc $H_2 \cong S_1$; in particular there is an arrow $1 \to 2$. Moreover, $\underline{\dim}\, P_1 = \underline{\dim}\, H_0 + \underline{\dim}\, H_2$.

(2) <u>top $\underset{\sim}{V}_2 \cong S_1$:</u> There is an exact sequence $0 \to \tilde{U}_0 \to P \to \underset{\sim}{V}_2 \to 0$ where $P$ is projective, and top $P \cong$ top $\underset{\sim}{V}_2 \cong$ soc $\tilde{U}_0$. We have now $\underline{\dim}\, P = \underline{\dim}\, \tilde{U}_0 + \underline{\dim}\, \underset{\sim}{V}_2 = \underline{\dim}\, U_0 + \underline{\dim}\, S_0 + \underline{\dim}\, V_2 + \underline{\dim}\, S_2 = \underline{\dim}\, H_2 + \underline{\dim}\, H_0 = \underline{\dim}\, P_1$. Since the Cartan matrix is non-singular, it follows that $P \cong P_1$; and therefore also soc $U_0 \cong S_1$.

We have now proved that the quiver contains

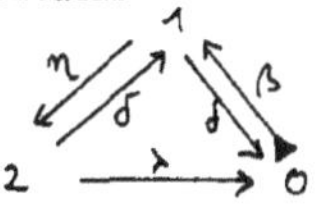


and no other arrows start or end at vertex 0 or 2.

(3) <u>No other arrow starts at 1:</u> We have $\underset{\sim}{V}_2 \cong \delta\Lambda$; since $\Omega\underset{\sim}{V}_2$ ($\cong \tilde{U}_0$) has a simple top $\cong S_0$, we see that $\delta\eta$ is not involved in any relation. Therefore the universal cover of $\Lambda/J^3$ contains the tree

$$0 \xrightarrow{\beta} 1 \xleftarrow{\delta} 2 \xrightarrow{\lambda} 0 \xleftarrow{\gamma} 1^* \xrightarrow{\eta} 2 \xrightarrow{\delta} 1 \xleftarrow{\beta} 0$$

Suppose there is another arrow, $\mu$ say, starting at 1, then $\mu$ may be attached to * in this tree. This is $\tilde{E}_7$, and $\Lambda$ is wild.

Assume for contradiction there is an arrow $\mu$ say ending at 1 with $e = s(\mu) \neq 2$ or 0. Then there is a path from 1 to e; hence there must be an arrow starting at 1 (or 0 or 2) but ending at some other vertex. This is not possible, by (3) (or earlier results in this proof). This shows that the quiver is of form $(3\mathcal{K})$.

Lemma VIII.6.3 has a dual version; this does not lead to new algebras since the quivers (3$\mathcal{F}$) and (3$\mathcal{K}$) are self-dual.

VIII.6.4 THEOREM *Let $\Lambda$ be of semidihedral type, with quiver (3$\mathcal{F}$). Then $\Lambda \cong KQ/I$ where I is defined by the relations*

$$\lambda\beta = (\eta\delta)^{k-1}\eta, \quad \beta\delta\eta = 0 = \eta\delta\lambda, \quad \delta\lambda\beta\delta = 0,$$

*where $k \geq 2$.*

Proof: We use VIII.6.2* where we obtained the structure of $P_0$. This implies that $\beta\delta\lambda\beta = 0 = \beta\delta\eta$ and also that $\langle (\beta\delta\lambda) \rangle = \mathrm{soc}(e_0\Lambda)$.

(1) $\langle (\delta\lambda\beta) \rangle = \mathrm{soc}(e_1\Lambda)$ and $\delta\lambda \in \mathrm{soc}_2\Lambda$: Since $\dim e_1\Lambda e_0 = \dim e_0\Lambda e_1 = 1$ and $0 \neq \delta\lambda \in e_1\Lambda e_0$, it follows that $e_1\Lambda e_0$ is spanned by $\delta\lambda$. On the other hand, $S_0$ occurs in $\mathrm{soc}_2(e_1\Lambda)$; and therefore we deduce that $0 \neq \delta\lambda J = \mathrm{soc}(e_1\Lambda)$.

(2) $\langle (\lambda\beta\delta) \rangle = \mathrm{soc}(e_2\Lambda)$ and $\lambda\beta \in \mathrm{soc}_2\Lambda$: We have that $(\delta\lambda\beta)\delta \in (\mathrm{soc}\ e_1\Lambda)J = 0$, hence $J(\lambda\beta\delta) = 0$ and $\lambda\beta\delta \in \mathrm{soc}(\Lambda e_2) = \mathrm{soc}(e_2\Lambda)$. Let $\psi$ be a symmetrizing linear form, then $\psi[\lambda\beta\delta] = \psi[\delta\lambda\beta] \neq 0$, by (1), and therefore $\lambda\beta\delta \neq 0$.

Let k be as large as possible such that $(\delta\eta)^k \neq 0$. Then $k \geq 1$; otherwise we would have that $0 = \langle (\delta\eta) \rangle = J\eta$ and $\eta \in \mathrm{soc}\ \Lambda \cap e_2\Lambda e_1 = 0$.

(3) $\mathrm{soc}\ e_1\Lambda = \langle (\delta\eta)^k \rangle$ : Since $(\delta\eta)^k\delta J = \langle (\delta\eta)^k\delta\lambda \rangle \subseteq J^{2k}(\mathrm{soc}_2\Lambda) = 0$, we have that $(\delta\eta)^k\delta$ lies in $\mathrm{soc}\ e_1\Lambda \cap e_1\Lambda e_2$, hence is zero. Consequently $(\delta\eta)^k J = 0$, that is, $\langle (\delta\eta)^k \rangle \subseteq \mathrm{soc}(e_1\Lambda)$; and equality holds since $(\delta\eta)^k \neq 0$.

(4) $\mathrm{soc}(e_2\Lambda) = \langle (\eta\delta)^k \rangle$ and $(\eta\delta)^{k-1}\eta \in \mathrm{soc}_2\Lambda$: By (3) we have that $\delta(\eta\delta)^k \in (\mathrm{soc}\ e_1\Lambda)J = 0$. Consequently $J(\eta\delta)^k = 0$, and $(\eta\delta)^k$ lies in $\mathrm{soc}(e_2\Lambda)$. If $\psi$ is a symmetrizing form then $\psi[(\eta\delta)^k] = \psi[(\delta\eta)^k] \neq 0$ and therefore $(\eta\delta)^k \neq 0$.

(5) We may assume that $\delta\lambda\beta = (\delta\eta)^k$: By (1) and (3) there is some $0 \neq c \in K$ such that $c(\delta\lambda\beta) = (\delta\eta)^k$. Without loss of generality $c = 1$; otherwise we replace $\lambda$ by $c\lambda$.

(6) $\lambda\beta = (\eta\delta)^{k-1}\eta$: By (2) and (4) we have that $\lambda\beta$ and $(\eta\delta)^{k-1}\eta$ are non-zero elements in $\mathrm{soc}_2(e_2\Lambda)e_1$. From the quiver we know that this space is 1-dimensional. Consequently $\lambda\beta = c(\eta\delta)^{k-1}\eta$ for some $0 \neq c \in K$. We deduce from (5) that $c = 1$.

We see from (6) that $k \geq 2$; otherwise $\eta$ would lie in $J^2$.

By (1) we have that $\eta\delta\lambda \in \mathrm{soc}\ \Lambda \cap e_2\Lambda e_0 = 0$. The remaining relations in the statement follow from the socle conditions. If $k = 2$ then the component of $S_0$ is $\cong \mathbb{Z}\tilde{D}_5$, this is easy to see directly; and $\Lambda$ is not of semidihedral type.

<u>VIII.6.5</u> THEOREM *Let $\Lambda$ be of semidihedral type, with quiver of the form $(\mathcal{SH})$. Then $\Lambda \cong KQ/I$ where I is defined by the reations*

$$\delta\lambda = (\gamma\beta)^{k-1}\gamma, \qquad \beta\delta\eta = 0 = \gamma\beta\delta,$$

$$\lambda\beta = (\eta\delta)^{s-1}\eta, \qquad \eta\gamma = 0,$$

*where $k,\ s \geq 2$.*

Proof: We see from VIII.2.0.1 that $S_0$ and $S_2$ must lie at the end of $\mathbb{Z}D_\infty$-components and satisfy IV.3.3(a), and $V_0 \cong S_2$ and $U_2 \cong S_0$. Moreover, $U_2$ lies at the end of the 3-tube (IV.6.4). Let $\eta\Lambda = \underset{\sim}{V}_2$ and $\lambda$ be an arrow $2 \to 0$ such that $\lambda J \subseteq \eta\Lambda$ ; then $\lambda J \subseteq soc_2(e_2\Lambda)$ but not in soc $\Lambda$. Moreover, we have

(1) <u>$\lambda\beta \in soc_2\Lambda$; $\langle (\lambda\beta\delta) \rangle = soc(e_2\Lambda)$ and $\lambda\beta\gamma = 0$.</u>

We have $\beta\Lambda = rad(e_2\Lambda)$. There is a submodule $\underset{\sim}{V}_0 = \mathcal{U}(S_2, S_0)$ $[\subseteq soc_2(e_0\Lambda)]$. Moreover, $U_0 \cong \beta\Lambda/\underset{\sim}{V}_0$ ,and $\Omega U_0 \cong \Omega^{-1}\underset{\sim}{V}_0 \cong \Omega^{-1}S_2 \cong \delta\Lambda$. Hence, considering a projective cover shows that:

(2) <u>We may choose $\delta$ such that $\beta\delta \in soc_2\Lambda$, hence $\beta\delta\eta = 0 = \gamma\beta\delta$.</u>

(3) <u>$soc(e_1\Lambda) = \langle (\delta\lambda\beta) \rangle$ and $\delta\lambda \in soc_2\Lambda$, hence $\eta\delta\lambda = 0$:</u> By (1) we know that $\lambda\beta \in soc_2\Lambda$ - soc $\Lambda$, so $0 \neq J\lambda\beta \subseteq$ soc $\Lambda$ and $J\lambda\beta = \Lambda(\delta\lambda\beta) =$ soc $\Lambda e_1 =$ soc $e_1\Lambda$.

(4) <u>$soc(e_0\Lambda) = \langle (\beta\delta\lambda) \rangle$:</u> Using (3) we have that $(\beta\delta\lambda)J = \beta(\delta\lambda\beta)\Lambda \subseteq \beta(soc\ \Lambda) = 0$. Let $\psi$ be a symmetrizing linear form, then $\psi[\beta\delta\lambda] = \psi[\delta\lambda\beta] \neq 0$, by (1), and therefore $\beta\delta\lambda \neq 0$.

Let k be as large as possible such that $(\beta\gamma)^k \neq 0$. We must have that $k \geq 1$: Otherwise we would have by (2), (4) that $e_0J$ is spanned by $\beta$, $\beta\delta$, $\beta\delta\lambda$, and $S_1$ would not occur in $soc_2(e_0\Lambda)$.

(5) <u>soc $e_0\Lambda = \langle (\beta\gamma)^k \rangle$:</u> We have that $(\beta\gamma)^k\beta J = (\beta\gamma)^k\beta\delta J \subseteq J^{2k}(soc_2\Lambda) = 0$, by (2). Hence $(\beta\gamma)^k\beta \in$ soc $\Lambda \cap e_0\Lambda e_1 = 0$ and then $(\beta\gamma)^k J = 0$. Similarly one shows

(6) <u>soc $e_1\Lambda = \langle (\gamma\beta)^k \rangle$.</u>

(7) <u>$\eta\gamma = 0$:</u> By (1), (2) and (5) the structure of $e_0\Lambda$ is given. In particular $1 = \dim e_0\Lambda e_2 = \dim e_2\Lambda e_0$. It follows that $e_2\Lambda$ is spanned by $\lambda$, and we deduce that $\eta\gamma = 0$.

Let s be as large as possible such that $(\eta\delta)^s \neq 0$. Then $s \geq 1$, for otherwise $\eta J = 0$, using (7), and $\eta \in$ soc $\Lambda \cap e_2\Lambda e_1 = 0$.

(8) <u>soc $e_2\Lambda = \langle (\eta\delta)^s \rangle$ and soc $e_1\Lambda = \langle (\delta\eta)^s \rangle$:</u> This is straightforward.

(9) <u>We may assume that $\lambda\beta = (\eta\delta)^{s-1}\eta$:</u> By (1) and (8) we know that the elements $\lambda\beta$

and $(\eta\delta)^{s-1}\eta$ lie in $[soc_2(e_2\Lambda)]e_1$ and are non-zero. This space is 1-dimensional, so $c(\lambda\beta) = (\eta\delta)^{s-1}\eta$ for some $0 \neq c \in K$. We may assume that $c = 1$.

(10) <u>We may assume that $\delta\lambda = (\gamma\beta)^{k-1}\gamma$ :</u> From (1) and (6) we obtain a basis for $e_1\Lambda$. Now $\delta\lambda \in e_1\Lambda e_0 \cap soc_2\Lambda$, so there is some $c \in K$ with $\delta\lambda = c(\gamma\beta)^{k-1}\gamma$. Using (9) we see that $c \neq 0$. We may assume that $c = 1$, otherwise we replace $\gamma$ (This does not change (9)).

We shall now prove that these are the only algebras where a simple module is of type d where U, V are defined and U or V is simple.

<u>VIII.6.6</u>  LEMMA  *Suppose $\Lambda$ is of semidihedral type having a simple module $S_0$ of type d such that IV.3.3(a) holds; and assume that $U_0$ or $V_0$ is simple. Then the quiver of $\Lambda$ is of the form $(\mathcal{SF})$ or $(\mathcal{SR})$ or $(\mathcal{SR}^{op})$.*

Proof: By IV.3.5 it suffices to consider one of the cases. Suppose $U_0 \cong S_i$, then by IV.3.5 it follows that $V_i \neq S_0$ (and $S_i$ satisfies IV.3.3(a) ). Note that $S_i \neq S_0$ since otherwise $S_0 \cong U_0 \cong V_0 \cong \tau S_0$ and $S_0$ does not lie in an infinite component.

Moreoover, we take $\tilde{U}_0 = \mathcal{U}(S_0, S_i) \cong \underset{\sim}{V}_i$. By IV.6.4 the modules $\tilde{U}_0$, $\underset{\sim}{V}_0$ and $\tilde{U}_i$ lie at the end of the 3-tube. This shows that

(1) top $V_0 \cong$ soc $U_i$.

We have that soc $V_0 \cong$ soc $H_0$ (see IV.3.4); also $H_0 \neq$ rad $Q$/soc $Q$ for $Q \neq P_0$, $Q$ indecomposable projective (by the hypothesis). It suffices to show that soc $H_0$ is simple; then we get the result from the dual of VIII.6.3. We may assume that $V_0$ is not simple; otherwise we are done.

(2) <u>There is a unique arrow $0 \overset{\alpha}{\to} i$:</u> . If there were another arrow then $S_i \subseteq$ top $V_0$ and then also $S_i \subseteq$ soc $U_i$. The quiver would contain

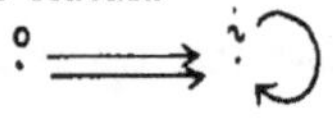

and $\Lambda$ would be wild, by I.10.8.

(2) <u>$H_0$ has a submodule $\mathcal{U}(S_i, S_k)$:</u>  Let $0 \to V_0 \overset{\lambda}{\to} H_0 \to U_0 \to 0$ be an AR-sequence. Choose a simple submodule $S_k \subseteq$ soc $V_0$; then by the AR-property we have that $H_0/\lambda(S_k) \cong V_0/S_k \oplus U_0$; and moreover $U_0 \cong S_i$ here.

By (1) we may now assume that $\alpha\Lambda/S_0 = \mathcal{U}(S_i, S_k)$ and that $V_0 \cap \alpha\Lambda/S_0 = S_k$.
(3) soc $H_0$ is simple: Let soc $V_0 = X \oplus S_k$. Then $H_0/X = V_0/X \oplus \mathcal{U}(S_i, S_k)$. By (1) we know that $S_i$ does not occur in top $V_0$, and therefore $H_0/X$ does not have a summand $\cong S_i$. Suppose $X \neq 0$, then by the AR-property $H_0/X \cong V_0/X \oplus U_0$ and $U_0 \cong S_i$, so there is a summand $\cong S_i$; a contradiction.

From now we always assume that whenever IV.3.3(a) holds then U and V are not simple.

## VIII.7 *Algebras where projectives have isomorphic hearts*

Assume that $\Lambda$ is an algebra of semidihedral type having non-isomorphic projective modules $P_1$ and $P_2$ such that rad $P_1$/soc $P_1 \cong$ rad $P_2$/soc $P_2 =: H$, and moreover H is indecomposable. Put $H_i :=$ rad $P_i$/ soc $P_i$ for i = 1 and 2. We will show that then the quiver of $\Lambda$ must be or the form (3$\mathcal{A}$) or (3$\mathcal{C}$); and then we will determine the possible algebras.

VIII.7.0 (a) With the above hypothesis, H lies at the end of a $\mathbb{Z}D_\infty$-component with three predecessors: We know that H has at least two predecessors, namely $\Omega S_1$ and $\Omega S_2$. Moreover, H is the only successor of $\Omega S_1$ and $\Omega S_2$. Now the statement follows from our hypothesis on the graph structure of $\Gamma_s(\Lambda)$.
(b) We assume also that H is not simple; otherwise, we are in the situation of VIII.5, and we have already seen there that the quiver is of the form (3$\mathcal{C}$).

VIII.7.1 LEMMA *With the hypothesis as above, we have that* $|top\ H_1| \leq 2$ *and* $|soc\ H_1| \leq 2$. *Moreover, if there is a loop at vertex 1 then the quiver of* $\Lambda$ *is of the form*

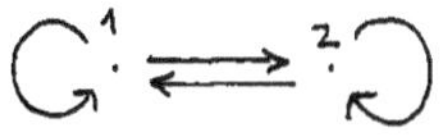


Proof: The hypothesis implies top $H_1 \cong$ top $H_2$ and soc $H_1 \cong$ soc $H_2$. If top $H_1$ would have length $\geq 3$ then the universal cover of $\Lambda/J^2$ would contain

$$\begin{array}{ccccccccccccccc} a & \leftarrow & 2 & \rightarrow & b & \leftarrow & 1 & \rightarrow & a & \leftarrow & 2 & \rightarrow & b & \leftarrow & 1 \\ & & & & & & \downarrow & & & & & & & & \\ & & & & & & c & & & & & & & & \end{array}$$

and $\Lambda$ would be wild (I 4.7 and I.9.5).

Now assume there is a loop at vertex 1 then $S_1 \subsetneq \text{top } H_1 \cong \text{top } H_2$, and then there is an arrow $2 \rightarrow 1$. This shows that $S_2$ occurs in soc $H_1$, and then $S_2 \subsetneq \text{soc } H_2$, and there is also a loop at vertex 2. Consequently $S_2 \subsetneq \text{top } H_2$ and $S_2 \subsetneq \text{top } H_1$, and by the first part of the proof, top $H_i \cong \text{soc } H_i \cong S_1 \oplus S_2$ for $i = 1, 2$. Since the quiver of $\Lambda$ is connected, the statement follows.

We study first the situation when there is a unique arrow starting and ending at vertex 1, and then also at vertex 2.

<u>VIII.7.2</u> LEMMA *Suppose the quiver of $\Lambda$ contains*

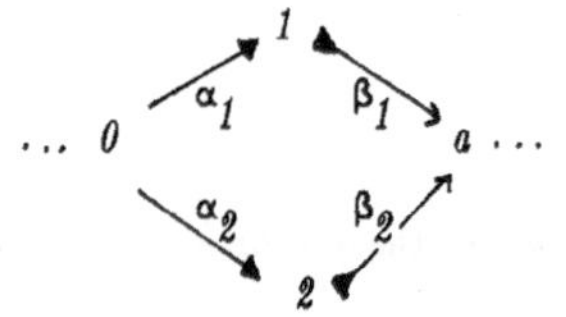


*with $H_1 \cong H_2$. Consider the module $X = \alpha_1\Lambda + \alpha_2\Lambda$; then*

*(a) $X$ lies at the end of some component, and moreover*

*(b) $\Omega^4 X \ncong X$; and if $\Omega^3 X \cong X$ then $S_a \cong S_0$.*

Proof: (1) <u>We may choose arrows satisfying $\alpha_1\beta_1 = \alpha_2\beta_2$:</u> Since $\alpha_1\Lambda \cong e_1\Lambda/S_1$, we deduce $\alpha_1\beta_1\Lambda \cong H_1$; and similarly $\alpha_2\beta_2\Lambda \cong H_2$. Any isomorphism $\alpha_1\beta_1\Lambda \rightarrow \alpha_2\beta_2\Lambda$ may be realized by left multiplication with a unit, see IV.1.9.1, and we get $u\alpha_1\beta_1 = \alpha_2\beta_2 v$ for u, v units. We replace $\alpha_1$ and $\beta_2$ by $e_0 u\alpha_1$ and $\beta_2 v e_a$.

Let $0 \rightarrow \tau X \xrightarrow{\mu} N \xrightarrow{\epsilon} X \rightarrow 0$ be the AR-sequence ending in X. We shall prove that N is indecomposable.

(2) <u>soc $\tau X \cong S_a$:</u> Recall that soc $\tau X \cong$ top $\Omega X$. We identify $\Omega X$ with $\{ (x,y) \in e_1\Lambda \oplus e_2\Lambda : \alpha_1 x - \alpha_2 y = 0 \}$, see I.6.5. By (1) we know that $(\beta_1, \beta_2)\Lambda \subsetneq \Omega X$; and equality holds since the lengths are the same. We see now that top $\Omega X \cong S_a$.

(3) <u>soc $N \cong$ soc $X \oplus$ soc $\tau X$:</u> Since $X$ is not simple, the inclusion soc $X \to X$ factors through $\epsilon$, and (3) follows.

(4) <u>$N$ is indecomposable:</u> Since soc $\tau X$ is simple, there is a summand of $N$, $N_1$ say, such that $\mu_1 : \tau X \to N_1$ is one-to-one where $\mu_1$ is the component of $\mu$ corresponding to $N_1$. Write $N = N_1 \oplus N_2$ and let $\epsilon_i = \epsilon|_{N_i}$. With $\mu_1$ is also $\epsilon_2$ one-to-one.

Assume for contradiction that $N_2$ is non-zero. Now, $X$ has only two maximal submodules, namely $\alpha_1\Lambda$ and $\alpha_2\Lambda$. Say $\epsilon_2$ maps $N_2$ into $\alpha_2\Lambda$; then since $\epsilon_2$ is irreducible, it follows that $N_2 \cong \alpha_2\Lambda \cong \Omega^{-1}S_2$. Since there is an irreducible map $\tau X \to \Omega^{-1}S_2$, we must have that $\tau X \cong H$ and therefore, by VIII.7.1 we deduce that $N \cong \Omega^{-1}S_1 \oplus \Omega^{-1}S_2 \oplus N'$ and $N' \neq 0$. This is not possible; by (2) and (3) we know that $|\mathrm{soc}\ N| = 2$.

(5) <u>$X$ is not of period dividing 4</u>: Assume for contradiction that $\Omega^4 X \cong X$; then there is an exact sequence $0 \to X \quad P_0 \to Q \to P_a \to P_1 \oplus P_2 \to X \to 0$ where $Q$ is projective; and it follows that the Cartan matrix is singular, a contradiction.

(6) <u>If $S_a \neq S_0$ then $X$ does not have period 3:</u> Suppose that $\Omega^3 X \cong X$; then there are exact sequences $0 \to X \to P_0 \to \Omega^2 X \to 0$ and $0 \to \Omega^2 X \to P_a \to \Omega X \to 0$. Now, we have that $\underline{\dim}\ \Omega X = \underline{\dim}\ X$, and it follows that $\underline{\dim}\ P_0 = \underline{\dim}\ \Omega^2 X + \underline{\dim}\ X = \underline{\dim}\ \Omega^2 X + \underline{\dim}\ \Omega X = \underline{\dim}\ P_a$, and $C$ is singular.

<u>VIII.7.3</u> **LEMMA** *Suppose $\Lambda$ satisfies the hypothesis in VIII.7.2; and assume also that $X$ is of type d. Then $S_0 \cong S_a$, and moreover in the quiver of $\Lambda$ there is a loop at vertex 0.*

Proof: The stable AR-quiver of $\Lambda$ contains

$$(*)\qquad \begin{array}{ccccc} \tau X & & & \overset{\lambda}{\nearrow} & X \\ & \searrow & & & \\ \tau Z & \longrightarrow & M & \xrightarrow{\lambda_1} & Z \end{array} \qquad \text{and} \qquad \begin{array}{ccccc} \Omega X & & & \nearrow & \Omega^{-1}X \\ & \searrow & & & \\ \Omega Z & \longrightarrow & \Omega^{-1}M & \longrightarrow & \Omega^{-1}Z \end{array}$$

(1) <u>$X$ and $Z$ do not have a projective predecessor:</u> Since top $X \neq$ soc $\tau X$, there is no projective $P$ with rad $P = \tau X$ and $P/\mathrm{soc}\ P \cong X$. Assume for contradiction that there is a projective $P_j$ such that $P_j/\ \mathrm{soc}\ P_j = \tau Z$. Then $X \cong U_j$, and there are arrows $j \to 1$, $j \to 2$. Consequently $j = 0$ and $M \cong H_0$. We claim that the map $\lambda$ splits: Identify $M$

with rad $P_0/S_0$ and denote by - residues modulo $S_0$. Then - induces a map $\psi$: $X \to M$. Since $S_1$ and $S_2$ occur with multiplicity 1 in top $M$, we have that $\lambda(\bar{\alpha}_1) = \alpha_1 u$ and $\lambda(\bar{\alpha}_2) = \alpha_2 v$ for units u, v. Hence $\lambda\psi$ is an isomorphism.

This is a contradiction to the fact that $\lambda$ is irreducible.

The structure of $\Omega X$ is "dual"to that of X, which gives also

(1*) <u>The modules $\Omega X$, $\Omega Z$ do not have a projective successor.</u>

(Hence in the diagram (*), each row corresponds to an AR-sequence.)

We consider now $\tau X$ and $\tau Z$ as submodules of M.

(2) <u>$\tau X \cap \tau Z \neq 0$:</u> Let $M_0 = \tau X \oplus \tau Z$ ; then $M_0$ is a proper submodule of M which contains the kernels of $\lambda$ and of $\lambda_1$. Let $X_1 = \alpha_1\Lambda$; then $X_1$ is a maximal submodule of X which is isomorphic to $\Omega^{-1}S_1$; we denote also by $M_1$ the module $\lambda^{-1}(X_1)$. Then $M_0 \subsetneq M_1$, hence $M_1$ is a maximal submodule of M and moreover $Z_1 := \lambda_1(M_1)$ is a maximal submodule of Z. We use the AR-property twice and deduce that

$$M_1 \cong \tau X \oplus X_1 \quad \text{and also} \quad M_1 \cong \tau Z \oplus Z_1.$$

Recall $X_1 \cong \Omega^{-1}S_1$; therefore $M_1$ is the direct sum of two indecomposable summands. Since $\tau Z \neq \tau X$, it follows that $\tau Z \cong X_1 \cong \Omega^{-1}S_1$. Now exactly the same arguments are valid if we start off with $\alpha_2\Lambda$ instead of $\alpha_1\Lambda$; and we deduce $\tau Z \cong \alpha_2\Lambda$. Therefore $\alpha_1\Lambda \cong \alpha_2\Lambda$, a contradiction.

We deduce from (1*) that Z is not simple. Therefore we have that soc $M \cong$ soc $Z \oplus$ soc $\tau Z$. In particular, soc Z and soc $\tau Z$ are simple. Since $0 \neq \tau Z \cap \tau X$ and soc $\tau X = S_a$, we deduce that

(3) <u>soc $\tau Z \cong S_a$ and soc $Z \cong S_0$</u>.

(4) <u>$S_a \cong S_0$:</u> We note first that the restriction of $\lambda$ to $\tau Z$ is onto: Since $\tau X \cap \tau X \neq 0$, the irreducible map $\tau X \oplus \tau Z \to M$ is not 1-1 hence is onto. Therefore $\tau X + \tau Z = M$, and then $X = \lambda(M) = \lambda(\tau Z)$.

Let $\omega \in \tau Z$ with $\omega = \omega e_1$ such that $\lambda(\omega) = \alpha_1$. Then $\omega\Lambda$ has a simple top $\cong S_1$, and $\lambda$ induces an epimorphism $\omega\Lambda \to \alpha_1\Lambda \cong \Omega^{-1}S_1$. Since $\tau Z$ does not have a projective summand, we must have that $\omega\Lambda \cong \Omega^{-1}S_1$. Therefore soc $\Omega^{-1}S_1 \cong S_0 =$ soc $\omega\Lambda \subsetneq$ soc $\tau Z \cong S_a$, and we are done.

(5) <u>There is a loop at vertex 0:</u> Recall that $S_0 \subsetneq \tau X \cap \tau Z$. By the AR-property we have that $M/S_0 \cong \tau X/S_0 \oplus X$ and also $M/S_0 \cong \tau Z/S_0 \oplus Z$. Since $X \neq Z$, it follows that $\tau X/S_0 \cong Z \oplus C$ for some module C. Now, soc $\tau X$ is simple, this allows to deduce that

$\tau Z$ has a submodule of the form $\mathcal{U}(S_0, S_0)$, and the statement follows.

VIII.7.4 **PROPOSITION** *Suppose $\Lambda$ satisfies the hypothesis in VIII.7.2. Then $S_0 \cong S_a$, and $\Lambda$ has three simple modules. Moreover, if there is no loop at vertex 0 then $S_0$ has period 3 and $\Omega S_0 \cong X$.*

Proof: We know from VIII.7.2 and 3 that in any case, $S_a \cong S_0$. If X is of type d then there is a loop at vertex 0. Since $\Lambda$ is not wild, it follows that the quiver is just (3C), and we are done. Assume now that X is not of type d; then X must lie in the 3-tube. Then $\Omega^2 X$ has top and socle simple and $\cong S_0$. If $\Omega^2 X \cong S_0$ then we are done.

We assume now that $\Omega^2 X$ is not simple, and that there is no loop at vertex 0, and that moreover $\Lambda$ has $\geq$ four simple modules, and we aim to get a contradiction.

(1) $S_0$ is not periodic: There is no space for $S_0$ in the 3-tube. Since $\underline{\mathrm{Hom}}_\Lambda(X, S_0) = 0$ but $\underline{\mathrm{Hom}}_\Lambda(\Omega^2 X, S_0) \neq 0$, the module $S_0$ does not have period 2 or 4 using (IV.1.1). By IV.1.4, $S_0$ does not have period 1.

(2) $\mathfrak{U}_0$ and $\underset{\sim}{V}_0$ are defined: $S_0$ is of type a or of type d. If it is of type d then $H_0 \neq H_j$ for $j \neq 0$, since 1, 2 are not joined to any other vertex of the quiver. Hence in any case, $\mathfrak{U}_0$ and $\underset{\sim}{V}_0$ are defined.

By IV.5.5 these modules lie at the end of some component.

(3) Either top $\underset{\sim}{V}_0$ or soc $\mathfrak{U}_0$ is not simple: Otherwise it would follow from IV.6.2 (and IV.5.5) that $\mathfrak{U}_0$ and $\underset{\sim}{V}_0$ lie at the end of a tube. None of the modules M at the end of the 3-tube has the property that M and $\Omega M$ has simple socles and tops; and therefore $\mathfrak{U}_0$ can only have period dividing 4. This is not possible by IV.6.6 since there is no loop attached to vertex 0.

(10) Three arrows start at vertex 0: We assume there are at least four simple modules, so there are at least three arrows starting at 0 (using also that $\Lambda$ is symmetric). Suppose there are more, then the universal cover of $\Lambda/J^2$ contains $\tilde{D}_4$, namely

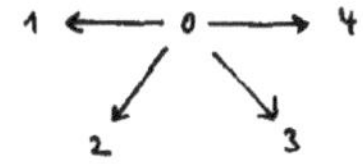


It follows that no other arrow can end at vertex 3 or 4 or 0; otherwise $\Lambda/J^2$ would be wild (I.9.3 or I.9.7); and by the hypothesis, VIII.7.2, the quiver must be of the

form

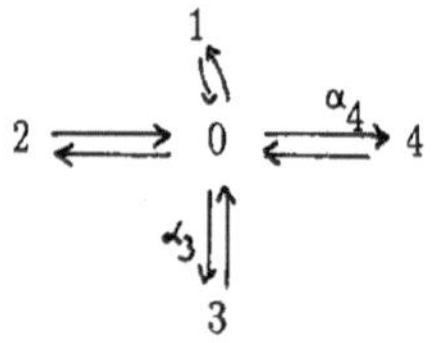


If $H_3 \cong H_4$, then we deduce from the first part of the proof that $\alpha_3\Lambda + \alpha_4\Lambda$ must lie in the 3-tube; but there is no space. Then the only possibility is that $S_3$ and $S_4$ are periodic, and then of period 4. It follows that no path of length 2 starting at 3 or 4 is involved in relations, and $\Lambda$ is wild (see the argument in I.10.9).

Similarly, three arrows end at vertex 0.

(5) $\underline{\alpha_1\Lambda \cap \alpha_2\Lambda \subset soc_2\Lambda}$: Recall that the intersection is $H_1$, and soc $H_1 \cong$ top $H_1 \cong S_0$. We assume there is no loop at vertex 0 and that $H_1$ is not simple (VIII.7.0(b)), and (5) holds.

It follows now that either $X = \underset{\sim}{V}_0$ or otherwise if $X = \langle U_0 \rangle$. In the first case we deduce that $\Omega^{-1}\underset{\sim}{V}_0$ has socle $\cong S_0$, that is, soc $U_0 \cong S_0$, and there is a loop at vertex 0, a contradiction. Hence $\langle U_0 \rangle = X$. But then we see that $(\beta_1, \beta_2) \in \Omega U_0$ - rad $\Omega U_0$, and $S_0$ occurs in top $\Omega U_0 \cong$ soc $V_0$, and there is a loop at vertex 0.

<u>VIII.7.5</u> **PROPOSITION** *Suppose the quiver of $\Lambda$ contains*

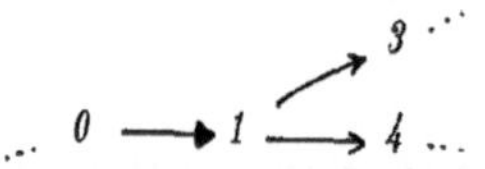


*and no other arrow starts at vertex 1. Then $H_1 \neq H_k$ for $k \neq 1$.*

Proof: Suppose that $H_1 \cong H_2$ where $S_1 \neq S_2$. Then by VIII.7.1(b) there is no loop at vertex 1 or 2, and the quiver contains

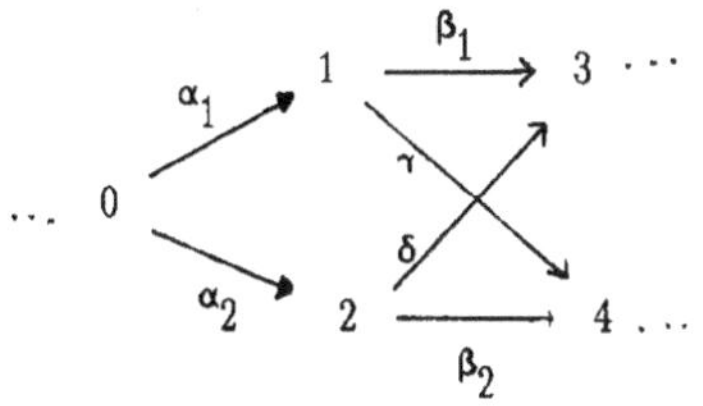


and no other arrow starts or ends at vertex 1, 2 or ends at 3, 4.

(1) <u>We may choose arrows satisfying $\alpha_1\beta_1 = \alpha_2\delta$ and $\alpha_1\gamma = \alpha_2\beta_2$:</u> Since $\alpha_1\Lambda \cong e_1\Lambda/S_1$, we have $H_1 \cong \alpha_1\beta_1\Lambda + \alpha_1\gamma\Lambda$. This is equal to $\alpha_1(\beta_1 + \gamma)\Lambda$ since $\beta_1$ and $\beta_2$ end at distinct vertices. Similarly $H_2 \cong \alpha_2(\beta_2 + \delta)\Lambda$. Any isomorphism may be realized by left-multiplication (IV.1.9.1). It follows that there is a unit $w \in e_0\Lambda e_0$ such that $w\alpha_1(\beta_1 + \gamma) = \alpha_2(\beta_2 + \delta)v$ where $\beta_2 v$ and $\delta v$ do not lie in $J^2$. Replace $\alpha_1$ by $w\alpha_1$ and $\beta_2$, $\delta$ by $\beta_2 v e_4$, $\delta v e_2$ respectively.

(2) <u>$H_3 \cong \operatorname{rad} Q/\operatorname{soc} Q$ for some projective $Q \neq P_3$</u>: We exclude all other possibilities. $S_3$ is non-periodic since $\operatorname{soc} H_3 \neq \operatorname{top} H_3$ (IV.1.5). By (1) and since $\Lambda$ is symmetric we have that $\operatorname{soc}_3(e_3\Lambda)$ has length 4 and is cyclic. In particular, $H_3$ is indecomposable; therefore $S_3$ can only lie at the end of a $\mathbb{Z}D_\infty$-component, and one of the possibilities in IV.3.3 holds.

Suppose $U_3$ and $V_3$ are defined. Let $T \subsetneq V_3$ be simple, then $H_3/T \cong V_3/T \oplus U_3$, by the AR-property (see IV.3.4.1), and hence $H_3/\operatorname{soc} H_3$ is a direct sum. On the other hand, $H_3/\operatorname{soc} H_3$ has a simple socle and is therefore indecomposable. Consequently $V_3/T = 0$ and $V_3$ is simple. Thus $V_3 \cong S_1$ or $S_2$ and then $U_1$ or $U_2$ is isomorphic to $S_3$, by IV.3.5. This means implicitly that $U_1$ or $U_2$ is defined, contrary to our hypothesis.

This shows that $U_3$, $V_3$ are not defined, and (2) holds by IV.3.3.

Then necessarily $H_3 \cong H_4$. If $|\operatorname{top} H_3| = 2$ then then quiver of $\Lambda$ contains

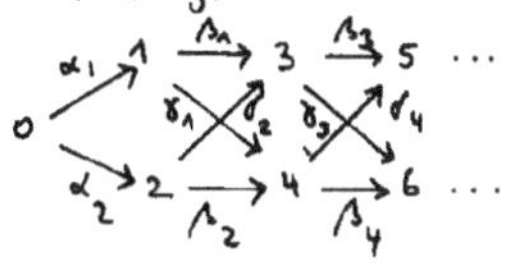

Otherwise it contains

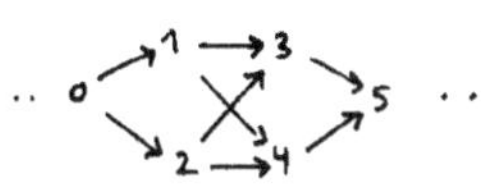

(where possibly $S_5 \cong S_0$.) To avoid too much notation we assume here $|\operatorname{top} H_3| = 2$, the other case is done with the same method; one simply ignores the appropriate arrows.

(3) <u>We may assume that, modulo $J^3$, and up to non-zero scalars $\beta_1\beta_3 \equiv \gamma_1\delta_4$ and $\delta_2\beta_3 \equiv$</u>

$\beta_2\delta_4$, and also $\beta_1\gamma_3 \equiv \gamma_1\beta_4$ and $\delta_2\gamma_3 = \beta_2\beta_4$: Recall $(\beta_1, \delta_2)\Lambda \cong e_3\Lambda/S_3$ and $(\gamma_1, \beta_2)\Lambda \cong e_4\Lambda/S_4$. Hence $H_2 \cong (\beta_1, \delta_2)y\Lambda$ where $y = (\beta_3 + \gamma_3)$, and $H_4 \cong (\gamma_1, \beta_2)z\Lambda$ with $z = (\delta_4 + \beta_4)$. Let $\Psi$: $H_3 \to H_4$ be an isomorphism. This may be lifted to an isomorphism of the injective hulls, that is of $e_1\Lambda \oplus e_2\Lambda$, and therefore $\Psi$ may be represented by a matrix

$$\begin{bmatrix} w_1 & \zeta \\ \rho & w_2 \end{bmatrix}$$

where $w_i$ is a unit in $e_i\Lambda e_i$ for $i = 1, 2$, and $\zeta$, $\rho$ lie in J. In fact, we may assume $\Psi[(\beta_1,\delta_2)y] = [(\gamma_1, \beta_2)z]$; otherwise we replace the arrows starting at vertex 4. Then we have, modulo $J^3$, that $w_1\beta_1(\beta_3 + \gamma_3) \equiv \gamma_1(\delta_4 + \beta_4)$ and $w_2\delta_2(\beta_3 + \gamma_3) \equiv \beta_2(\delta_4 + \beta_4)$.

Now consider the module $M = \mathcal{U}(S_1, S_3)$. By (3) we have that $\Omega M \cong \gamma_1\Lambda$; and also $\Omega^{-1}M$ has a simple socle, by (1). By IV.4.2 and IV.4.3 we know that $\gamma_1\Lambda$ and then also M lie at the end of some AR-components.

(4) M must lie in a tube: Suppose not; then it is of type d. Let $X = \Omega^{-1}M$. Then the stable AR-quiver around X is of the form

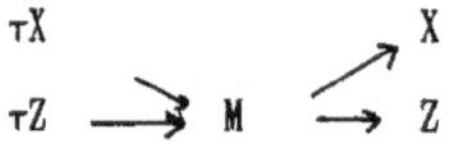


We may apply IV.6.5 and deduce that either $Z \cong \Omega S_1$ or $\tau Z \cong \Omega^{-1}S_3$. Suppose $Z \cong \Omega S_1$. Then $\tau^{-1}M \cong H_1 \cong H_2$, and $X \cong \Omega S_2$ and M is simple, a contradiction. Similarly if $\tau Z \cong \Omega^{-1}S_3$ then it follows since $H_3 \cong H_4$ that $\tau X \cong \Omega^{-1}S_4$ which is also not the case.

(5) M must lie at the end of the 3-tube: Suppose not; then $\Omega^4 M \cong M$ and there is an exact sequence $0 \to M \to P_3 \to P_2 \to P_4 \to P_1 \to M \to 0$, and $\underline{\dim}\,[P_1 \oplus P_2] = \underline{\dim}\,[P_4 \oplus P_3]$, and the Cartan matrix is singular.

The same arguments apply for the other modules of length 2 with tops $S_1$ or $S_2$ and socles $S_3$ or $S_4$. There are four modules of this form, but there is only one 3-tube, and we have derived a contradiction.

VIII.7.6 **PROPOSITION** *The quiver does not contain*

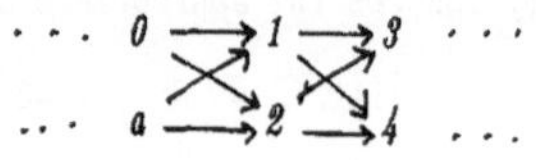

*where* $H_1 \cong H_2$ *(and* $S_0$, $S_a \neq S_3$, $S_4$*).*

Proof: Suppose this occurs. The simple modules $S_0$ and $S_a$ are distinct; otherwise there would be two double arrows, and $\Lambda/J^2$ would be wild, by the dual of I.10.8(iii), and then $\Lambda$ as well (I.4.7). Similarly $S_3 \neq S_4$. Moreover, since $\Lambda$ is not wild there is no loop at any of the vertice 0, a, 3, 4, by I.10.8.

(1) $H_3 \cong$ rad $Q$/soc $Q$ for $Q$ indecomposable projective, $Q \neq P_3$: Suppose not; then also $H_4 \neq$ rad $Q$/soc $Q$ for $Q$ indecomposable projective and $Q \neq P_4$. Then since $S_3$ and $S_4$ are non-periodic it follows that $U_i$ and $V_i$ are defined, for i = 3 and 4. Moreover, they lie at the end of the 3-tube, by IV.6.6. Consequently, either $\mathfrak{U}_4 \cong \underset{\sim}{V}_3$ or $\mathfrak{U}_3 \cong \underset{\sim}{V}_4$. In both cases, there are arrows between vertces 3 and 4, a contradiction.

Then necessarily $H_3 \cong H_4$.

The arguments in VIII.7.5 may be used here as well. We see that the modules of length two with top $S_1$ or $S_2$ and socle $S_3$ or $S_4$ all lie at the end of the 3-tube. There are four of them but $\Lambda$ has ony one 3-tube, a contradiction.

VIII.7.7 PROPOSITION *There is no algebra of semidihedral type with quiver*

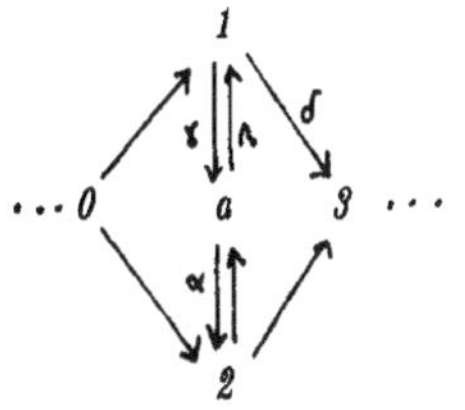


*and* $H_1 \cong H_2$.

Proof: Suppose this occurs. Consider the simple module $S_a$. Since $S_0 \neq S_3$ it is not possible here that $H_a \cong$ rad $Q$/soc $Q$ for $Q$ indecomposable projective and $Q \neq P_a$.

(1) $S_a$ is not periodic: Suppose that $S_a$ is periodic. If $\Omega^4 S_a \cong S_a$ then by IV.1.6 all paths of length 2 starting or ending at vertex a are independent. It follows that the universal cover of an appropriate factor algebra of $\Lambda$ contains the tree

$$\begin{array}{ccccccccccccccc} 2 & \leftarrow & 0 & \rightarrow & 1 & \leftarrow & a & \rightarrow & 2 & \leftarrow & 0 & \rightarrow & 1 & \leftarrow & a \\ & & & & & & & & \downarrow & & & & & & \\ & & & & & & & & a & & & & & & \end{array}$$

This is wild (I.5.9), and $\Lambda$ is wild (I.4.7).

Now suppose that $\Omega^3 S_a \cong S_a$. Then top $\Omega S_a \cong S_a$. To avoid that the universal cover of a factor algebra contains the above wild tree there must be a commutativity relation modulo $J^3$ of the form

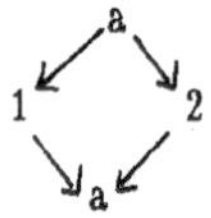


But then $\Lambda$ is still wild, since there is some ideal I of $\Lambda$ such that $\Lambda/I$ has a universal cover which contains a quiver I.9.6.

It follows now from IV.6.6 that $S_a$ is of type d, and $\tilde{U}_a$, $\underset{\sim}{V}_a$ are defined and lie at the end of the 3-tube. ($U_a$ and $V_a$ cannot be simple, for otherwise $U_1$ or $U_2$ would be defined.)

Say top $U_a \cong S_1$. Suppose the arrows are $a \xrightarrow{\beta} 1$, $1 \xrightarrow{\gamma} a$ and $1 \xrightarrow{\delta} 3$.

(2) <u>One of $\beta\gamma$, $\beta\delta$ must be involved in some relation:</u> Suppose not; then the universal cover of an appropriate factor algebra of $\Lambda/J^3$ contains the tree

$$\begin{array}{ccccccccccccccc} & & & & & & a & & & & & & & & \\ & & & & & & \downarrow\beta & & & & & & & & \\ 3 & \leftarrow & 2 & \rightarrow & a & \xleftarrow{\gamma} & 1 & \xrightarrow{\delta} & 3 & \leftarrow & 2 & \rightarrow & a & \leftarrow & 1 \end{array}$$

without relations, and $\Lambda$ is wild.

We identify now $U_a$ with $\beta\Lambda/\beta\Lambda \cap \alpha\Lambda$ where $\alpha$ is the other arrow starting at vertex a; then $\Omega U_a$ may be taken as $\{ x \in e_1\Lambda : \beta x \in \alpha\Lambda \}$. Say $\beta\delta$ occurs in some relation, then $\beta\delta = \beta y + \alpha z$ where $y \in J$, and $z \in \Lambda$. It follows that $(\delta - y)\Lambda \subsetneq \Omega U_a - \text{rad}\, \Omega U_a$. Hence $S_3$ occurs in top $\Omega U_a$. But top $\Omega U_a \cong \text{soc}\, V_a$, so there is an arrow $3 \rightarrow a$, a contradiction. Similarly, if $\beta\gamma$ is involved in some relation it follows that there is a loop at vertex a which is not the case.

This completes the proof of VIII.7.7.

<u>VIII.7.8</u> **PROPOSITION** *There is no algebra of semidihedral type whose quiver contains* $\ldots a \rightleftarrows 1 \rightleftarrows b \ldots$ *with* $H_1 \cong H_2$ *for* $S_1 \neq S_2$.

Proof: Suppose this occurs. Then the quiver of $\Lambda$ contains

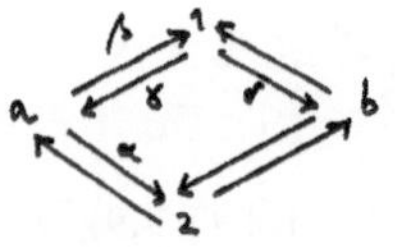

This must be the whole quiver, since otherwise $\Lambda$ would be wild (use I.10.8 for $\Lambda/J^2$).
(1) <u>We must have that $H_a \cong H_b$</u>: The argument in (1) of VIII.7.7 may be applied here as well and show that $S_a$ and $S_b$ can not be periodic. So if (1) does not hold then by IV.6.6, $S_a$ and $S_b$ are of type d, and $U_a$, $V_a$ and $U_b$, $V_b$ are defined. Moreover none of these modules can be simple, since otherwise $U_1$ or $U_2$ would also be defined.
Say top $U_a \cong S_1$; and let $\alpha$, $\beta$ be the arrow starting at vertex a such that $U_a \cong \beta\Lambda/\alpha\Lambda \cap \beta\Lambda$, and let $1 \xrightarrow{\gamma} a$ and $1 \xrightarrow{\delta} b$ be the arrows in the quiver. Then one of $\beta\gamma$, $\beta\delta$ must be involved in some relation, since otherwise $\Lambda$ would be wild (I.10.9). As in VIII.7.7 one shows now that either $S_a$ or $S_b$ occurs in soc $V_a$, this is not the case from the quiver.

Now we will determine the algebras where projectives have isomorphic hearts.

<u>VIII.7.9</u> PROPOSITION *Let $\Lambda$ be of semidihedral type, with quiver (3A), and assume that $H_1 \cong H_2$. Then $\Lambda \cong KQ/I$ where I is defined by the relations*

$$\gamma\beta = \delta\eta, \quad (\beta\gamma)^k\beta = 0, \quad (\eta\delta)^k\eta = 0, \quad (\gamma\beta)^k\gamma = 0 = (\gamma\beta)^k\delta,$$
$$(\beta\gamma)^{k-1}\beta\delta = 0 \quad \text{and} \quad (\eta\delta)^{k-1}\eta\gamma = 0.$$

*where $k \geq 2$.*

Proof: (1) <u>We may assume that $\gamma\beta = \delta\eta$:</u> We have $\gamma\Lambda \cong e_2\Lambda/S_2$ and consequently $\gamma\beta\Lambda \cong H_2$; and the same argument shows that $\delta\eta\Lambda \cong H_1$. By IV.1, there is a unit x of $e_0\Lambda e_0$ satisfying $\delta\eta\Lambda = x\gamma\beta\Lambda$; and we may replace $\gamma$ by $x\gamma$ without changing the situation.

Let k be as large as possible such that $(\gamma\beta)^k \neq 0$.
(2) <u>$\langle (\gamma\beta)^k \rangle = \text{soc } e_0\Lambda$:</u> Since $(\gamma\beta)^k\gamma\beta = 0$, we have that $(\gamma\beta)^k\gamma J = 0$ and therefore $(\gamma\beta)^k\gamma \in \text{soc } \Lambda \cap e_0\Lambda e_2 = 0$. Moreover, $(\gamma\beta)^k\delta\eta = (\gamma\beta)^k\gamma\beta = 0$, hence $(\gamma\beta)^k\delta J = 0$ and $(\gamma\beta)^k\delta \in \text{soc } \Lambda \cap e_0\Lambda e_1 = 0$. This shows that $(\gamma\beta)^k J = 0$. On the other hand, soc $e_0\Lambda$ is simple, and (2) follows.
(3) <u>$\langle (\beta\gamma)^k \rangle = \text{soc } e_2\Lambda$, and $\langle (\eta\delta)^k \rangle = \text{soc } e_1\Lambda$:</u> We deduce from (2) that $(\beta\gamma)^k J = \langle (\beta\gamma)^k\beta \rangle = 0$, therefore $(\beta\gamma)^k$ lies in soc $\Lambda$. Let $\psi$ be a symmetrizing linear form of $\Lambda$, then $\psi[(\beta\gamma)^k] = \psi[(\gamma\beta)^k] \neq 0$ and then $(\beta\gamma)^k \neq 0$. The other statement is clear because $(\gamma\beta)^k = (\delta\eta)^k$.

(4) <u>$k > 2$</u>: Otherwise, $\Lambda/\mathrm{soc}\,\Lambda$ would be special biserial, by (1), (2) and (3), and $\Lambda$ would not be of semidihedral type (II,7.1 and II.7.2).

Now it is straightforward to determine the remaining relations, see also [$E_5$].

<u>VIII.7.10</u> PROPOSITION Assume that $\Lambda$ is an algebra of semidihedral type, with quiver (3C). Then $\Lambda \cong KQ/I$ where I is defined by the relations

$$(\beta\gamma)^{k-1}\beta\delta = 0 = (\eta\delta)^{k-1}\eta\gamma, \quad \eta\rho = 0 = \rho\gamma, \quad \rho\delta = 0 = \beta\gamma,$$

$$(\beta\gamma)^{k}\beta = 0 = (\eta\delta)^{k}\eta = \rho^{s+1},$$

where $s \geq 2$ and $k \geq 1$.

Proof: By VIII.5.2 we know that $H_1 \cong H_2$. Assume first that $H_1$ is simple; hence $\cong S_0$.

Then we know that $S_0$ has three predecessors in $\Gamma_s(\Lambda)$; and we obtain the statement directly from VIII.5.1. Hence we assume now that $H_1$ is not simple.

(1) <u>Without loss of generality, $\gamma\beta = \delta\eta$:</u> From the shape of the quiver one sees that $H_1 \cong \gamma\beta\Lambda$ and $H_2 \cong \delta\eta\Lambda$ By the hypothesis, there is an isomorphism $\psi$: $\gamma\beta\Lambda \to \delta\eta\Lambda$. We may lift $\psi$ to an isomorphism of the injective hulls and deduce that there are units w, v in $e_0\Lambda e_0$ such that $w\gamma\beta v = \delta\eta$. Replace $\gamma$ by $w\gamma$ and $\beta$ by $\beta v$.

(2) <u>$H_1 \cong \beta\Lambda/S_1$ and $H_2 = \eta\Lambda/S_2$</u>: Actually, by (1) we see that there is a well-defined isomorphism which takes $\beta x + S_1$ to $\eta x + S_2$.

Let $X = \gamma\Lambda + \delta\Lambda$. This is the module we studied in VIII.7.2 and VIII.7.3. We proved that X lies at the end of its component, and $\Omega^4 X \neq X$. Here we can say more.

(3) <u>X is not of type d:</u> Suppose not; then, by the results from VIII.7.3, the AR-quiver around X is of the form

$$\begin{array}{ccccc} \tau X & & & \overset{\lambda}{\nearrow} & X \\ & \searrow & & & \\ \tau Z & \longrightarrow & M & \overset{\kappa}{\longrightarrow} & Z \end{array}$$

We have proved that X, Z do not have projective predecessors; and Z is not simple, hence $\mathrm{soc}\,Z \cong S_0 \cong \mathrm{soc}\,\tau Z$. Moreover, $\tau X \cap \tau Z \neq 0$. This means that

(*) the restriction of $\lambda$ to $\tau Z$ is not one-to-one.

On the other hand, $\lambda|\tau Z$ is onto ( since the map $[\lambda, \kappa]: M \to X \oplus Z$ is irreducible and must here be onto).

Choose $w = we_1 \in \tau Z$ with $\lambda(w) = \gamma$. Then $\lambda$ induces an epimorphism $\tilde{\lambda}$: $w\Lambda \to \gamma\Lambda$. But $\gamma\Lambda \cong \Omega^{-1}S_1$, and $\tau Z$ does not have a projective summand. Consequently $\lambda$ is an isomorphism. In particular $\tilde{\lambda}|_{\mathrm{soc}\ w\Lambda}$ is 1-1. But soc $\tau Z$ is simple, therefore $\lambda|_{\tau Z}$ is 1-1, a contradiction to (*). It follows now that

(4) <u>$X$ lies at the end of the 3-tube, hence top $\Omega^2X \cong$ soc $\Omega^2X \cong S_0$.</u>

We study now relations for the algebra. Identify $\Omega X$ with $(\beta,\eta)\Lambda$, then we have:

(5) <u>$\beta\gamma$ is not involved in any relation:</u> Suppose, for a contradiction, that $\gamma\beta = \beta y$ for some element $y \in e_0J^2$. Then, by (2), we have that $\eta\gamma = \eta y + \eta w$ where $\eta w$ lies in soc $e_2\Lambda$. Using the isomorphism in (2) again, we have $\beta w \in$ soc $e_1\Lambda \cap e_1\Lambda e_2$, hence $\beta w = 0$. Consequently $(\gamma - y - w)\Lambda \subsetneq \Omega^2X$ - rad $\Omega^2X$ and $S_1$ occurs in top $\Omega^2X$, a contradiction, see (4). At the same time we also see that $\eta\gamma$ is independent; and $\beta\delta$ and $\eta\delta$ are also independent.

(6) <u>$\beta\rho$ is dependent:</u> Suppose not, then $\eta\rho$ is also independent, using the isomorphism in (2). Then there is an ideal I conaining $J^3$ such that $\Lambda/I$ is generated by zero relations, and its universal cover contains

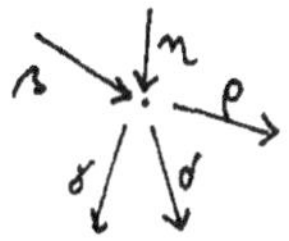

and $\Lambda$ is wild, see I.9.7, I.4.7, a contradiction.

We see now from the quiver that $\beta\rho$ and $\eta\rho$ lie in $J^3$, similarly $\rho\gamma$ and $\rho\delta$.

(7) <u>The elements $\rho\gamma$, $\rho\delta$, $\beta\rho$ and $\eta\rho$ are zero:</u> By (6), the ideal generated by $\rho$ is spanned by powers of $\rho$, and (7) is clear.

Let s be as large as possible such that $\rho^s \neq 0$. Then $\rho\Lambda = \mathrm{span}\,\{\rho, \rho^2, \ldots, \rho^s\}$.

(8) <u>$H_0$ is a direct sum:</u> We have $\mathrm{rad}(e_0\Lambda) = \rho\Lambda + (\gamma\Lambda + \delta\Lambda)$. So it suffices to show that $\rho\Lambda \cap (\gamma\Lambda + \delta\Lambda) \subsetneq$ soc $\Lambda$. If $w$ lies in the intersection then $\rho w = \rho(\gamma x + \delta y) = 0$ and also $\beta w = \beta\rho z = 0$ and $\eta w = \eta\rho z = 0$. Hence $Jw = 0$ and $w \in$ soc $\Lambda$.

Now it is straightforward to derive the remaining relations.

## VIII.8 *End vertices*

VIII.8.1 Exploiting the work done so far, we will in the following tacitly assume the following:

(a) Distinct projectives have non-isomorphic hearts (VIII.7).

(b) Suppose S is of type d, then U and V are not simple (VIII.6).

(c) Simple modules do not have three predecessors (VIII.5).

Then we have some consequences:

(1) Simple modules for end vertices are periodic. This follows from IV.3.7, by (b).

(2) If S is of type d or a, then soc H $\cong$ soc U $\oplus$ soc V; and top H $\cong$ top U $\oplus$ top V.

Our first aim is to show that if the quiver has an end vertex then $\Lambda$ has at most three simple modules. We start with some preparations.

VIII.8.2 LEMMA *Suppose the quiver of* $\Lambda$ *contains* 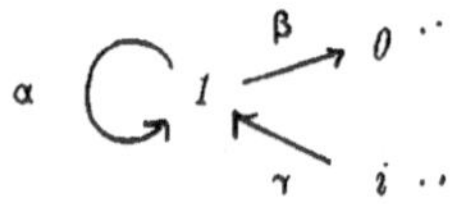

*such that no other arrow starts or ends at vertex 1, and that* $\alpha\beta$ *is involved in some relation. Assume also that* $S_1$ *is non-periodic. Then* $S_1$ *is of type a; and moreover* $\alpha\Lambda = \mathcal{U}(S_1, S_1, \ldots, S_1)$. *If* $S_i \neq S_0$ *then* $\alpha\Lambda$ *lies at the end of the* $\mathcal{I}$*-tube.*

Proof: Assume that $S_1$ is not periodic and not of type d. Then (with the conventions of VIII.8,1) $S_1$ is of type d, and there is an AR-sequence $0 \to V_1 \to H_1 \to U_1 \to 0$.

Case 1 $U_1 = \alpha\Lambda/\alpha\Lambda \cap \beta\Lambda$ and $\underset{\sim}{V}_1 = \beta\Lambda$. Since $\alpha\beta$ is dependent, it follows that $U_1$ is spanned by the cosets of $\alpha, \alpha^2, \alpha^3, \ldots$ . Now consider a projective cover $\pi$: $P_1 \to U_1$. Then $\pi(\beta) = \pi(\beta)e_0 \in U_1e_0 = 0$ and $\beta\Lambda \subset \ker \pi \cong \Omega U_1$. Clearly $\beta\Lambda \not\subseteq \mathrm{rad}^2(\ker \pi)$, and since $\Omega U_1$ has a simple top, it follows that $\beta\Lambda = \Omega U_1$. On the other hand, $\beta\Lambda = V_1$ which lies in a tube, and $\Omega U_1$ is periodic; a contradiction.

Case 2 $U_1 = \beta\Lambda/\beta\Lambda \cap \alpha\Lambda$ and $\underset{\sim}{V}_1 = \alpha\Lambda$. Since $\alpha\beta$ is dependent, we deduce that $V_1$ is spanned by $\alpha, \alpha^2, \ldots$, and $\underset{\sim}{V}_1e_0 = 0$. Now consider a projective cover $\pi$: $P_1 \to \underset{\sim}{V}_1$; we have that $\pi(\beta) = \pi(\beta)e_0 = 0$ and so $\beta\Lambda \subsetneq \ker \pi$. On the other hand, $|\beta\Lambda| = |\langle U_1\rangle| \gneq |\Omega V_1| = |\ker \pi|$, a contradiction.

Now we know that $H_1 \cong U_1 \oplus V_1$, and, say, $\underset{\sim}{V}_1 = \alpha\Lambda$; and then the structure of $\underset{\sim}{V}_1$ follows since $\alpha\beta$ is dependent. Moreover we have that $\underset{\sim}{V}_1 \cong V_1$. Assume that $\Omega^4 V_1 \cong V_1$; then there is an exact sequence $0 \to \underset{\sim}{V}_1 \to P_1 \to P_i \to P_0 \to P_1 \to \underset{\sim}{V}_1 \to 0$. By exactness and since the Cartan matrix is non-singular it follows that $P_0 \cong P_i$ and $S_0 \cong S_i$.

VIII.8.3 LEMMA *Suppose the quiver of* $\Lambda$ *contains*

$$\cdots \quad a \underset{\gamma}{\overset{\beta}{\rightleftarrows}} 0 \underset{\eta}{\overset{\delta}{\rightleftarrows}} b \quad \cdots$$

*and no other vertex starts or ends at 0, and suppose* $S_0$ *does not have period 3. Then each of* $\gamma\beta$, $\delta\eta$ *is independent.*

Proof: If $S_0$ has period dividing 4 then the statement follows from IV.1.6. Otherwise, $S_0$ is of type d, and $U_0$ and $\underset{\sim}{V}_0$ lie at the end of the 3-tube by IV.6.6. Say $U_0 \cong \gamma\Lambda/\gamma\Lambda \cap \delta\Lambda$ and $\underset{\sim}{V}_0 = \delta\Lambda$. We identify $\Omega U_0$ with $\{ x \in P_a : \gamma x \in \delta\Lambda \}$ and $\Omega\underset{\sim}{V}_0$ with $R_\delta$.

Suppose $\gamma\beta = \delta x + \gamma y$ for $x \in J$, $y \in J^2$. Then $(\beta - y)\Lambda \subseteq \Omega U_0$ but does not lie in rad $\Omega U_0$; and $S_0$ occurs in top $\Omega U_0$. But top $\Omega U_0 \cong$ soc $\Omega^2 U_0 \cong$ soc $V_0$, and there is a loop at vertex 0, a contradiction.

Suppose $\delta\eta = \gamma x + \delta y$ where $x \in J$ and $y \in J^2$. Then $\eta\Lambda \subseteq \Omega\underset{\sim}{V}_0$ but does not lie in rad $\Omega\underset{\sim}{V}_0$, and $S_0$ occurs in top $\Omega\underset{\sim}{V}_0$. But top $\Omega\underset{\sim}{V}_0 \cong$ soc $\Omega^2\underset{\sim}{V}_0 \cong$ soc $U_0$, and there is a loop at vertex 0, a contradiction.

Now we deal with the remaining quivers having few arrows. We say that the vertex e of $Q$ is an *end vertex* if e is only joined to one other vertex, f say, and does not have a loop.

VIII.8.4 If e is an end vertex and $\Lambda$ is a tame symmetric algebra with $\geq$ three simple modules then the quiver around e is of the form

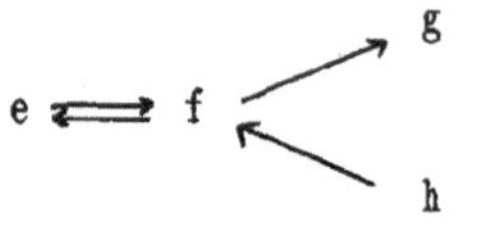


(possibly g = h).

Clearly, since $\Lambda$ is symmetric there is a subquiver of the stated form. If there were another arrow between e and f then $\Lambda$ would be wild (I.10.8).

VIII.8.5 PROPOSITION *Suppose $\Lambda$ is an algebra of semidihedral type whose quiver has an end vertex. Then $\Lambda$ has at most three simple modules.*

Proof: Suppose $\Lambda$ has at least 3 simple modules; let 1 be an end vertex and suppose $Q$ contains

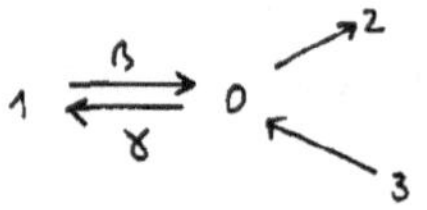


(1) We may assume that $S_1$ is periodic: If $H_1 \cong \mathrm{rad}\, Q/\mathrm{soc}\, Q$ for $Q$ projective $\neq P_1$ then $Q \cong P_2$ or $P_3$. Then we have the situation of VIII.7.2, and we are done. So we may assume that $H_1 \ncong \mathrm{rad}\, Q/\mathrm{soc}\, Q$ for $Q$ projective, $\neq P_1$. We have that $H_1$ is indecomposable. Suppose that $S_1$ is not periodic; then it must be of type d; and $U_1$ and $V_1$ are defined. Since soc $H_1$ and top $H_1$ are simple, it follows that $U_1 \cong S_0 \cong V_1$ and $\tau U_1 \cong V_1 \cong U_1$, a contradiction.

This implies now that most paths of length 2 starting or ending at vertex 1 are not involved in relations, see IV.1.6 and IV.1.7.

(2) If two arrows $\neq \gamma$ start at vertex 0 then $S_1$ has period 3: Let $\delta$, $\mu$ be distinct arrows $\neq \gamma$ startin at 0. Suppose $S_1$ does not have period 3; then top $\Omega^2 S_1 \cong S_0$, see (IV.1), and $\beta\gamma$, $\beta\delta$, $\beta\mu$ are not involved in any relation. Consequently the universal cover of an appropriate quotient of $\Lambda/J^3$ contains

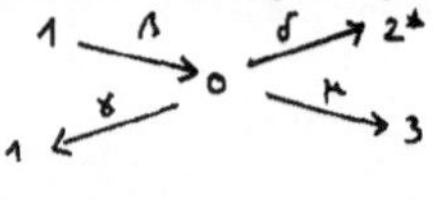

with no relations. If 2 is not an end vertex then attach an arrow $i \xrightarrow{\eta} 2$ at * where $\eta \neq \delta$, and we see that $\Lambda$ is wild (I.9.3, I.4.7). Similarly, 3 must be an end vertex. At least one of $S_2$, $S_3$ has period, say $S_2$. Then the universal cover of an appropriate quotient algebra of $\Lambda$ contains the tree without relations

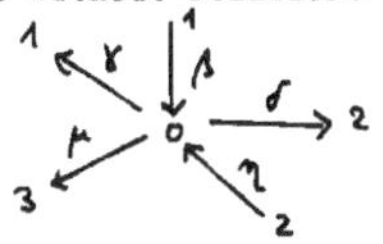

and $\Lambda$ is wild (I.9.7, I.4.7).

<u>Case 1</u> $\Lambda$ has an end vertex 1 where $S_1$ is of period 4.

Then by (2) and the dual of (2), the quiver is of the form

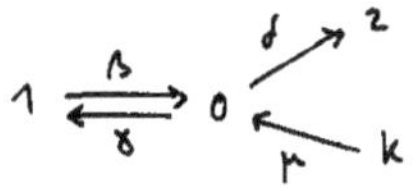

and

(*) no other arrow starts or ends at vertex 0.

Then none of $\delta$ or $\mu$ is a loop since otherwise there could not be a path from 2 to k.

<u>Case 1a</u> $S_0$ is periodic. Then it must have period dividing 4, by IV.1.3. Moreover, by IV.1.5 there is also an arrow $2 \rightarrow 0$ and an arrow $0 \rightarrow k$. The hypothesis implies that $2 = k$.

We assume that $\Lambda$ has at least three simple modules, and 1 is an end vertex. So there must be an arrow $2 \xrightarrow{\zeta} 3$ for a new vertex 3. Then $\delta\zeta$ is not involved in any relation since $S_0$ has period 4, see IV.1.6. It follws now that the universal cover of an approprate factor algebra of $\Lambda$ contains the tree without relations

$$\begin{array}{ccccccccccccccc} 0 & \leftarrow & . & \leftarrow & 0 & \rightarrow & 2 & \rightarrow & 0 & \leftarrow & . & \leftarrow & 0 & \rightarrow & . \\ & & & & & & \downarrow\zeta & & & & & & & & \end{array}$$

and $\Lambda$ is wild (I.9.5, I.4.7).

<u>Case 1b</u> $S_0$ is non-periodic. Then $U_0$ and $\underset{\sim}{V}_0$ are defined. From the hypothesis on the arrows at 0, we have that the tops and socles of $U_0$ and $V_0$ are simple. Therefore, by IV.6.6 we have that $U_0$ and $\underset{\sim}{V}_0$ lie at the end of the 3-tube. By IV.1.2 (with $S_1$ and $U_0$) we deduce that top $\underset{\sim}{V}_0 \neq S_1 \neq$ soc $U_0$. It follows that top $U_0 \cong S_1 \cong$ soc $V_0$, and then $\underline{\dim}\ V_0 = \underline{\dim}\ U_0$ since $\tau U_0 \cong V_0$. Consequently $\underline{\dim}\ P_2 = \underline{\dim}\ \underset{\sim}{V}_0 + \underline{\dim}\ \Omega\underset{\sim}{V}_0 + \underline{\dim}\ U_0 + \underline{\dim}\ \Omega\underset{\sim}{V}_0 = \underline{\dim}\ P_k$, and $P_2 \cong P_k$. Since $\Lambda$ has $\geq$ three simple modules, there must be an arrow $2 \xrightarrow{\xi} 3$. Then $\delta\xi$ is not involved in any relation, also $\delta\eta$ is independent, and $\gamma\beta$ as well (since $S_0$ does not occur in top $\Omega\underset{\sim}{V}_0$). It follows now as in case 1a that $\Lambda$ is wild.

<u>Case 2</u> No end vertex of $\Lambda$ has period 4. Take an end vertex such that $S_1$ has period 3, let $Q$ be

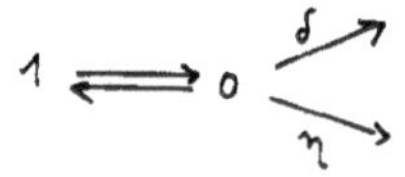

Then $\Lambda$ has no other end vertices since there is only one 3-tube.

(3) <u>At most one arrow $\neq \gamma$ starts at 0:</u> Suppose not, let $\delta$ and $\mu$ be distinct arrows $\neq$ $\gamma$ starting at 0. Since 1 is the only end vertex, the universal cover of an appropriate factor algebra of $\Lambda/J^3$ contains a tree with no relations

$$\begin{array}{c} \qquad\qquad\qquad\qquad\qquad 1 \\ \qquad\qquad\qquad\qquad\qquad \downarrow \\ \cdot \leftarrow \cdot \rightarrow \cdot \xleftarrow{\eta} 0 \xrightarrow{\delta} \cdot \leftarrow \cdot \rightarrow \cdot \leftarrow \cdot \end{array}$$

and $\Lambda$ is wild (I.9.5 and I.4.7).

Dually, only one arrow $\neq \beta$ ends at vertex 1. Hence the quiver around 1 is of the form

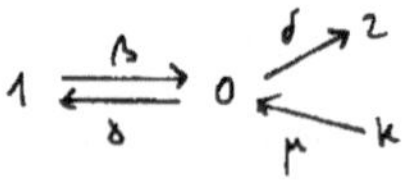

and no other arrow starts or ends at 0. As in case 1, none of $\delta$, $\mu$ are loops, and we may assume that the verices 2 and k are distinct. Also, $\underset{\sim}{U}_0$ and $\underset{\sim}{V}_0$ are defined and lie at the end of the 3-tube. Hence $\underset{\sim}{U}_0 \cong \Omega S_1$ and $\underset{\sim}{V}_0 \cong \Omega^{-1} S_1$, and it follows that $\underline{\dim}\, U_0 = \underline{\dim}\, V_0$, and moreover top $U_0 \cong S_2$ and soc $V_0 \cong S_k$. On the other hand, $\Omega^2 U_0 \cong V_0$. We deduce $\underline{\dim}\, P_2 = \underline{\dim}\, U_0 + \underline{\dim}\, \Omega U_0 = \underline{\dim}\, V_0 + \underline{\dim}\, \Omega U_0 = \underline{\dim}\, P_k$, and the Cartan matrix is singular. This completes the proof of VIII.8.5.

<u>VIII.8.6</u> LEMMA *Suppose the quiver of $\Lambda$ contains*

*where only one arrow starts at vertex 0. Then $\Lambda$ has only wo simple modules.*

Proof: Assume there is another simple module, $S_2$ say. There must be an arrow $i \to 2$ with $i = 0$ or 1 since $\Lambda$ is connected and symmetric, and therefore by the hypothesis there is an arrow $1 \to 2$. Then the quiver contains $0 \leftrightarrows 1 \to 2$, and $\Lambda$ is wild by I.10.8.

We shall now determine the algebras where distinct hearts are non-isomorphic and whose quiver is of the form (3A):

VIII.8.7 **PROPOSITION** *Let $\Lambda$ be of semidihedral type, with quiver (3A), and assume that $H_1 \neq H_2$. Then $\Lambda \cong KQ/I$ where I is given by the relations*

$$\beta\gamma = 0, \quad \delta\eta\delta = (\gamma\beta\delta\eta)^{k-1}\gamma\beta\delta, \quad \eta\delta\eta = (\eta\gamma\beta\delta)^{k-1}\eta\gamma\beta,$$

*where $k \geq 1$. That is, $\Lambda$ belongs to family $SD(3A)_1$.*

Proof: By IV.3.7 we know that the simple modules $S_1$ and $S_2$ are both periodic. Therefore, by VIII.2.1, the simple module $S_0$ must be non-periodic. Consequently, by IV.6.6, $S_0$ lies at the end of a component of type $\mathbb{Z}D_\infty$, with one predecessor. Moreover, IV.3.3(a) holds.

We choose the notation such that top $U_0 \cong S_1$ and top $V_0 \cong S_2$. Then $\underset{\sim}{V}_0 \cong \gamma\Lambda$ and $\langle U_0\rangle \cong \delta\Lambda$. By IV.6.4 we have that $\gamma\Lambda$ lies in a 3-tube. We also know here that $\gamma\Lambda \cong e_2\Lambda/S_2$, so the modules at the end of the 3-tube are $S_2$, $\gamma\Lambda$ and $\beta\Lambda$. In particular, we may assume that $\beta\gamma = 0$. It follows also that $\underset{\sim}{U}_0 \cong \beta\Lambda$ and hence

(2) <u>soc $U_0 \cong S_2$ and soc $V_0 \cong S_1$</u>.

(3) <u>The socle and top of $\Omega U_0$ is $\cong S_1$, and **dim** $U_0$ = **dim** $V_0$</u>:
We know that $\Omega^2 U_0 \cong V_0$, and that top $V_0 \cong S_1 \cong$ top $U_0$. Hence there is an exact sequence $0 \to V_0 \to P_1 \to P_1 \to U_0 \to 0$. This implies (3).

(4) <u>The Cartan matrix of $\Lambda$ is of the form</u>

$$\begin{pmatrix} 4k & 2k & 2k \\ 2k & k+2 & k \\ 2k & k & k+1 \end{pmatrix}$$

Let $\underline{\dim}\, U_0 = [r, k, l]$; then $\underline{\dim}\, P_0 = [2r, 2k, 2l]$ and $\underline{\dim}\, P_2 = [r, k, l+1]$. Moreover $\delta\Lambda \cong e_1\Lambda/S_1$ and $\underline{\dim}\, \delta\Lambda = \underline{\dim} \langle U_0\rangle = [r, k+1, l]$. Consequently

$$C = \begin{pmatrix} 2r & 2k & 2l \\ r & k+2 & l \\ r & k & l+1 \end{pmatrix}$$

Since C is symmetric, we have that $r = 2k$ and $k = l$.

(5) <u>$\Omega U_0 = \mathcal{U}(S_1, S_0, S_1)$:</u> This follows from (3) and (4).

(6) <u>$R_\eta + R_\beta = R_\eta + \gamma\Lambda = \Omega[\mathcal{U}(S_0, S_1)]$:</u> We have that $R_\eta \cap R_\beta$ = soc $e_0\Lambda$. Moreover, we know the composition factors and see that $\underline{\dim}\, e_0\Lambda - \underline{\dim}\, [R_\eta + R_\beta] = \underline{\dim}\, S_0 + \underline{\dim}$

$S_1$.

(7) <u>We may assume that $\delta\eta\delta \in soc_2(\gamma\Lambda)$:</u> Let $\omega \in e_0\Lambda e_0$ such that $R_\eta = \omega\Lambda$. Now, $\delta\eta \in top(\omega\Lambda + \gamma\Lambda)$ but $\delta\eta \notin \gamma\Lambda$. So $\omega = c(\delta\eta) + \gamma x + \delta y$ where $0 \neq c \in K$ and $x \in J$ and $y \in J^2$. Then we may assume that $c = 1$. Now, $\omega\delta = 0$ and therefore $\delta\eta\delta + \delta y\delta = \gamma x\delta \in \delta\Lambda \cap \gamma\Lambda \subseteq soc_2(\gamma\Lambda)$.

Now we are able to determine the relations for $\Lambda$. We replace $\eta$ by $\eta + y$, then we have (8) <u>$\delta\eta\delta = \gamma x\delta \in soc_2(\gamma\Lambda)$.</u> The rest is routine, we omit details; see also [$E_5$].

Now we shall study algebras of semidihedral type with quiver (3*B*).

<u>VIII.8.8.1</u> *Some consequences of known facts*

(1) The simple module $S_2$ is periodic, see VIII.8.1. This implies:

(2) *The simple module $S_0$ cannot have period 3*:

Suppose $S_0$ has period 3. Since there is only one 3-tube, in which there is not enough space at the end for $S_0$ and $S_2$ and also $\Omega^{\pm 1}S_0$, $\Omega^{\pm 1}S_2$, and since the period is not 1 (IV.1.4), it follows that $S_0$ and $S_2$ have coprime period, which contradicts IV.1.2.

(3) The elements $\eta\gamma$, $\beta\delta$ do not occur in any relation. This follows directly from IV.1.6 and IV.1.7.

(4) The elements $\delta\eta$ and $\gamma\beta$ do not occur in any relation, see VIII.8.3 and (1) above.

(5) *There is a relation involving $\alpha\beta$ and a relation involving $\gamma\alpha$:*

Suppose $\alpha\beta$ is independent. Let $\Lambda' = e\Lambda e$ where $e = e_1 + e_0$. also By (3) and (4) we see that the universal cover of $\Lambda'/J(\Lambda')^3$ contains the tree without relations

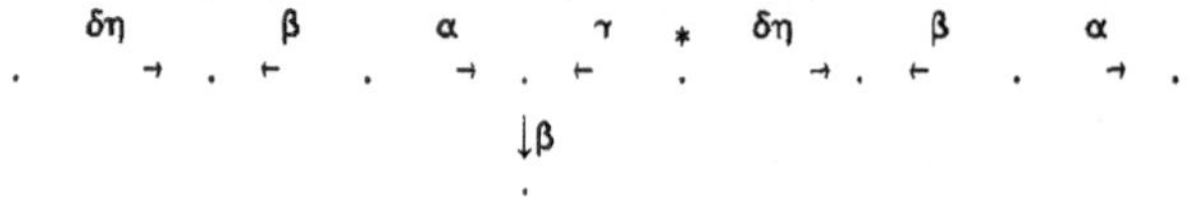

and $\Lambda$ is wild (I.9.5 and I.4.7). Similarly, if $\alpha\gamma$ were independent then $\Lambda$ would be wild. This is seen by taking the tree

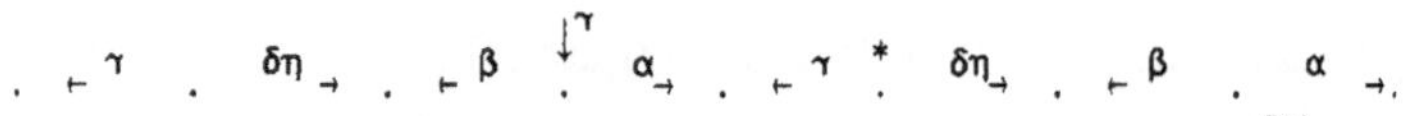

(6) The elements $\eta\delta\eta$ and $\delta\eta\delta$ are dependent: Consider similar trees $\cong \tilde{\tilde{E}}_7$ as in (5) where $\eta$ or $\delta$ is attached to *, without the vertical $\beta$- and $\gamma$-arrow.

VIII.8.8.2 LEMMA *$S_0$ is not periodic.*

Proof: Suppose $S_0$ is periodic. Since $S_1$ is always periodic, we know that $S_1$ is non-periodic, and then by VIII.8.2, the module $S_1$ has two predecessors, and $\alpha\Lambda = \mathcal{U}(S_1, S_1, \ldots, S_1)$, and $\alpha\Lambda e_0 = 0$.

(1) $\alpha\beta = 0 = \gamma\alpha$: We have $\alpha\beta \in \alpha\Lambda e_0 = 0$. Consider the homomorphism $\varphi: \alpha\Lambda \to e_0\Lambda$ given by $\varphi(x) = \gamma x$. Since $soc(e_0\Lambda) \cong S_0$ and $(\alpha\Lambda)e_0 = 0$, $\varphi$ must be zero.

By VIII.8.8.1, part (2) we know that $S_0$ has period 4, and then top $\Omega^2 S_0 \cong S_1 \oplus S_2$. Consider a projective cover $\pi: P_1 \oplus P_2 \to \gamma\Lambda + \delta\Lambda$ given by $\pi(x, y) = \gamma x + \delta y$. We identify ker $\pi$ with $\Omega^2 S_0$. Let $W = (\alpha, 0)\Lambda$ ; then we have by (1) that $W \subseteq \ker \pi$ and clearly $W \not\subseteq rad(\ker \pi)$; and moreover $W \cong V_1$. Dually, there is an epimorphism $\pi^*$ : $\Omega^{-2}S_0 \to V_1$ such that the restriction of $\pi^*$ to soc $\Omega^{-2}(S_0)$ is non-zero. Now $\Omega^{-2}S_0 \cong \Omega^2 S_0$, and its socle is multiplicity-free. So it follows that $\pi^*|$soc $W$ is non-zero and then $\pi^*$ is a split epimorphism, a contradiction.

VIII.8.8.3 THEOREM *Assume $\Lambda$ is of semidihedral type with quiver (3B), and assume that $S_2$ has period 4. Then $\Lambda \cong KQ/I$ where I is the ideal defined by*

$$\alpha\beta = 0 = \gamma\alpha = \beta\gamma$$

$$\alpha^s = (\beta\delta\eta\gamma)^k$$

$$\eta\delta\eta = (\eta\gamma\beta\delta)^{k-1}\eta\gamma\beta \quad \text{and} \quad \delta\eta\delta = (\gamma\beta\delta\eta)^{k-1}\gamma\beta\delta$$

*where $k \geq 1$ and $s \geq 2$. That is, $\Lambda$ belongs to family $SD(3B_1)$.*

Proof: By VIII.8.8.2 and IV.6.6 we know that $S_0$ satisfies IV.3.3(a), and that $\underset{\sim}{U}_0$ and $\underset{\sim}{V}_0$ lie at the end of the 3-tube.

(1) $\underset{\sim}{V}_0 \cong \gamma\Lambda$, and $\underset{\sim}{U}_0 \cong \delta\Lambda/\delta\Lambda \cap \gamma\Lambda$: We have $\underline{Hom}_\Lambda(\underset{\sim}{U}_0, S_2) = 0$. By IV.1.2 it follows that $\underline{Hom}_\Lambda(\underset{\sim}{V}_0, S_2) = 0$ as well, and then (1) must hold.

(2) $\Omega\underset{\sim}{V}_0$ is not simple: We identify $\underset{\sim}{V}_0$ with $\gamma\Lambda$ and $\Omega\underset{\sim}{V}_0$ with $R_\gamma$. If $R_\gamma$ were simple then $R_\gamma = soc(e_1\Lambda)$, and $\gamma$ would be the only arrow terminating at vertex 1, a contradiction.

Hence there is no simple module at the end of the 3-tube.

(3) $S_1$ is not periodic: Suppose it is; then it must have period 4. However, $\underset{\sim}{U}_0$ has

period 3 and $\underline{\mathrm{Hom}}_\Lambda(\mathrm{U}_0, \mathrm{S}_1) = 0$; consequently by IV.1.2 we deduce that $0 = \underline{\mathrm{Hom}}_\Lambda(\Omega \mathrm{U}_0, \mathrm{S}_1) \cong \underline{\mathrm{Hom}}_\Lambda(\underset{\sim}{\mathrm{V}}_0, \mathrm{S}_1)$. This is a contradiction since top $\underset{\sim}{\mathrm{V}}_0 \cong \mathrm{S}_1$.
Now we obtain using VIII.8.8.1(5) that $\underset{\sim}{\mathrm{V}}_1 = \alpha\Lambda \cong \mathcal{U}(\mathrm{S}_1, \mathrm{S}_1, \ldots, \mathrm{S}_1) \cong \mathrm{V}_1$, and $\underset{\sim}{\mathrm{U}}_1 \cong \beta\Lambda$.

(4) $\underline{\alpha\beta = 0 = \gamma\alpha}$: Clearly $\alpha\beta \in \alpha\Lambda e_0 = 0$. Let $\varphi: \alpha\Lambda \to e_0\Lambda$ be the $\Lambda$-homomorphism $\varphi(x) = \gamma x$. Then $\varphi = 0$ since $\alpha\Lambda e_0 = 0$ but $\mathrm{soc}(e_0\Lambda) \cong \mathrm{S}_0$.

We have taken $\gamma\Lambda = \underset{\sim}{\mathrm{V}}_0$; then (4) implies that $\alpha\Lambda \subseteq \Omega\underset{\sim}{\mathrm{V}}_0$. Since $\alpha\Lambda$ is not contained in rad $\Omega\underset{\sim}{\mathrm{V}}_0$ and top $\Omega\underset{\sim}{\mathrm{V}}_0$ is simple it follows that $\alpha\Lambda \cong \Omega\underset{\sim}{\mathrm{V}}_0$. Consequently

(5) The modules at the end of the 3-tube are $\underset{\sim}{\mathrm{V}}_0 \cong \widetilde{\mathrm{U}}_1$, $\Omega\underset{\sim}{\mathrm{V}}_0 \cong \widetilde{\mathrm{V}}_1 \cong \underset{\sim}{\mathrm{V}}_1$ and $\Omega\widetilde{\mathrm{V}}_1 \cong \underset{\sim}{\mathrm{U}}_1 \cong \widetilde{\mathrm{U}}_0 \cong \beta\Lambda$.

(6) $\underline{\beta\gamma = 0}$: This follows since we identify $\Omega(\beta\Lambda)$ and $\gamma\Lambda$.
By VIII.8.8.1 we know that there is a relation involving $\eta\delta\eta$; on the other hand $\eta\delta$ is independent. We are now able to determine the relations.

Let k be the integer such that $(\eta\gamma\beta\delta)^k \neq 0$ but $(\eta\gamma\beta\delta)^{k+1} = 0$. Then $\langle (\eta\gamma\beta\delta)^k \rangle = e_2 J^{4k}$, and it follows that soc $e_2\Lambda = \langle (\eta\gamma\beta\delta)^k \rangle$. Hence a basis of $e_2 J$ adapted to the Loewy series is $\{ \eta, \eta\delta, \eta\gamma, \eta\gamma\beta, \eta\gamma\beta\delta, \eta\gamma\beta\delta\eta, \ldots, (\eta\gamma\beta\delta)^k \}$.

(7) $\underline{\text{We may assume that } \eta\delta\eta = (\eta\gamma\beta\delta)^{k-1}\eta\gamma\beta}$: There is a uniserial module $\mathcal{U}(\mathrm{S}_2, \mathrm{S}_0, \mathrm{S}_2)$ [ take $e_2\Lambda(\eta\gamma)\Lambda$ ]. This mus be isomorphic to a submodule $\omega\Lambda$ say of $e_2\Lambda$; then $\omega$ lies in $\mathrm{soc}_3(e_2\Lambda)$. By considering the Loewy series it follows that $\omega\Lambda = \eta\delta\Lambda$; and then $\eta\delta\eta$ is a non-zero element of $\mathrm{soc}_2(e_2\Lambda)e_0$. Now (7) follows.
Now it is straightforward to obtain the remaining relations; using for example III.16. We omit details.

<u>VIII.8.8.4</u> **THEOREM** *Assume that $\Lambda$ is of semidihedral type, with quiver (3B), and assume that $S_2$ has period 3. Then $\Lambda \cong KQ/I$ where I is the ideal defined by the relations*

$$\eta\delta = 0, \qquad \beta\gamma = \alpha^{s-1},$$
$$\gamma\alpha = (\delta\eta\gamma\beta)^{k-1}\delta\eta\gamma \quad \text{and} \quad \alpha\beta = (\beta\delta\eta\gamma)^{k-1}\beta\delta\eta,$$

*where $k \geq 1$ and $s \geq 3$. That is, $\Lambda$ is a member of family $SD(3B)_2$.*

Proof: By VIII.8.8.2 and IV.6.6 we know that $\mathrm{S}_0$ satisfies IV.3.3(a), and that $\mathrm{U}_0$ and

$\underset{\sim}{V}_0$ lie at the end of the 3-tube. On the other hand, $S_2$ lies also at the end of the 3-tube. It follows that $\Omega\underset{\sim}{V}_0 \cong S_2$ and $\Omega S_2 \cong U_0$; consequently

(1) $\underline{\underset{\sim}{V}_0 \cong \delta\Lambda \text{ and } U_0 \cong \gamma\Lambda/\gamma\Lambda \cap \delta\Lambda;\ \text{soc}\ U_0 \cong S_2 \text{ and soc}\ V_0 \cong S_1}$.

Moreover, by IV.1.8(b) we may assume

(3) $\underline{\eta\delta = 0}$.

(2) $\underline{\dim}\ U_0 = \underline{\dim}\ V_0$: Since top $U_0 \cong$ soc $V_0$ $[\cong S_1]$ and $\Omega^2 U_0 \cong V_0$, there is an exact sequence $0 \to V_0 \to P_1 \to P_1 \to U_0 \to 0$.

(4) $\underline{\text{We may assume that } \gamma\alpha \in \text{soc}_2(\delta\Lambda) \text{ and } \Omega U_0 \cong \alpha\Lambda}$: By VIII.8.8.2, we know that $\gamma\alpha = \gamma x + \delta y$ for $x, y \in J^2$ [and $x = e_2 x e_2$]. Then $\gamma(\alpha - x) \in \gamma\Lambda \cap \delta\Lambda \subset \text{soc}_2(\delta\Lambda)$, and we may replace $\alpha$ by $\alpha - x$.

We identify $\Omega U_0$ with $\{ z \in e_2\Lambda\colon \gamma z \in \delta\Lambda \}$ and obtain that $\alpha\Lambda \subseteq \Omega U_0$; clearly $\alpha$ does not lie in rad $\Omega U_0$. Now equality follows from the fact that top $\Omega U_0$ is simple. In particular, $\alpha\Lambda$ is non-periodic. This implies that $S_1$ is periodic: For, otherwise we would have by VIII.8.2 that $S_1$ is of type a, and by IV.6.2 the module $\alpha\Lambda$ would lie at the end of some tube.

(5) $\underline{\gamma\alpha^2 = 0 \text{ and soc } e_0\Lambda = \langle(\gamma\alpha\beta)\rangle}$: By (4) we have that $\gamma\alpha^2$ lies in soc $e_0\Lambda \cap e_0\Lambda e_1$, hence is 0. Then $\gamma\alpha J = \langle (\gamma\alpha\beta) \rangle = \text{soc}\ e_0\Lambda$.

(6) $\underline{\text{soc}(e_1\Lambda) = \langle(\beta\gamma\alpha)\rangle}$: We have $\beta\gamma\alpha J = \langle \beta\gamma\alpha^2, \beta\gamma\alpha\beta \rangle = 0$. On the other hand, let $\psi$ be a symmetrizing linear form of $\Lambda$, then $\psi[\beta\gamma\alpha] = \psi[\gamma\alpha\beta] \neq 0$ since $\gamma\alpha\beta$ spans a non-zero ideal, and $\gamma\alpha\beta \neq 0$.

(7) $\underline{\beta\gamma \text{ is involved in some relation}}$ : Suppose not. Then $\beta\delta\eta$ is dependent, for otherwise a factor algebra of $\Lambda$ would have a covering which contains the tree without relations

$$\overset{\beta}{\leftarrow} \cdot \overset{\alpha}{\rightarrow} \cdot \overset{\gamma}{\leftarrow} \underset{\cdot}{\overset{\beta}{\downarrow}} \overset{\delta}{\rightarrow} \cdot \overset{\eta}{\rightarrow} \cdot \overset{\beta}{\leftarrow} \cdot \overset{\alpha}{\rightarrow} \cdot$$

and $\Lambda$ would be wild (I.9.5 and I.4.7).

It follows that $e_1 J$ has a basis $\{ \alpha, \alpha^2, \ldots, \beta, \beta\gamma, \beta\delta, \beta\gamma\beta, (\beta\gamma)^2, \ldots \}$. In particular $\dim e_1\Lambda e_2 = 1$. Hence using (3) we also have that $1 = \dim (e_2\Lambda/S_2)e_1 = \dim \underset{\sim}{V}_0 e_1 = \dim(V_0)e_1 = \dim (U_0)e_1$ and therefore $\dim e_0\Lambda e_1 = 2$ and $\dim e_1\Lambda e_0 = 2$. Consequently $e_1\Lambda e_0 = \text{span}\ \{ \beta, \beta\gamma\beta \}$. On the other hand, $\alpha\beta$ lies in $e_1 J^2 e_0$, and therefore $\alpha\beta = c(\beta\gamma\beta)$ for some $c \in K$. Let $\alpha' = \alpha - c(\beta\gamma)$, then $\alpha'$ is an arrow $\equiv \alpha$, and $\alpha'\beta = 0$. Then, by (6), we see that $\alpha'\Lambda \cap \beta\Lambda \subseteq \text{soc}\ e_1\Lambda$, and $H_1$ is a direct sum.

This is a contradiction; we have already proved that $S_1$ is periodic, and therefore $H_1$ must be indecomposable.

It is now straightforward to determine the relations, we will not give details.

## VIII.9 *Quivers with enough arrows.*

We have now considered all possible quivers which contain some vertex f with at which only one arrow starts or ends, see VIII.2.0. We assume now that $|s^{-1}(f)| \geq 2$ and $|e^{-1}(f)| \geq 2$ for each vertex f of the quiver. Since $\Lambda$ is not wild, it follows then that $|s^{-1}(f)| = |e^{-1}(f)| = 2$ for all f, see I.10.8. We will frequently use I.10.9.

VIII.9.1 We summarize now some known properties we shall use:

(a) By the results of VIII.7 we always assume that hearts of distinct projectives are non-isomorphic. This leaves then the following options:

(b) A simple module $S_0$ may be periodic; then soc $H_0 \cong$ top $H_0$.

(c) If a simple module is non-periodic then $U_0$ and $\underset{\sim}{V}_0$ are defined and have simple socles and tops, and they lie at ends of tubes (IV.5.5, IV.6.2). Moreover

(d) Suppose $S_0$ is non-periodic. If there is no loop at vertex 0 then $S_0$ is of type d, and $U_0$ and $\underset{\sim}{V}_0$ lie at the end of the 3-tube (IV.6.6).

To determine the possible quivers we shall use the following result:

VIII.9.2 LEMMA *Suppose $\Lambda$ is an algebra of semidihedral type whose quiver has enough arrows, Let $S_0$ be a non-periodic simple module. Then there is an arrow $i \to j$ in the quiver of $\Lambda$ in each of the folowing cases:*

*(i) $S_0$ is of type d, top $U_0 \cong S_i$ and soc $V_0 \cong S_j$.*

*(ii) top $\underset{\sim}{V}_0 \cong S_i$ and soc $\Omega\underset{\sim}{V}_0 \cong S_j$.*

*(iii) soc $U_0 \cong S_j$ and top $\Omega^{-1}U_0 \cong S_i$.*

Proof: (i) Suppose $\alpha$, $\delta$ are the arrows starting at verex 0, with $U_0 = \alpha\Lambda/\alpha\Lambda \cap \delta\Lambda$. We identify $\Omega U_0$ with $\{ x \in e_i\Lambda : \alpha x \in \delta\Lambda \}$. There are two arrows starting at i, $\beta$ and $\gamma$ say. By I.10.9 one of $\alpha\beta$, $\alpha\gamma$ must be involved in some relation. Say $\alpha\beta = \alpha y + \delta z$ with $y \in J^2$ and $z \in J$. Then $(\beta - z)\Lambda$ is contained in $\Omega U_0$ but not in rad $\Omega U_0$. Since top $\Omega U_0 \cong$ soc $V_0 \cong S_j$ which is simple, it follows that $(\beta - z)\Lambda = \Omega U_0$ and moreover $\beta$ ends at vertex j. Parts (ii) and (iii) are proved similarly.

<u>VIII.9.3</u> LEMMA *Suppose the quiver of* $\Lambda$ *contains*

$$\cdots\ 1 \underset{\gamma}{\overset{\beta}{\rightleftarrows}} 0 \underset{\eta}{\overset{\delta}{\rightleftarrows}} 2\ \cdots$$

*Then* $\Lambda$ *has three simple modules.*

Proof: Assume first that $S_0$ is periodic, of period 4. Then if $\omega$ is a path of length 2 which starts at 0 but does not end at 1 or 2 then $\omega$ is not involved in any relation. Dually, any path of length 2 which ends at 0 but does not start at 1 or 2 is also not involved in any relation. Hence the universal cover of an appropriate factor algebra of $\Lambda$ contains lines of arbitrary length

$$\cdot \leftarrow 0 \rightarrow \cdot \rightarrow 0 \leftarrow \cdot \leftarrow 0 \rightarrow \cdot \rightarrow 0 \leftarrow \ \cdots\ .$$

Assume for contradiction that $\Lambda$ has more than three simple modules. Then since the quiver is connected (and $\Lambda$ is not wild), there is an arrow $2 \xrightarrow{\rho} 3$, say. We deduce that $\delta\rho$ is not involved in any relation, and therefore the universal cover contains the wild tree $\tilde{E}_7$, namely

$$\begin{array}{ccccccccccccc} 0 & \leftarrow & \cdot & \leftarrow & 0 & \xrightarrow{\delta} & \cdot & \xleftarrow{\eta} & 0 & \leftarrow & \cdot & \leftarrow & 0 & \rightarrow & \cdot \\ & & & & & & \downarrow{\scriptstyle\rho} & & & & & & & & \\ & & & & & & \cdot & & & & & & & & \end{array}$$

and $\Lambda$ is wild (I.9.5, I.4.7). Hence if $S_0$ has period 4 then the statement holds.

The next possibility is that $S_0$ is of type d. Then (assuming VIII.9.1) we have that $U_0$ and $V_0$ are defined as in IV.3.3(a). We can continue as in the first part of the proof, provided we show that $\delta\eta$ and $\gamma\beta$ are not involved in any relations.

Say $U_0 = \delta\Lambda/\delta\Lambda \cap \gamma\Lambda$ and $\underset{\sim}{V}_0 = \gamma\Lambda$. We take $\Omega U_0 = \{ x \in e_2\Lambda :\ \delta x \in \gamma\Lambda \}$, with top isomorphic to $S_1$; hence ther can be no relation in which $\delta\eta$ occurs. Moreover, $V_0$ lies in the 3-tube here. So top $\Omega V_0 \cong$ soc $U_0 \cong S_1$ or $S_2$, and it does not contain $S_0$. Hence no relation exists in which $\gamma\beta$ occurs.

We are left to consider the case when $S_0$ has period 3. Then $S_1$ and $S_2$ are non-periodic, by IV.1.2. Then $\tilde{U}_i$ and $\underset{\sim}{V}_i$ are defined, and these modules lie at ends of tubes. They have simple socles and tops and are not simple, so they are distinct from the modules at the end of the 3-tube. Then by IV.6.6 we deduce that there is a loop at vertex 1 and one at vertex 2. Since $\Lambda$ is not wild, the quiver must be of the form ($3\mathcal{D}$), and $\Lambda$ has three simple modules. This completes the proof.

(Actually, the last case does not occur.)

<u>VIII.9.4</u> LEMMA *The quiver of $\Lambda$ does not contain a vertex 0 and*

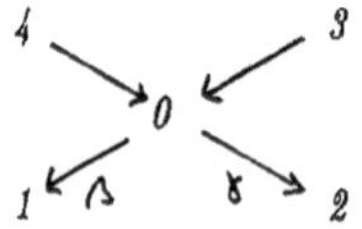


*where 1, 2, 3 and 4 are distinct vertices.*

Proof: Suppose this occurs; then in particular, there is no loop at vertex 0; moreover the simple modules $S_0, S_1, \ldots, S_4$ are non-periodic. With the convention VIII.9.1(a) we have that $\tilde{U}_i$ and $\underset{\sim}{V}_i$ are defined for $i = 0, 1, \ldots, 4$. Moreover $S_0$ is of type d, and $\tilde{U}_0$ and $\underset{\sim}{V}_0$ must lie at the end of the 3-tube. Say $\underset{\sim}{V}_0 = \beta\Lambda$ and $U_0 = \gamma\Lambda/\gamma\Lambda \cap \beta\Lambda$; assume that soc $V_0 \cong S_3$ and soc $U_0 \cong S_4$. Then the module $\Omega\underset{\sim}{V}_0$ has top $\cong S_4$ and socle $\cong S_1$. We see now from the socles and tops that none of the modules at the end of the 3-tube can be isomorphic to $\tilde{U}_2$ or $\underset{\sim}{V}_2$.

It follows now from VIII.9.1(d) that there is a loop at vertex 2, call it $\epsilon$. Let $\delta$ be the other arrow starting at 2. We may assume that $U_0 = \gamma\Lambda/\gamma\Lambda \cap \beta\Lambda$; and since $\Omega^2 U_0 \cong V_0$, we know that top $\Omega U_0 \cong S_3$. One of $\gamma\delta$, $\gamma\epsilon$ is involved in some relation, by I.10.9. If it were $\epsilon$ then we would have that $(\epsilon - x)\Lambda = \Omega U_0$ for some $x \in J^2$, and then top $\Omega U_0 \cong S_2$, a contradiction. Hence $\gamma\delta$ is involved in some relation, and $U_0$ is spanned by the cosets of $\gamma, \gamma\epsilon, \gamma\epsilon^2, \ldots$ . It follows that $U_0 e_4 = 0$. This is not possible since soc $U_0 \cong S_4$.

<u>VIII.9.5</u> LEMMA *The quiver of Λ does not contain*

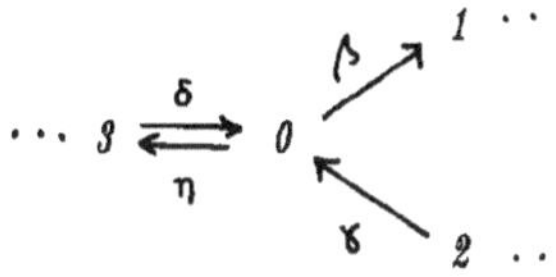


*with $S_1$, $S_2$ and $S_3$ distinct.*

Proof: Suppose this occurs. Then $S_0$, $S_1$ and $S_2$ are non-periodic. Moreover $S_0$ is of type d, and $\underset{\sim}{U}_0$ and $\underset{\sim}{V}_0$ are defined and lie at the end of the 3-tube.

(1) <u>It is not possible that top $U_0 \cong S_3 \cong$ soc $V_0$:</u> Suppose this happens. Since $\Omega^2 U_0 \cong V_0$, it follows that $\underline{\dim}\, U_0 = \underline{\dim}\, V_0$; and then $\underline{\dim}\, P_1 = \underline{\dim}\, \underset{\sim}{V}_0 + \underline{\dim}\, \Omega\underset{\sim}{V}_0 = \underline{\dim}\, \underset{\sim}{U}_0 + \underline{\dim}\, \Omega\underset{\sim}{V}_0 = \underline{\dim}\, P_2$, and the Cartan matrix is singular.

Say top $U_0 \cong S_3$, top $V_0 \cong S_1$ and soc $V_0 \cong S_2$ ; then soc $U_0 = S_3$. (The other case is dual.) By VIII.9.2, there is an arrow $3 \to 2$ in the quiver of Λ.

(2) <u>The modules $\underset{\sim}{U}_3$ and $\underset{\sim}{V}_3$ lie at the end of the 3-tube:</u> None of the arrows starting at 3 is a loop. By VIII.9.1(d) we may assume that there is no arrow $2 \to 3$ in the quiver. Then $S_3$ is not periodic and then of type d; and (2) follows, using VIII.9.1.

By comparing tops and socles we see that the only possibility is that $\underset{\sim}{U}_3 = \Omega\underset{\sim}{V}_0$ and $\underset{\sim}{V}_3 = \underset{\sim}{U}_0$; then

(3) $\underline{\dim}\, P_1 = \underline{\dim}\, \underset{\sim}{U}_3 + \underline{\dim}\, \underset{\sim}{V}_0$.

In particular also top $\underset{\sim}{V}_3 \cong S_0$ and top $U_3 = S_2$; and soc $U_3 \cong S_1$ and therefore soc $V_3 \cong S_0$.

Since $\Omega^2 U_0 \cong V_0$ and $\Omega^2 U_3 \cong V_3$ there are exact sequences $0 \to V_0 \to P_2 \to P_3 \to U_0 \to 0$ and $0 \to V_3 \to P_0 \to P_2 \to U_3 \to 0$. Hence we deduce that $0 = \underline{\dim}\, [U_0 \oplus U_3] - \underline{\dim}\, P_3 + \underline{\dim}\, P_0 - \underline{\dim}[V_0 \oplus V_3]$ consequently $\underline{\dim}\, \underset{\sim}{U}_0 = \underline{\dim}\, \underset{\sim}{U}_3$ and then by (3) we see that $\underline{\dim}\, P_1 = \underline{\dim}\, \underset{\sim}{U}_0 + \underline{\dim}\, \underset{\sim}{V}_0 = \underline{\dim}\, P_0$, and the Cartan matrix is singular, a contradiction.

<u>VIII.9.6</u> LEMMA *The quiver of Λ does not contain a double arrow if Λ. has ≥ two simple modules.*

Proof: By VIII.4.1 and 2 we may assume that $\Lambda$ has at least three simple modules. Suppose $\beta_1, \beta_2 : 0 \to 1$ are arrows in $Q$. These are the only arrows starting at 0 or ending at 1. If there are also two arrows $1 \Rightarrow 0$ then there are no more arrows, and $\Lambda$ has two simple modules. Assume therefore that there is no double arrow $1 \Rightarrow 0$. Then soc $H_i \neq$ top $H_i$ for i = 0, 1, and hence $S_0$ and $S_1$ are non-periodic. Moreover, there is no loop attached to vertex 0 or 1. Consequently, by VIII.9.1(d), $S_0$ and $S_1$ are of type d and $U_i$, $\underset{\sim}{V}_i$ lie at the end of the 3-tube for i = 0, 1. So some of them must be isomorphic, say $\underset{\sim}{V}_0 \cong U_1$. Then we have that $\Omega\underset{\sim}{V}_1 \cong U_0$, and hence top $V_1 \cong$ soc $U_0 \cong S_j$ say, and the quiver contains a triangle

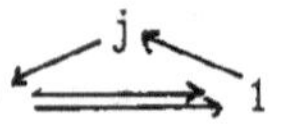


Let $S_m$ = soc $V_0$. By VIII.9.2 there must be an arrow $1 \to m$; and then top $U_1 \cong S_m$. Since $\Omega^2 U_0 \cong V_0$ and $\Omega^2 U_1 \cong V_1$, there are exact sequences $0 \to V_0 \to P_m \to P_1 \to U_0 \to 0$ and $0 \to V_1 \to P_0 \to P_m \to U_1 \to 0$. Hence $\underline{\dim}$ $[U_0 \oplus U_1]$ - $\underline{\dim}$ $P_1$ + $\underline{\dim}$ $P_0$ - $\underline{\dim}$ $[V_0 \oplus V_1] = 0$, which implies $\underline{\dim}$ $\underset{\sim}{V}_1$ = $\underline{\dim}$ $U_0$ . But then, remembering that $V_0 \cong U_1$, we deduce $\underline{\dim}$ $P_0$ = $\underline{\dim}$ $P_1$, and the Cartan matrix is singular.

<u>VIII.9.7</u> **LEMMA** *The quiver of $\Lambda$ does not contain*

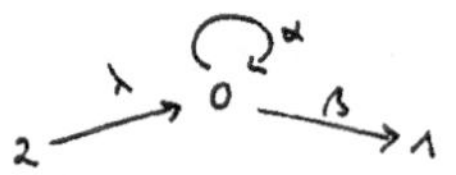


*with $S_2 \neq S_1$.*

Proof: Suppose this occurs. By VIII.9.6, there are no double arrows; and we deduce from VIII.9.4 and 5 that there can only be loops at vertices 1 and 2. Inductively, we see that the quiver can only be of the form

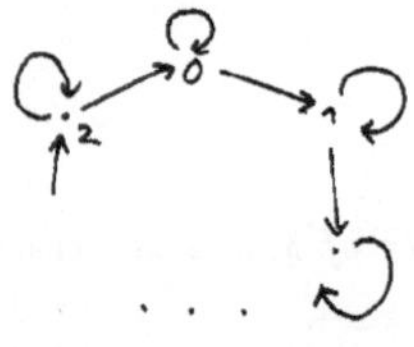


None of the simple modules is periodic.

(1) <u>If $S_0$ is of type d then top $U_0 \cong S_0$ and soc $V_0 \cong S_0$, and $U_0$ lies in the 3-tube:</u>
Suppose that top $U_0 \cong S_1$. Then by VIII.9.2 there must be an arrow $1 \to i$ where $S_i$ = soc $V_0$. Then soc $V_0 \cong S_2$ and soc $U_0 \cong S_0$. Let $\epsilon$ be the loop at vertex 1; then $U_0$ is spanned by the cosets of $\beta$, $\beta\epsilon$, $\beta\epsilon^2$, ..., and $U_0e_0 = 0$. This is a contradiction since soc $U_0 \cong S_0$.
Hence top $U_0 \cong S_0$; and then we deduce using VIII.9.2 that soc $V_0 \cong S_0$. The rest follows from VIII.9.2. Similarly if $S_0$ is of type a then also $U_0$ [say] lies in the 3-tube.

Hence to each simple module $S_i$, we have a module $U_i$ which lies at the end of the 3-tube and therefore $\Lambda$ can only have three simple modules.

We will now show that this is not possible after all. Considering the modules at the end of the 3-tube we see that for any loop $\epsilon$ of the quiver, $\epsilon^2$ is involved in some relation. We may then write down a basis of $\Lambda$ which consists of subwords of powers of the cyclic words where the cycle involves all arrows. One sees that the Cartan matrix is singular.

Now it remains to determine the algebras of semidihedral type whose quiver is of the form ($3\mathcal{D}$) or ($3\mathcal{K}$).

<u>VIII.9.8</u> *The simple module $S_0$ does not have period 3:*
For example, consider the quiver ($3\mathcal{D}$). If the statement is not true then each of $\delta\xi$, $\gamma\alpha$ and $\alpha\beta$ is independent; and without loss of generality $\gamma\beta = \delta\eta$, by IV.1.8. Then the universal cover of an appropriate factor algebra of $\Lambda/J^3$ involves the quiver

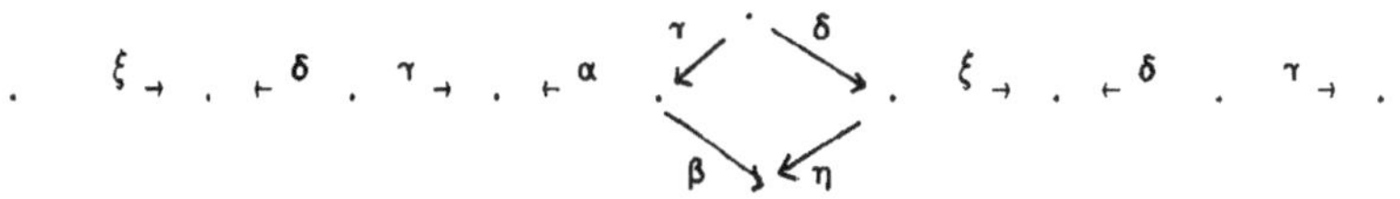


where the only relation is the commutative square, which is wild (I.9.7), and then $\Lambda$ is wild, by I.4.7. A similar argument holds for every simple module in case the quiver is of the form ($3\mathcal{K}$).

<u>VIII.9.9</u> LEMMA *Suppose the quiver of $\Lambda$ is of the form ($3\mathcal{D}$). Then there is no relation involving any of $\beta\delta$ or $\eta\gamma$, and of $\delta\eta$ or $\gamma\beta$. Moreover, each of the elements $\delta\xi$, $\xi\eta$, $\gamma\alpha$, $\alpha\beta$, $\beta\gamma$ is independent.*

Proof: (a) Suppose that $\beta\delta = \beta x + \alpha y$ where $x, y \in J$; then $x, y \in J^2$ and $e_1\Lambda e_2 \subseteq e_1J^3e_2$. Every path in the quiver from vertex 1 to 2 must involve $\beta\delta$, it follows therefore that $e_1\Lambda e_2 = 0$. Since $\Lambda$ is symmetric, also $e_2\Lambda e_1 = 0$. Then we have that $\gamma\Lambda \cap \delta\Lambda = \text{soc } e_0\Lambda$: If $\gamma x = \delta y$ then $\beta\gamma x = \beta\delta y \in e_1\Lambda e_2 y = 0$ and $\eta\gamma x \in e_2\Lambda e_1 x = 0$ and $J\gamma x = 0$. That is, $H_0$ is a direct sum. This is a contradiction to IV.6.6.
Similarly one shows that $\eta\gamma$ does not occur in any relation.

The statement for $\delta\eta$, $\gamma\beta$ is clear by IV.1.6 if $S_0$ has period 4. Otherwise, by I.10.9 and IV.6.6, we know that $S_0$ is of type d, and $\underset{\sim}{U}_0$ and $\underset{\sim}{V}_0$ lie at the end of the 3-tube. Say $U_0 = \delta\Lambda/\delta\Lambda \cap \eta\Lambda$ and $\underset{\sim}{V}_0 = \gamma\Lambda$. We take $\Omega U_0 = \{x \in e_2\Lambda\colon \delta x \text{ lies in } \eta\Lambda\}$ and $\Omega\underset{\sim}{V}_0 = R_\gamma$.

Suppose $\delta\eta = \delta x + \gamma y$ for some $x \in J^2$ and $y \in J$. Then $\eta - x$ lies in $\Omega U_0$ but not in rad $\Omega U_0$. Therefore $S_0$ occurs in top $\Omega U_0$ and hence in soc $V_0$, and there is a loop at vertex 0, a contradiction.

Now assume $\gamma\beta = \gamma x + \delta y$ for $x \in J$ and $y \in J^2$; then $\beta - x \in \Omega\underset{\sim}{V}_0 - \text{rad } \Omega\underset{\sim}{V}_0$ and $S_0$ occurs in top $\Omega\underset{\sim}{V}_0$ which is isomorphic to soc $\Omega^2\underset{\sim}{V}_0$, that is to soc $\underset{\sim}{U}_0$, and there is a loop at vertex 0, a contradiction. The rest follows now from I.10.9.

<u>VIII.9.10</u> LEMMA *Suppose the quiver of $\Lambda$ is of the form (3K). Then there is no relation involving a closed path in the quiver of length 2.*

This is proved by the argument of VIII.9.9.

Now we will first finish the study of algebras with quiver (3*D*).

<u>VIII.9.11</u> LEMMA *$S_1$ and $S_2$ are not both periodic.*

Proof: Assume for contradiction that $S_1$ and $S_2$ are both periodic. Then $S_0$ is non-periodic, by VIII.2.1. It follows now from IV.6.6 that $S_0$ is of type d, and $\underset{\sim}{U}_0$ and $\underset{\sim}{V}_0$ lie at the end of the 3-tube. Consequently all modules at the end of the 3-tube have simple socles and tops; hence none of them can be isomorphic to $\Omega S_2$ or $\Omega S_1$. Therefore $S_1$ and $S_2$ must have period 4. We apply now IV.1.1 with $\underset{\sim}{U}_0$ and $S_1$, $S_2$. Since $\underline{\text{Hom}}_\Lambda(\underset{\sim}{U}_0, S_i) = 0$ for $i = 1,2$ it follows that $\underline{\text{Hom}}_\Lambda(\underset{\sim}{V}_0, S_i) = 0$ as well for $i$ =

1, 2, This is a contradiction since $\underset{\sim}{V}_0$ is isomorphic to $\gamma\Lambda$ or $\delta\Lambda$.

<u>VIII.9.12</u> **THEOREM** *Assume that $\Lambda$ is of semidihedral type with quiver of the form (SD). Then $\Lambda \cong KQ/I$ where I is the ideal defined by the relations*

$$0 = \delta\xi = \xi\eta = \eta\delta,$$
$$\beta\gamma = \alpha^{s-1},$$
$$\gamma\alpha = (\delta\eta\gamma\beta)^{k-1}\delta\eta\gamma \quad \text{and} \quad \alpha\beta = (\beta\delta\eta\gamma)^{k-1}\beta\delta\eta,$$
$$\xi^t = (\eta\gamma\beta\delta)^k,$$

*where $k \geq 1$, $s \geq 3$ and $t \geq 2$.*

Proof: By VIII.9.11 we may choose the notation such that $S_2$ is non-periodic. Then by VIII.8.2 we know that $S_2$ is of type a, and $\xi\Lambda = \mathcal{U}(S_2, S_2, \ldots, S_2) = \underset{\sim}{V}_2$ say. It follows then that $\xi\eta = 0$ since $\xi\Lambda e_0 = 0$, and also $\delta\xi = 0$, since $\xi$ must lie in the kernel of the homomorphism $x \to \delta x$. Moreover $\tilde{V}_2 \cong \underset{\sim}{V}_2$.

(1) <u>$S_0$ is not periodic:</u> Suppose this is false. Then by VIII.9.8 we have hat $\Omega^4 S_0 \cong S_0$. We identify $\Omega^2 S_0$ with $\{ (x,y) \in e_1\Lambda \oplus e_2\Lambda\colon \gamma x + \delta y = 0 \}$. Since $\delta\xi = 0$, we have an embedding $j\colon \xi\Lambda = \underset{\sim}{V}_2 \to \Omega^2 S_0$; and clearly $j(\underset{\sim}{V}_2)$ does not lie in rad $\Omega^2 S_0$.

Dually, since also $\xi\eta = 0$, there is an epimorphism $\rho\colon \Omega^{-2} S_0 \to \underset{\sim}{V}_2$ such that the restriction of $\rho$ to soc $\Omega^{-2} S_0$ does not vanish. Now $\Omega^{-2} S_0 \cong \Omega^2 S_0$, and $S_1$ occurs with multiplicity 1 in soc $\Omega^2 S_0$. It follows that $\rho\, j$ is non-zero on soc $\underset{\sim}{V}_2$, and then since soc $\underset{\sim}{V}_2$ is simple, $\rho\, j$ is an isomorphism. Hence $\Omega^2 S_0 \cong \underset{\sim}{V}_2$ and its top is simple, a contradiction.

Now we see from IV.6.6 that $S_0$ is of type d, and the modules $\tilde{U}_0$ and $\underset{\sim}{V}_0$ lie at the end of the 3-tube.

(2) <u>$\underset{\sim}{V}_0 = \delta\Lambda$ and $U_0 = \gamma\Lambda/\gamma\Lambda \cap \delta\Lambda$:</u> Suppose this is not so; then $U_0 = \delta\Lambda/\,\delta\Lambda \cap \gamma\Lambda$; we identify $\Omega U_0$ with $\{ x \in e_2\Lambda\colon \delta x \in \gamma\Lambda \}$. Since $\delta\xi = 0$ we have that $\xi\Lambda \subseteq \Omega U_0$, and $\xi$ does not lie in rad $\Omega U_0$. But $\Omega U_0$ has a simple top, hence $\Omega_0 = \xi\Lambda$. This is a contradiction since $\xi\Lambda$ is periodic (IV.6.2) but $\Omega U_0$ is not.

Now, $\xi\Lambda = \Omega(\delta\Lambda)$ that is $\delta\Lambda \cong \tilde{V}_2$, and the modules at the end of the 3-tube are therefore $\underset{\sim}{V}_0 = \delta\Lambda \cong \tilde{U}_2$, $\xi\Lambda = \underset{\sim}{V}_2 \cong \tilde{V}_2$, and $\underset{\sim}{U}_2 = \eta\Lambda$. We deduce that $\Omega\underset{\sim}{U}_2 \cong \underset{\sim}{V}_0$; in particular

(3) <u>We may assume that $\eta\delta = 0$.</u>

(4) $\underline{S_1 \text{ is periodic, of period 4:}}$ Suppose that $S_1$ is not periodic, then by VIII.8.3 we have that $H_1 \cong U_1 \oplus V_1$, and $V_1 \cong \alpha\Lambda \cong U(S_1, S_1, \ldots, S_1)$. In particular $\gamma\alpha = 0$, and then $\alpha\Lambda \subsetneq \Omega U_0$ but not in rad $\Omega U_0$. Since top $\Omega U_0$ is simple it follows that $\Omega U_0 = \alpha\Lambda$. This is a contradiction since $\underset{\sim}{V}_1$ is periodic but $\Omega U_0$ is non-periodic. Clearly, the period is 4 since there is no space at the end of the 3-tube.

(5) $\underline{\text{We may assume that } \gamma\alpha \in \text{soc}_2\Lambda - \text{soc } \Lambda}$: By VIII.9.9 we know that $\gamma\alpha$ is dependent. Say $\gamma\alpha = \gamma x + \delta y$ where $x, y \in J^2$. Then $\gamma(\alpha - x) = \delta y$ and this element lies in $\text{soc}_2(\delta\Lambda)$ we may relplace $\alpha$ by $\alpha - x$. If $\gamma\alpha$ were in soc $\Lambda$ then $\gamma\alpha = 0$ and $R_\gamma = \Omega U_0$. This is not true.

From (5) and the fact that $\Lambda$ is symmetric we deduce

(6) $\underline{0 \neq \beta\gamma\alpha \in \text{soc}(e_1\Lambda) \text{ and } 0 \neq \gamma\alpha\beta \in \text{soc}(e_0\Lambda); \text{ moreover } \eta\gamma\alpha = 0 = \gamma\alpha^2.}$

Sinc $\gamma\alpha^2 = 0$ we also have that $\alpha^2\beta = 0$, and then $J\alpha\beta \subseteq$ soc $\Lambda$ and

(7) $\underline{\alpha\beta \in \text{soc}_2\Lambda.}$

Recall that $\beta\gamma$ is dependent, see VIII.9.9; therefore we deduce now the relations.

Let s be the integer with $\alpha^s \neq 0$ but $\alpha^{s+1} = 0$. Then $\langle \alpha^s \rangle = \text{soc}(e_1\Lambda)$.

(8) $\underline{\text{We may assume that } \beta\gamma = \alpha^{s-1}:}$ Write $\beta\gamma = \alpha x + \beta y$ where $y \in J^2$ and $\alpha x \in J^2 e_0$. Since $\alpha\beta \in \text{soc}_2\Lambda$ we have that $\alpha x = \Sigma c_i\alpha^i$ where $c_i \in K$. Now $\alpha\beta\gamma \in$ soc $\Lambda$, therefore all $c_i$ are 0 except possibly the last one. So $\beta\gamma = c\alpha^{s-1} + \beta y$. Replace $\gamma$ by $\gamma - y$. We must have that $c \neq 0$ since otherwise $\alpha\beta\gamma$ would be 0. Then, without loss of generality $c = 1$. The rest is straightforward.

Now we shall determine the algebras whose quiver is of the form ($3\mathcal{K}$).

VIII.9.13 THEOREM *Assume that* $\Lambda$ *is of semidihedral type, with quiver of the form* ($3\mathcal{K}$). *Then* $\Lambda \cong KQ/I$ *where I is the ideal defined by the relations*

$$\kappa\eta = \eta\gamma = \gamma\kappa = 0,$$

$$\delta\lambda = (\gamma\beta)^{a-1}\gamma, \quad \beta\delta = (\kappa\lambda)^{b-1}\kappa \text{ and } \lambda\beta = (\eta\delta)^{c-1}\eta,$$

*where* $a, b, c \geq 2$.

Proof: By VIII.9.10 and I.10.9 we have that any path $i \to j$ in the quiver of length two with $i \neq j$ must be dependent, and then the element lies in $J^3$, since for any given i, j there is only one such path. Therefore we are already able to write down a vector space basis of $\Lambda$. To determine precise relations we will study some modules.

At least one of the simple modules, $S_0$ say, is non-periodic (VIII.2.1). Then by IV.6.6, $S_0$ lies at the end of a $\mathbb{Z}D_\infty$-component, with one predecessor, and IV.3.3(a) holds. Moreover, $U_0$ and $\underset{\sim}{V}_0$ must lie at the end of the 3-tube.

(1) $S_2$ is not periodic: The modules at the end of the 3-tube have simple socles and tops. Therefore none of them can be isomorphic to $\Omega S_2$, and $S_2$ does not have period 3. Since $\underline{\mathrm{Hom}}_\Lambda(\underset{\sim}{V}_0, S_2) \neq 0 = \underline{\mathrm{Hom}}_\Lambda(U_0, S_2)$ we know that $S_2$ does not have period 4 (by IV.1.1. Thus (1) is established.

Now we apply IV.6.6 again and deduce that that $S_2$ lies also at the end of a $\mathbb{Z}D_\infty$-component, and IV.3.3(a) holds for $S_2$. Moreover, $U_2$ and $\underset{\sim}{V}_2$ lie at the end of the 3-tube.

(2) $U_2 \cong \underset{\sim}{V}_0$ and $\Omega\underset{\sim}{V}_2 \cong U_0$; in particular soc $U_2 \cong S_0$ and soc $V_2 \cong S_1$: We have that $U_2 \neq U_0$ by comparing tops, and $U_2 \neq \Omega\underset{\sim}{V}_0$, by comparing socles.

(3) soc $U_0 \cong S_1 \cong$ top $V_2$, soc $V_0 \cong S_2$ and top $U_2 \cong S_0$: By (2) we have that $\Omega\underset{\sim}{V}_2 \cong U_0$. Now from the quiver we know top $V_2 = S_0$ or $S_1$ and soc $U_0 \cong S_1$ or $S_2$. This proves the first statement, and the rest follows.

(4) $S_1$ is not periodic: This is seen by the argument in (1), using the fact that $\underline{\mathrm{Hom}}_\Lambda(\underset{\sim}{V}_2, S_1) \neq 0$ but $\underline{\mathrm{Hom}}_\Lambda(U_2, S_1) = 0$.

It follows again from IV.6.6 that $S_1$ lies also at the end of a $\mathbb{Z}D_\infty$-component, and IV.3.3(a) holds for $S_1$. The modules $U_1$ and $\underset{\sim}{V}_1$ also lie at the end of the 3-tube; and the only possibility is that

(5) $U_1 \cong \underset{\sim}{V}_2$ and $\underset{\sim}{V}_1 \cong U_0$; in particular soc $U_1 \cong S_2$, soc $V_1 \cong S_0$ and top $V_1 \cong S_0$, top $U_1 \cong S_2$.

Now we shall determine the relations.

(6) We may assume that $\kappa\eta = \eta\gamma = \gamma\kappa = 0$: First, we choose $\kappa$ such that $\underset{\sim}{V}_0 = \kappa\Lambda$; and we identify $\Omega\underset{\sim}{V}_0$ with $R_\kappa$. By (2), we may choose $\eta$ such that $\underset{\sim}{V}_2 = \eta\Lambda$ and $\kappa\eta = 0$. Using similarly (5) we may take $\underset{\sim}{V}_1 = \gamma\Lambda$ and $\eta\gamma = 0$. Now $R_\gamma \cong \kappa\Lambda$; and any isomorphism may be lifted to an isomorphism of the injective hulls and is therefore given by left-multiplication with a unit. That is, $R_\gamma = w\kappa\Lambda$ where $w$ is a unit in $e_0\Lambda e_0$. We replace $\kappa$ by $w\kappa$ without effecting the other relations.

Now we choose arrows $\beta$, $\delta$, $\lambda$ such that $U_0 = \beta\Lambda/\beta\Lambda \cap \kappa\Lambda$, $U_1 = \delta\Lambda/\delta\Lambda \cap \gamma\Lambda$ and $U_2 = \lambda\Lambda/\lambda\Lambda \cap \eta\Lambda$ and moreover $\beta\Lambda \cap \kappa\Lambda$ and also $\delta\Lambda \cap \gamma\Lambda$ and $\lambda\Lambda \cap \eta\Lambda$ are contained in

$soc_2\Lambda$. As we already noticed at the beginning, the elements $\beta\delta$, $\delta\lambda$, $\lambda\beta$ lie in $J^3$. The rest is straightforward, for details see [$E_5$].

<u>VIII.11</u> *On the converse* Only in some special cases it is at present possible to prove that the algebras listed are in fact of semidihedral type.

This is established for the local algebras, by [$CB_3$] and II.10. Also, whenever $\Lambda$ is of semidihedral type and $\Lambda$/soc $\Lambda$ is Morita equivalent to some block then $\Lambda$ is of semidihedral type, see VIII.11.

Otherwise, there are some small algebras where A. Skowroński has proved with covering methods that the converse holds.

# *IX Centres, blocks, decomposition numbers*

## IX *On the centres of algebras of dihedral, semidihedral and quaternion type*

For many of the algebras in our list, we will now determine the centres, in particular its dimensions. This will be used later fore the study of blocks.

We assume in this chapter that $\Lambda$ is basic; this is no restriction, by I.2.2. Let $Z(X)$ denote the centre of the ideal X. We have that $Z(\Lambda) = K.1_\Lambda \oplus Z(J(\Lambda))$; so it is enough to study the centre of $J(\Lambda)$. We will see that there is a certain uniformity. This seems to be connected with the isomorphism types of $\Lambda/\mathrm{soc}_2\Lambda$; and it might therefore be partly explained by I.3.9 and the fact that the algebras modulo $\mathrm{soc}_2\Lambda$ coincide for the different types.

IX.1.1 PROPOSITION *Suppose $\Lambda$ is tame, local and symmetric. If $\Lambda$ is not commutative then $\dim \Lambda = 4k$ and $\dim Z(\Lambda) = k + 3$.*

Proof: We apply III.1. If $\Lambda$ is not commutative then $\Lambda \cong K\langle X, Y\rangle / I$ where in any case, I is of the form

$[(XY)^k - (YX)^k,\ X^2 - r(YX)^{k-1}Y - s(XY)^k,\ Y^2 - t(XY)^{k-1}X - u(XY)^k]$ for appropriate $r, s, t, u \in K$; and $\dim \Lambda = 4k \geq 8$. Moreover, $\mathrm{soc}_2\ \Lambda = \mathrm{span}\ \{(XY)^{k-1}X, (YX)^{k-1}Y, (XY)^k\}$. By I.3.9, this space is contained in $Z(\Lambda)$. Using this, it is an easy calculation to show that $Z(\Lambda)$ consists of the elements of the form

$a_0 1 + \sum_{i=1}^{k-1} a_i[(XY)^i + (YX)^i] + b_1(XY)^{k-1}X + b_2(YX)^{k-1}Y + c(XY)^k$ with $a_i$, $b_i$ and $c \in K$ arbitrary. Hence $\dim Z(\Lambda) = k+3$.

IX.1.2 LEMMA *Let $\Lambda$ be an algebra of dihedral or semidihedral or quaternion type, with Cartan matrix of the form*

$$\begin{bmatrix} 4k & 2k \\ 2k & k+r \end{bmatrix} \quad \text{with } r \geq 1.$$

*Then $\dim Z(\Lambda) = k + r + 2$.*

Proof: A general element of $Z(J\Lambda)$ is of the form $x = x_0 + x_1$ where $x_i \in e_i\Lambda e_i$;

actually $x_i$ is central in $e_i\Lambda e_i$. Using this and I.3.9 it is straightforward to verify the formulae given below; we will not give details.

(i) Assume first that the quiver of $\Lambda$ is of the form $(2\mathcal{A})$. If $\Lambda$ belongs to the family $D(2\mathcal{A})$ of $SD(2\mathcal{A})_2$ then the parameter r occuring in C is 1. On the other hand, the general element of $Z(J\Lambda)$ is given by

$$x_0 = \sum_{i=1}^{k-1} a_i[(\alpha\beta\gamma)^i + (\beta\gamma\alpha)^i] + u(\beta\gamma\alpha)^{k-1}\beta\gamma + a'(\alpha\beta\gamma)^k$$

$$x_1 = \sum_{i=1}^{k-1} a_i(\gamma\alpha\beta)^{k-1} + b'(\gamma\alpha\beta)^k$$

where $a_i$, u, a', b' are arbitrary elements of K. Hence $\dim Z(\Lambda) = k+3$.

Now assume that $\Lambda$ belongs to either $SD(2\mathcal{A})_1$ or $Q(2\mathcal{A})$. Then the parameter r in C is 2. On the other hand, x belongs to $Z(\Lambda)$ if and only if

$$x_0 = \sum_{i=1}^{k-1} a_i[(\alpha\beta\gamma)^i + (\beta\gamma\alpha)^i] + u(\alpha\beta\gamma)^{k-1}\alpha + v(\beta\gamma\alpha)^{k-1}\beta\gamma + a'(\alpha\beta\gamma)^k$$

$$x_1 = \sum_{i=1}^{k-1} a_i(\gamma\alpha\beta)^i + u(\gamma\beta) + b'(\gamma\alpha\beta)^k$$

for $a_i$, u, v, a', b' in K. Therefore $\dim Z(\Lambda) = k + 4$.

(ii) Now assume that the quiver of $\Lambda$ is of the form $(2\mathcal{B})$. Then consider the subalgebra $\Lambda_0$ of $\Lambda$ obtained from $\Lambda$ by removing the loop $\eta$; that is, $\Lambda_0 = \mathrm{alg}_K\{e_0, e_1, \alpha, \beta, \gamma\}$. By inspection one sees that $\Lambda_0$ is also an algebra belonging to our list, with quiver of the form $(2\mathcal{A})$.

If $\Lambda$ belongs to $D(2\mathcal{B})$ or $SD(2\mathcal{B})_1$ then $\eta$ is central in $\Lambda$, and $Z(\Lambda) = Z(\Lambda_0) \oplus \mathrm{span}\{\eta, \eta^2, \ldots, \eta^{s-1}\}$. Otherwise, the element $\eta' = \eta + (\alpha\beta\gamma)^{k-1}\alpha$ is central in $\Lambda$, and $Z(\Lambda) = Z(\Lambda_0) \oplus \mathrm{span}\{\eta', \eta^2, \eta^3, \ldots, \eta^{s-1}\}$. Hence $\dim Z(\Lambda) = \dim Z(\Lambda_0) + (s-1)$ which gives the statement.

<u>IX.I.3</u> LEMMA *Let $\Lambda$ be an algebra of dihedral or semidihedral or quaternion type, with Cartan matrix of the form*

$$\begin{bmatrix} 4k & 2k & 2k \\ 2k & k+s & k \\ 2k & k & k+t \end{bmatrix}.$$

*Then* $\dim Z(\Lambda) = k + s + t + 1$.

Proof: A general element of $Z(J\Lambda)$ is of the form $x = x_0 + x_1 + x_2$ where $x_i$ is central in $e_iJe_i$. Using this and I.3.9 the calculations needed for the proofs below are straightforward, and we omit details.

If $\Lambda$ has such a Cartan matrix then the quiver is of the form ($3\mathcal{A}$) or ($3\mathcal{B}$) or ($3\mathcal{D}$).

(i) Assume first the quiver of $\Lambda$ is of the form ($3\mathcal{A}$). For $\Lambda$ belonging to $D(3\mathcal{A})$, x lies in $Z(\Lambda)$ if and only if

$$D(3\mathcal{A}) \quad x_0 = \sum_{i=1}^{k-1} a_i[(\gamma\beta\delta\eta)^i + (\delta\eta\gamma\beta)^i] + a'(\gamma\beta\delta\eta)^k;$$

$$x_1 = \sum_{i=1}^{k-1} a_i(\beta\delta\eta\gamma)^i + b'(\beta\delta\eta\gamma)^k \quad \text{and} \quad x_2 = \sum_{i=1}^{k-1} a_i(\eta\gamma\beta\delta)^i + c'(\eta\gamma\beta\delta)^k,$$

where $a_i$, $a'$, $b'$, $c'$ are arbitrary elements of K. Hence $\dim Z(\Lambda) = k + 3$; and on the other hand, $s = t = 1$.

Now consider an algebra belonging to $SD(3\mathcal{A})_1$. Here $x \in Z(J\Lambda)$ if and only $x_1$ is as for $D(3\mathcal{A})$, and

$$SD(3\mathcal{A})_1 \quad x_0 = \sum_{i=1}^{k-1} a_i[(\gamma\beta\delta\eta)^i + (\delta\eta\gamma\beta)^i] + u(\gamma\beta\delta\eta)^{k-1}\gamma\beta + a'(\gamma\beta\delta\eta)^k;$$

$$x_2 = \sum_{i=1}^{k-1} a_i(\eta\gamma\beta\delta)^i + u(\eta\delta) + c'(\eta\gamma\beta\delta)^k,$$

where $a_i$, u, $a'$, $b'$ and $c'$ are arbitrary $\in$ K. Hence $\dim Z(\Lambda) = k + 4$, as required.

Now assume that $\Lambda$ belongs to $Q(3\mathcal{A})_2$. Then a general element of $Z(J\Lambda)$ is given as follows:

$$Q(3\mathcal{A})_2 \quad x_0 = \sum_{i=1}^{k-1} a_i[(\gamma\beta\delta\eta)^i + (\delta\eta\gamma\beta)^i] + u(\gamma\beta\delta\eta)^{k-1}\gamma\beta + v(\delta\eta\gamma\beta)^{k-1}\delta\eta + a'(\gamma\beta\delta\eta)^k;$$

$$x_1 = \sum_{i=1}^{k-1} a_i(\beta\delta\eta\gamma)^i + v(\beta\gamma) + b'(\beta\delta\eta\gamma)^k \quad \text{and} \quad x_2 \text{ as in } SD(3\mathcal{A})_2;$$

where $a_i$, u, v, $a'$, $b'$, $c'$ are arbitrary in K. Hence $\dim Z(\Lambda) = k+5$; and on the other hand $s = t = 2$.

(ii) Now assume the quiver is of the form ($3\mathcal{B}$). Let $\Lambda_0$ be the subagebra of $\Lambda$ obtained from $\Lambda$ by removing the loop $\alpha$. By inspection we see that $\Lambda_0$ is also an algebra belonging to our list, with quiver of the form ($3\mathcal{A}$). If $\Lambda$ belongs to $D(3\mathcal{B})_1$ or $SD(3\mathcal{B})_1$ then $\alpha$ is central in $\Lambda$, and $Z(\Lambda) = Z(\Lambda_0) \oplus \text{span}\ \{\alpha, \alpha^2, \ldots, \alpha^{s-1}\}$. Otherwise, $\Lambda$ belongs to $SD(3\mathcal{B})_2$ or $Q(3\mathcal{B})$, and $Z(\Lambda) = Z(\Lambda_0) \oplus \text{span}\ \{\alpha', \alpha^2, \alpha^3, \ldots, \alpha^{s-1}\}$ where $\alpha' = \alpha + (\delta\eta\gamma\beta)^{k-1}\delta\eta$. In all cases, we obtain the stated formula for dim $Z(\Lambda)$.

(iii) Now assume that the quiver of $\Lambda$ is of the form ($3\mathcal{D}$). Let $\Lambda_0$ be the subalgebra of $\Lambda$ obtained from $\Lambda$ by removing the loop $\xi$. Then, by inspection, $\Lambda_0$ is an algebra belonging to our list, with quiver (3$\mathcal{B}$). The method used in (ii) may be used again and gives a similar answer.

<u>IX.1.4</u> **LEMMA** *Let $\Lambda$ be an algebra of dihedral or semidihedral or quaternion type, with Cartan matrix of the form*

$$\begin{bmatrix} a+b & a & b \\ a & a+c & c \\ b & c & b+c \end{bmatrix}.$$

*Then* $\dim Z(\Lambda) = a + b + c + 1$.

Proof: A general element of $Z(J\Lambda)$ is of the form $x = x_0 + x_1 + x_2$ where $x_i$ is central in $e_i\Lambda e_i$. Using this and I.3.9 the calculations below are straightforward; we omit details.

If $\Lambda$ has such a Cartan matrix then the quiver of $\Lambda$ is of the form (3$\mathcal{F}$), (3$\mathcal{H}$) of (3$\mathcal{K}$), or $\Lambda$ belongs to the family $Q(3\mathcal{A})_1$.

(i) Assume first that the quiver is of the form (3$\mathcal{F}$); then $\Lambda$ belongs to the family SD(3$\mathcal{F}$), and x is given by

$$x_0 = a'(\beta\delta\lambda), \quad x_1 = \sum_{i=1}^{k-1} a_i(\delta\eta)^i + b'(\delta\eta)^k \text{ and } x_2 = \sum_{i=1}^{k-1} a_i(\eta\delta)^i + c'(\eta\delta)^k$$

where $a_i$ are arbitrary $\in K$. Hence $\dim Z(\Lambda) = k+3$, as required.

(ii) Now assume that the quiver of $\Lambda$ is of the form (3$\mathcal{H}$); then $\Lambda$ belongs to the family SD(3$\mathcal{H}$), and x is given by

$$x_0 = \sum_{i=1}^{k-1} a_i(\beta\gamma)^i + a'(\beta\gamma)^k, \quad x_1 = \sum_{i=1}^{k-1} a_i(\gamma\beta)^i + \sum_{i=1}^{s-1} b_i(\eta\delta)^i + b'(\gamma\beta)^k \text{ and}$$

$$x_2 = \sum_{i=1}^{s-1} b_i(\eta\delta)^i + c'(\eta\delta)^k,$$

where $a_i$, $b_i$ and a', b', c' are arbitrary elements of K. Hence $\dim Z(\Lambda) = k + s + 2$.

(iii) Consider now algebras with quiver (3$\mathcal{K}$); these occur for all three types of algebras. In all cases, we find that x belongs to $Z(J\Lambda)$ if and only if

$$x_0 = \sum_{i=1}^{a-1} a_i(\beta\gamma)^i + \sum_{i=1}^{b-1} b_i(\kappa\lambda)^i + a'(\beta\gamma)^a;$$

$$x_1 = \sum_{i=1}^{a-1} a_i(\gamma\beta)^i + \sum_{i=1}^{c-1} c_i(\delta\eta)^i + b'(\gamma\beta)^a \text{ and}$$

$$x_2 = \sum_{i=1}^{b-1} b_i(\lambda\kappa)^i + \sum_{i=1}^{c-1} c_i(\eta\delta)^i + c'(\lambda\kappa)^b,$$

with $a_i$, $b_i$ , $c_i$ and a', b', c' arbitrary in K. In particular, $\dim Z(\Lambda) = a+b+c+1$.

(iv) Now suppose $\Lambda$ belongs to $Q(3\mathcal{A})_1$. Here x lies in $Z(J\Lambda)$ if and only if

$$x_0 = \sum_{i=1}^{k-1} a_i(\gamma\beta)^i + \sum_{i=1}^{s-1} c_i(\eta\delta)^i + a'(\gamma\beta)^k,$$

$$x_1 = \sum_{i=1}^{k-1} a_i(\beta\gamma)^i + b'(\beta\gamma)^k \text{ and}$$

$$x_2 = \sum_{i=1}^{k-1} c_i(\delta\eta)^i + c'(\delta\eta)^k$$

were $a_i$, $c_i$, $a'$, $b'$ and $c'$ are arbitrary elements of K. In particular $\dim Z(\Lambda) = k + s + 2$, as required.

IX.1.5 LEMMA *Suppose $\Lambda$ is an agebra of dihedral or semidihedral or quaternion type, with Cartan matrix of the form*

$$\begin{bmatrix} k+s & k & k \\ k & k+1 & k-1 \\ k & k-1 & k+1 \end{bmatrix} \quad \text{where } k,\ s \geq 1.$$

*Then* $\dim Z(\Lambda) = k + s + 2$.

Proof: A general element of $Z(J\Lambda)$ is of the form $x = x_0 + x_1 + x_2$ where $x_i$ is central in $e_i\Lambda e_i$. If $\Lambda$ has such a Cartan matrix then the quiver of $\Lambda$ is of the form $(3\mathcal{C})$; or $\Lambda$ belongs to the family $SD(3\mathcal{A})_2$.

Assume first that $\Lambda$ belongs to the family $SD(3\mathcal{C})_1$. Then x is central in $\Lambda$ if and only if $x_0 = \sum_{i=1}^{s} a_i\rho^i$, $x_1 = b'(\beta\gamma)$ and $x_2 = c'(\eta\delta)$ where $a_i$, $b'$ and $c'$ are arbitrary in K. Hence $\dim Z(\Lambda) = s + 3$.

Now consider an algebra $\Lambda$ which belongs to family $SD(3\mathcal{C})_2$. Here $x \in Z(\Lambda)$ if and only if

$$x_0 = \sum_{i=1}^{k-1} b_i(\gamma\beta)^i + \sum_{i=1}^{s-1} a_i\rho^i + a'(\gamma\beta)^k, \quad x_1 = \sum_{i=1}^{k-1} b_i(\beta\gamma)^i + b'(\beta\gamma)^k \text{ and}$$

$$x_2 = \sum_{i=1}^{k-1} b_i(\eta\delta)^i + c'(\eta\delta)^k$$

where $a_i$, $b_i$ and $a'$., $b'$, $c'$ are arbitrary elements in K. Therefore $\dim Z(\Lambda) = k + s + 2$.

Now let $\Lambda$ be an algebra as in $Q(3\mathcal{C})$. Then a general element of $Z(J\Lambda)$ is given by

$$x_0 = \sum_{i=1}^{k-1} b_i(\gamma\beta)^i + \sum_{i=1}^{s-1} a_i\rho^i + a'(\gamma\beta)^k, \quad x_1 = \sum_{i=1}^{k-1} b_i(\beta\gamma)^i + b'(\beta\gamma)^k \text{ and}$$

$$x_2 = \sum_{i=1}^{k-2} b_i(\eta\delta)^i + (b_{k-2} + a_1 a)(\eta\delta)^{k-1} + c'(\eta\delta)^k$$

where $a_i$, $b_i$ and $a'$, $b'$, $c'$ are arbitrary elements of K. Hence $\dim Z(\Lambda) = k + s + 2$.

Now suppose $\Lambda$ belongs to the family $SD(3\mathcal{A})_2$; here s = 1. Then $x \in Z(\Lambda)$ if and only if $x_0 = \sum_{i=1}^{k-1} a_i(\gamma\beta)^i + a'(\gamma\beta)^k$, $x_1 = \sum_{i=1}^{k-1} a_i(\beta\gamma)^i + b'(\beta\gamma)^k$ and

$$x_2 = \sum_{i=1}^{k-1} a_i(\eta\delta)^i + c'(\eta\delta)^k$$ with $a_i$, $a'$, $b'$, $c'$ arbitrary in K. Hence

$\dim Z(\Lambda) = k+3$.

Now we have determined the centres for all algebras except for the ones of dihedral type where two 3-tubes are permuted by $\Omega$ and the few families with complicated relations.

## IX.2 *Cartan matrices and decomposition numbers for tame blocks*

Our aim now is to determine which of the algebras in our list are Morita equivalent to blocks. In particular, we will obtain the Cartan matrices for tame blocks on the way. Then we will determine the decomposition numbers for these blocks, and also some generalized decomposition matrices.

Concerning block theory, apart from the formulae for k(B) derived in V.5 (due to $[B_2]$, [O]) which are summarized below in IX.2.3, we will essentially only use the following general principles; see V.1.6 for details:

### IX.2.1 SOME GENERAL FACTS FROM BLOCK THEORY

Suppose B is an arbitrary p-block with Cartan matrix C and decomposition matrix D. Recall that k(B) denotes the number of ordinary irreducible characters of B, and $\ell(B)$ is the number of irreducible Brauer characters, which is equal to the number of simple B-modules.

(1) The determinant of C is a power of p. The highest elementary divisor of C is the order of the defect group, and it occurs with multiplicity 1.

(2) The entries of D are non-negative integers; moreover D does not have a row consisting of zeros only.

(3) The matrix D has $\ell(B)$ columns and k(B) rows, and $D^tD = C$.

(4) $k(B) = \dim Z(B)$ where Z(B) is the centre of B.

<u>IX.2.2</u> **LEMMA** *Suppose $B$ is a p-block with defect group $\delta(B)$ and $B \sim_M \Lambda$ where $\Lambda$ is one of the algebras in the list. Then $p = 2$, and moreover:*

*(i) If $\Lambda$ is of dihedral type then $\delta(B)$ is dihedral.*

*(ii) If $\Lambda$ is of semidihedral type and assuming $\Lambda$ is tame, then $\delta(B)$ is semidihedral.*

*(iii) If $\Lambda$ is of quaternion type then $\delta(B)$ is quaternion.*

Therefore, we will for the rest of this chapter always assume that char $K = 2$.

Proof: Recall that B and $\Lambda$ have the same Cartan matrix and the same Auslander-Reiten quiver. In all cases, except possibly III.1(a), we find that the determinant of the Cartan matrix is even; therefore $p = 2$ by IX.2.1(1). Assume $\Lambda$ is as in III.1(a); then $\Lambda$ is special biserial and hence tame (II.3.1 and I.4.3.1). Therefore by [BD] here $p = 2$ as well.

Suppose now that $\Lambda$ is of dihedral type; then $\Lambda$ is tame (VI.10.1); and hence by [BD], a defect group can only be dihedral or semidihedral or generalized quaternion. Comparing the stable AR-quivers, using V.4 shows that $\delta(B)$ can only be dihedral.

Suppose $B \sim_M \Lambda$ where $\Lambda$ is of semidihedral type. If B is tame then $\delta(B)$ is restricted according to [BD]. Again, comparing the AR-quivers shows that $\delta(B)$ can only be semidihedral.

Suppose $\Lambda$ is of quaternion type; then the simple modules are periodic. For blocks of infinite type this can only happen when the defect groups are quaternion, by cohomological considerations, see for example [Be].

<u>IX.2.3</u> *Results on $k(B)$, proved in [V.5].* Suppose $\delta(B)$ has order $2^n$.

| $\delta(B)$ | $k(B)$ |
|---|---|
| dihedral | $2^{n-2} + 3$ |
| semidihedral | either $\ell(B) + 2^{n-2} + 1$ and $\ell(B) \geq 2$, |
| | or $\ell(B) + 2^{n-2} + 2$ and $\ell(B) \leq 2$. |
| quaternion | $\ell(B) + 2^{n-2} + 2$. |

## IX.3 *One simple module*

Let $B$ be a tame block with one simple module, and $\delta(B)$ of order $2^n$. Assume also that $\Lambda$ is a basic algebra which is tame local and symmetric. Then $\Lambda$ is one of the algebras in III.1. The information we need to determine whether it is possible that $\Lambda \sim_M B$ is as follows:

| Algebra $\Lambda$ | dim $\Lambda$ | dim $Z(\Lambda)$ | |
|---|---|---|---|
| III.1(a) | n+m | n+m | [here n+m > 4]. |
| III.1(b), (b') | 4 | 4 | |
| III.1(c) to (e') | 4k | k+3 | |

No algebra of the family III.1(a) occurs as a block:

IX.3.1 **LEMMA** *Suppose $\Lambda \sim_M B$. Then one of the following holds:*
*(i) $\Lambda$ is commutative of dimension 4, and $\delta(B)$ is a Klein 4-group.*
*(ii) $\Lambda$ belongs to one of the families in (c) to (e') and $k = 2^{n-2}$.*

Proof: Since the algebra $\Lambda$ is local, we have that dim $\Lambda$ is equal to the only Cartan number which is equal to $|\delta(B)|$; and dim $Z(\Lambda)$ = k(B) (IX.2.1(1) and (4)).
(i) Assume first that $\Lambda$ is commutative; then $\Lambda$ is one of the algebras in III.1 (a) to (b'). In particular, $\Lambda$ is of dihedral type (VI.10.2), and hence D must be dihedral, by IX.2.2. Therefore we know by V.5.10 that $k(B) = 2^{n-2}+3$; and this is equal to $2^n$ only for n = 2.
(ii) Otherwise, $4k = |\delta(B)| = 2^n$.

IX.3.2 We deduce from these considerations the value for k(B). Of course, one could also calculate it by group theory.

Consider the families (c) to (e'). From the representation theory, we know that blocks with dihedral [semidihedral, quaternion ] defect groups belong to (c), (d) or (e) respectively. In these algebras there are scalars appearing in the socle relations. It needs more work to decide which scalars are possible for blocks.

(c) The information from the representation theory of algebras is not sufficient to determine the decomposition numbers. According to Brauer, the decomposition matrix is

(*) $$D^t = [1\ 1\ 1\ 1\ 2\ \dots\ 2]$$

On the other hand, for example if $n = 5$ then $k(B) = 11$, and the arithmetic would also allow the solution $D^t = [4\ 2\ 2\ 1\ 1\ \dots\ 1]$.

Concerning Brauer's method, he proves that $k_0(B)$, the number of characters of B of height zero (see V.1), is 4 [$B_2$]. Therefore the number of decomposition numbers which is odd must be 4; and this implies immediately, by IX.2.1(3), that D must satisfy (*).

## IX.4 *Two simple modules*

Let $\Lambda$ be an algebra in our list with two simple modules, and suppose $\Lambda \sim_M B$ for some tame block, with $\delta(B)$ of order $2^n$.

In [$E_8$], we have determined the possible Cartan matrices for such blocks, and we have calculated the decomposition matrices. We will here give a brief outline; the detailed results may be found in the tables at the end; and for the proofs we refer to [$E_8$].

By studying the list of algebras, we see that only two shapes of Cartan matrices occur, namely

$$\begin{bmatrix} 4k & 2k \\ 2k & k+r \end{bmatrix} \quad \text{and} \quad \begin{bmatrix} s+2 & s \\ s & s+2 \end{bmatrix}, \text{ and also } \begin{bmatrix} 4 & 2 \\ 2 & t+1 \end{bmatrix}.$$

In the first case, the dimension of $Z(\Lambda)$ is always $k + r + 2$; this has been proved here in IX.1.3. The possible Cartan matrices are then given as follows:

<u>IX.4.1</u> *Suppose B is any 2-block with defect group of order $2^n$.*

*(i) If the Cartan matrix of B is of the form* $\begin{bmatrix} 4k & 2k \\ 2k & k+r \end{bmatrix}$ *and if* $\dim Z(B) = k + r + 2 \equiv 2 \bmod 4$ *then* $\{k, r\} = \{2^{n-2}, 1\}$ *or* $\{2^{n-2}, 2\}$.

*(ii) If the Cartan matrix of B is of the form* $\begin{bmatrix} s+2 & s \\ s & s+2 \end{bmatrix}$ *then* $s = 2^{n-2} - 1$.

*(iii) There is no block with* $k(B) = 2^{n-2} + 4$ *and Cartan matrix* $\begin{bmatrix} 4 & 2 \\ 2 & t+1 \end{bmatrix}$.

In $[E_9]$ the last Cartan matrix, which occurs for algebras of quaternion type only, was omitted. We will therefore give a proof of (iii).

The determinant of C is 4t; for a block t must be a 2-power. If t were 1 then both elementary divisors of C would be 2, a contradiction to IX.2.1(1). Hence $t \geq 2$, and the elementary divisors of C are 1 and 4t. Consequently $t = 2^{n-2}$, by IX.2.1(1). It is now easy to see that there is no matrix D of size $(2^{n-2}+4)$ by 2 satisfying also IX.2.1(2), with $D^tD = C$, and hence no such block exists.

In total, this gives five different Cartan matrices. Suppose B is a block with Cartan matrix as in (ii) and $k(B) \geq 2^{n-2} + 3$. Then there is only a solution for a decomposition matrix when $k(B) = 2^{n-2} + 3$, and then it is unique.

Consider blocks with Cartan matrices as in (i). If one of k, r is 1 then there is a unique solution for a matrix D satisfying IX.2.1(2) and (3). If one of k, r is 2 and if in addition the condition[**], given below, is satisfied then there is also a unique solution for a matrix D satisfying IX.2.1(2), (3).

[**] Let $P_0$ be the first indecomposable projective. Then $P_0$ does not have a submodule whose composition factors are $rS_0$ with $r \geq 3$. Therefore the decomposition matrix does not have rows whose sum is [r 0].

We remark that IX.4.1 and the calculations on decomposition numbers are more general and depend only on arithmetic hypothesis. We do not know whether these are satisfied for other blocks.

It is clear that the decomposition numbers give directly the relations satisfied by the characters on elements of odd order, as obtained in $[B_3]$, [O]. Moreover, the general fact that at least one ordinary character and also one Brauer character in the block is of height zero, implies also the results in $[B_3]$, [O] on the heights of the characters. The answer is that in the matrices given in the tables, the first four characters are of height zero, and the last row corresponds th characters of height 1.

## IX.5 *Cartan matrices and blocks, three simple modules*

Now we shall determine which of the algebras in the list having three simple modules can be Morita equivalent to blocks. In particular, we obtain all possible Cartan matrices for tame blocks with three simple modules.

By studying the list of algebras we observe that the most common type of Cartan matrix is

$$\begin{bmatrix} 4k & 2k & 2k \\ 2k & k+s & k \\ 2k & k & k+t \end{bmatrix}.$$

It occurs for $D(3\mathcal{A})_1$, $SD(3\mathcal{A})_1$, $Q(3\mathcal{A})_2$, $D(3\mathcal{B})_1$, $SD(3\mathcal{B})_{1,2}$, $Q(3\mathcal{B})$ and $D(3\mathcal{D})_1$, $SD(3\mathcal{D})$ and $Q(3\mathcal{D})$. In IX.1.3 we proved that always $\dim Z(\Lambda) = k + s + t + 1$. The shape of the Cartan matrix and $\dim Z(\Lambda)$ determine already necessary conditions for blocks amongst these algebras.

<u>IX.5.1</u> LEMMA *Assume that $B$ is a 2-block whose Cartan matrix is of the form*

$$\begin{bmatrix} 4k & 2k & 2k \\ 2k & k+s & k \\ 2k & k & k+t \end{bmatrix}$$

*Suppose also that $\dim Z(B) = k + s + t + 1$ and $2^{n-2}+3 \leq k(B) \leq 2^{n-2}+5$. Then one of the following holds:*

*(i) $k(B) = 2^{n-2}+3$ and $\{k, s, t\} = \{1, 1, 2^{n-2}\}$.*

*(ii) $k(B) = 2^{n-2}+4$ and $\{k, s, t\} = \{1, 2, 2^{n-2}\}$.*

*(iii) $k(B) = 2^{n-2}+5$ and $\{k, s, t\} = \{2, 2, 2^{n-2}\}$.*

Proof: We have that $\det C = 4kst$. For a block we must have that $4kst$ is a power of 2. Moreover, $k + s + t + 1 = k(B)$. We shall also need the elementary divisors of $C$; denote them by $d_1$, $d_2$, $d_3$, such that $d_i$ divides $d_{i+1}$; then $d_3 = |\delta(B)|$, by IX.2.1(1). We have that $d_1 = \text{g.c.d.}\ \{k, s, t\}$ and $d_1d_2 = \text{g.c.d.}\{\text{2-minors of } C\} = \text{g.c.d.}\{2ks, 2kt, ks + kt + st\}$, hence $d_1d_2 = (kst)/2^{n-2}$.

<u>Case 1</u> $k(B) = 2^{n-2}+3$. Then $k+s+t = 2^{n-2}+2$; and since $k$, $s$, $t$ are powers of 2, the only solutions for $\{k, s, t\}$ are $\{1, 1, 2^{n-2}\}$ or $\{2^{n-3}, 2^{n-3}, 2\}$. Assume for

contradiction that the second possibility occurs, with $n > 3$; then $d_1d_2 = 2^{n-3}$. On the other hand, $ks + kt + st = 2^{n-1} + 2^{2(n-3)}$ and g.c.d.$\{ 2ks, 2kt \} = 2^{n-1}$. We deduce that $d_1d_2 = \min \{ 2^{n-1}, 2^{2(n-3)} \}$, and hence $n = 3$, a contradiction.

<u>Case 2</u> $k(B) = 2^{n-2}+4$. Then $k + s + t = 2^{n-2}+ 3$, and it follows immediately that (ii) must hold.

<u>Case 3</u> $k(B) = 2^{n-2}+ 5$. Here we have the possibilities that $\{k, s, t\}$ is $\{ 2, 2, 2^{n-2}\}$ or $\{ 4, 2^{n-3}, 2^{n-3}\}$. Assume for contradiction the second solution occurs when $n \neq 4$. Then we have that $d_1d_2 = 2^{n-2}$. On the other hand, $ks + kt + st = 2^n + 2^{2(n-3)}$ and g.c.d.$\{ 2ks, 2kt \} = 2^n$. Hence $d_1d_2 = \min \{ 2^n, 2^{2(n-3)} \}$. Since $n-2 \neq n$ here, this is not possible.

<u>IX.5.1.1</u> This result shows that we should study seven cases in finding decomposition matrices. We will list them now and give references.

| k | s | t | Decompositon matrix | | k(B) |
|---|---|---|---|---|---|
| $2^{n-2}$ | 1 | 1 | IX.6.1(i) | , IX.6.4(i) | $2^{n-2}+3$ |
| 1 | $2^{n-2}$ | 1 | -"- | IX.6.6(i) | -"- |
| | | | | | |
| $2^{n-2}$ | 2 | 1 | IX.6.1(ii) | , IX.6.4(ii) | $2^{n-2}+4$ |
| 1 | $2^{n-2}$ | 2 | -"- | , IX.6.6(ii) | -"- |
| 2 | $2^{n-2}$ | 1 | IX.6.2 | , IX.6.7(ii) | -"- |
| | | | | | |
| $2^{n-2}$ | 2 | 2 | IX.6.3 | , IX.6.5 | $2^{n-2}+5$ |
| 2 | $2^{n-2}$ | 2 | -"- | , IX.6.7 | -"- . |

The shape of Cartan matrix we will deal with now occurs for the quivers $(3\mathcal{F})$, $(3\mathcal{H})$ and $(3\mathcal{K})$, and for the algebra Qu. $(3\mathcal{A})_1$. In IX.1.4 we have seen that always $\dim Z(\Lambda) = a + b + c + 1$.

IX.5.2 LEMMA *Assume that B is a block with defect group of order $2^n$ whose Cartan matrix is of the form*

$$\begin{bmatrix} a+b & a & b \\ a & a+c & c \\ b & c & b+c \end{bmatrix}$$

*Assume also that* $\dim Z(B) = 1 + a + b + c$ *and* $2^{n-2}+3 \leq k(B) \leq 2^{n-2}+5$. *Then one of the following holds:*

*(i)* $k(B) = 2^{n-2}+3$ *and* $\{a, b, c\} = \{1, 1, 2^{n-2}\}$.

*(ii)* $k(B) = 2^{n-2}+4$ *and* $\{a, b, c\} = \{1, 2, 2^{n-2}\}$.

*(iii)* $k(B) = 2^{n-2}+5$ *and* $\{a, b, c\} = \{2, 2, 2^{n-2}\}$.

Proof: We have that $\det C = 4abc$. For a block we must have, by IX.2.1(1) that a, b, c are powers of 2. Moreover, $1 + a + b + c = k(B)$.

We shall need the elementary divisors of C; they are $d_1 = \text{g.c.d.}\{a\ b,\ c\}$, $d_1d_2 = \text{g.c.d.}\{ab + ac + bc, ac + bc - ab, ab + bc - ac, ab + ac - bc\}$, and $d_1d_2d_3 = \det C$. Since also $d_3 = 2^n$, we have that $d_1d_2 = (abc)/2^{n-2}$.

Case 1 $k(B) = 2^{n-2}+3$. Then $a + b + c = 2^{n-2}+2$; and the possibilities for {a, b, c} are (only) $\{1, 1, 2^{n-2}\}$ or $\{2, 2^{n-3}, 2^{n-3}\}$. Assume that the second solution occurs, and $n > 3$. Then we have that $d_1d_2 = 2^{n-3}$. On the other hand, $d_1d_2 = \text{g.c.d.}\{2^{2(n-3)}, 2^{n-1} \pm 2^{2(n-3)}\}$, and it follows that $n = 3$.

Case 2 $k(B) = 2^{n-2}+4$. Then $a + b + c = 2^{n-2}+3$ and it follows that (ii) must hold.

Case 3 $k(B) = 2^{n-2}+5$. Then the possibilities for {a, b, c} are $\{2, 2, 2^{n-2}\}$ and $\{2^{n-3}, 2^{n-3}, 4\}$. Suppose the second alternative occurs; then $d_1d_2 = 2^{n-2}$. On the other hand, $d_1d_2 = \text{g.c.d.}\{2^n \pm 2^{2(n-3)}, 2^{2(n-3)}\} = \min\{2^n, 2^{2(n-3)}\}$. It follows that $n = 4$.

IX.5.2.1 By symmetry of notation, we may assume in IX.5.2 that $a \leq b \leq c$; therefore we will have to study three possible cases for decomposition matrices. The references are as follows:

| Block | decompositon matrix |
|---|---|
| $a = b = 1,\ c = 2^{n-2}$ | IX.6.1(i), IX.6.8 |
| $a = 1,\ b = 2,\ c = 2^{n-2}$ | IX.6.2, IX.6.9 |
| $a = b = 2,\ c = 2^{n-2}$ | IX.6.3, IX.6.8. |

We will now investigate Cartan matrices which occur for algebras with quiver (*3C*). For these, dim $Z(\Lambda) = k+s+2$, by IX.1.5.

IX.5.3 LEMMA *Suppose B is a block with defect group of order $2^n$ with Cartan matrix*

$$\begin{bmatrix} k+1 & k-1 & k \\ k-1 & k+1 & k \\ k & k & k+s \end{bmatrix}$$

*where $k,\ s \geq 1$. Assume also that dim $Z(B) = k+s+2$ and $2^{n-2}+4 \leq k(B) \leq 2^{n-2}+5$. Then $\{k,\ s\} = \{2,\ 2^{n-2}\}$ and $n > 3$.*

Proof: The Cartan matrix C has determinant 4ks; hence p = 2 and ks is a power of 2. Moreover, $k + s + 2 = k(B)$.

Suppose it is possible that $k(B) = 2^{n-2} + 5$. Then $\{s,\ k\} = \{1,\ 2^{n-2}+2\}$, and since s and k are powers of 2, we must have $n-2 = 1$ and $\{s,\ k\} = \{1,\ 4\}$. In both cases, the first two elementary divisors of C are odd and therefore 1; and we deduce that the highest elementary divisor is det $C = 16 = 2|\delta(B)|$. This is a contradiction to IX.2.1(1).

Hence $k(B) = 2^{n-2}+4$ and consequently $\{k,\ s\} = \{2,\ 2^{n-2}\}$. Suppose $n = 3$; consider the elementary divisors $d_1$, $d_2$, $d_3$ for C. We have $d_1 = 1$ and $d_2 = 4$, therefore $d_3 = ks = 4 = |\delta(B)|$, but $|\delta(B)| = 2^n$, a contradiction.

IX.5.3.1 The decomposition matrices for IX.5.3 are determined in IX.6.2 ($k = 2$, $s = 2^{n-2}$); one takes $C = \tilde{C}$ and all signs for D equal to 1; and in IX.6.10 ($k = 2^{n-2}$, $s = 2$).

There are some more Cartan matrices which occur for algebras of dihedral type. We will not study them in generality since the algebras in question cannot arise as blocks:

<u>IX.5.4</u> **LEMMA** *Suppose that Λ is an algebra of dihedral type which belongs to one of the families* $D.(3\mathcal{A})_2$, $D.(3\mathcal{B})_2$, $D.(3\mathcal{D})_2$, $D.(3\mathcal{C})$, $D.(3\mathcal{Q})$ *or* $D.(3\mathcal{R})$. *Then Λ is not Morita equivalent to a block.*

Proof: Suppose $\Lambda \sim_M B$ for some block B; then B must be of dihedral type, by IX.2.2. We know from V.4 that B has two 3-tubes, and they are fixed by $\Omega$; on the other hand, for any of the listed algebras, the 3-tubes are permuted by $\Omega$.

## IX.6 *Decomposition numbers for tame blocks with three simple modules*

In order to determine the decomposition matrix D, we will essentially only study solutions of equations $D^tD = C$, for the possible Cartan matrices C. It is convenient to reduce these to a few diophantine equations, which can be solved easily.

As a by-product, this gives also the generalized decomposition matrices $D^J$ where J is a central involution of the defect group.

We start by studying the appropriate diophantine equations: Given certain 3×3-matrices $\mathbb{C}$ over $\mathbb{Z}$, we wish to find all k×3- matrices $\mathbb{D}$ with entries over $\mathbb{Z}$ satisfying

$$\mathbb{D}^t\mathbb{D} = \mathbb{C},$$

for various values of k. We also assume that $\mathbb{D}$ does not have a row consisting of zeros only.

<u>IX.6.1</u> LEMMA *Suppose that either*

(i) $\mathfrak{C} = \begin{bmatrix} 2 & 1 & 0 \\ 1 & 2^{n-2}+1 & 1 \\ 0 & 1 & 2 \end{bmatrix}$ or (ii) $\mathfrak{C} = \begin{bmatrix} 3 & 1 & -1 \\ 1 & 2^{n-2}+1 & 1 \\ -1 & 1 & 3 \end{bmatrix}$

*and* $k = 2^{n-2}+3$ *and* $k = 2^{n-2}+4$.

*Then* $\mathfrak{D}$ *is of the form either*

(i) $\begin{bmatrix} \pm 1 & 1 & 0 \\ \pm 1 & 0 & 0 \\ \pm 0 & 1 & 1 \\ \pm 0 & 0 & 1 \\ \pm 0 & 1 & 0 \end{bmatrix} (2^{n-2}-1)\text{-times}$ or (ii) $\begin{bmatrix} \pm 1 & 0 & -1 \\ \pm 1 & 1 & 0 \\ \pm 1 & 0 & 0 \\ \pm 0 & 1 & 1 \\ \pm 0 & 0 & 1 \\ \pm 0 & 1 & 0 \end{bmatrix} (2^{n-2}-1)\text{times}$

*respectively.*

Proof: Let $[u_i]^t$, $[v_i]^t$ and $[w_i]^t$ be the columns of $\mathfrak{D}$, and denote by $c_{ij}$ the entries of $\mathfrak{C}$.

(i) We may assume that $u_1$ and $u_2$ are ±1 and $u_i = 0$ otherwise. Then $\pm v_1 \pm v_2 = 1$ and hence $v_1^2 + v_2^2 \geq 1$.

(1) <u>$w_1$ and $w_2$ are 0:</u> Suppose not, then $\pm w_1 \pm w_2 = 0$ and $w_1^2 + w_2^2 = 2$. Then $w_i = 0$ for $i \geq 3$. Since $\mathfrak{D}$ does not have a row of zeros, it follows that $v_i \neq 0$ for $i \geq 3$ and then $c_{11} - 1 \geq \sum_{i\geq 3} v_i^2 \geq 2^{n-2} + 1$, a contradiction.

We may now choose the notation such that $w_3 = \pm 1$ and $w_4 = \pm 1$, and then $v_3^2 + v_4^2 \geq 1$. Also, $v_r \neq 0$ for $r \geq 5$ since $\mathfrak{D}$ does not have a row of zeros. Consequently $2^{n-2}-1 \leq \sum_{r\geq 5} v_r^2 \leq c_{11}-2$, and it follows that equality holds. This shows that $v_r = \pm 1$ for $r \geq 5$, and also only one of $v_1$, $v_2$ and one of $v_3$, $v_4$ is ± 1; and the statement follows.

(ii) We may assume that $v_1$, $u_2$, $u_3$ are ± 1 and $u_i = 0$ otherwise. Then $\sum_{i=1}^{3} \pm v_i = 1$ and hence $\sum_{i=1}^{3} v_i^2 \geq 1$. Similarly $\sum_{i=1}^{3} \pm w_i = -1$ and hence $\sum_{i=1}^{3} w_i^2 \geq 1$.

(2) <u>One of $w_1$, $w_2$, $w_3$ is $\mp$ 1 and the others are 0:</u>

Suppose not; since $|w_i| \leq 1$ we must have $|w_i| = 1$ for $i \leq 3$ and then $w_r = 0$ for $r \geq 4$. Since $\mathfrak{D}$ does not have a row of zeros, we have $v_r \neq 0$ for $r \geq 4$, and then $c_{11}-1 \geq \sum_{r\geq 4} v_r^2 \geq 2^{n-2} + 1$, a contradiction.

Now we may choose the notation such that $w_1 = -u_1$ and $w_2 = w_3 = 0$ and $w_4$, $w_5$ are $\pm 1$. We have now that $\pm v_1 \pm v_2 \pm v_3 = 1$ and $\mp v_1 \pm v_4 \pm v_5 = 1$. It follows that at least two of $v_1, \ldots, v_5$ are $\neq 0$. Moreover, we have again that $v_r \neq 0$ for $r \geq 6$. Therefore we deduce $c_{11}-2 \geq \sum_{r\geq 2} v_r^2 \geq 2^{n-2}-1$, and equality holds. This shows that $v_r = \pm 1$ for $r \geq 6$; and also that precisely two of $v_1, \ldots, v_5$ are $\neq 0$, and the rest is clear.

By similar arguments one shows

IX.6.2 LEMMA *Suppose that* $\mathcal{C}$ *is of the form*

$$\begin{bmatrix} 3 & 2 & 1 \\ 2 & 2^{n-2}+2 & 2 \\ 1 & 2 & 3 \end{bmatrix}$$

*and* $k = 2^{n-2}+4$. *Then* $\mathcal{D}$ *is given by*

$$\begin{matrix} \pm \\ \pm \\ \pm \\ \pm \\ \pm \\ \pm \end{matrix}\begin{bmatrix} 1 & 1 & 1 \\ 1 & 1 & 0 \\ 1 & 0 & 0 \\ 0 & 1 & 1 \\ 0 & 0 & 1 \\ 0 & 1 & 0 \end{bmatrix} \begin{matrix} \\ \\ \\ \\ \\ (2^{n-2}-1)\text{-times.} \end{matrix}$$

IX.6.3 LEMMA *Suppose that* $\mathcal{C}$ *is of the form*

$$\begin{bmatrix} 4 & 2 & 0 \\ 2 & 2^{n-2}+2 & 2 \\ 0 & 2 & 4 \end{bmatrix}$$

*and that* $k = 2^{n-2}+5$. *Then* $\mathcal{D}$ *is of the form*

$$\begin{matrix} \pm \\ \pm \\ \pm \\ \pm \\ \pm \\ \pm \\ \pm \end{matrix}\begin{bmatrix} 1 & 0 & 0 \\ 1 & 1 & 0 \\ 1 & 1 & 1 \\ 1 & 0 & -1 \\ 0 & 1 & 1 \\ 0 & 0 & 1 \\ 0 & 1 & 0 \end{bmatrix} \begin{matrix} \\ \\ \\ \\ \\ \\ (2^{n-2}-1)\text{-times.} \end{matrix}$$

Proof: Let $[u_i]^t$, $[v_i]^t$ and $[w_i]^t$ be the columns of $\mathbb{D}$, and denote by $c_{ij}$ the entries of $\mathbb{C}$.

(1) Four of the $u_i$ are $\pm 1$: Suppose not, then $u_1 = \pm 2$ (say), and $u_i = 0$ otherwise; moreover $w_1 = 0$. Then $c_{01} = \pm 2v_1 = 2$ and $v_1 = \pm 1$. At most four of the $w_i$ are $\neq 0$; so we may assume that $w_r \neq 0$ for $r \geq 6$. Since $\mathbb{D}$ does not have a row of zeros, it follows that $v_r \neq 0$ for $r \geq 6$. We deduce that $2^{n-2} - 1 \leq \sum_{r\geq 6} v_r^2 \leq c_{11} - 1$, and equality holds. This shows that $v_r = \pm 1$ for $r \geq 6$, and $v_2, \ldots v_5$ are 0. But then $c_{12} = 0$, a contradiction.

By symmetry, we also get that four or the $w_i$ are $\pm 1$. We may choose the notation such that $u_i = \pm 1$ for $i \leq 4$, and $u_i = 0$ otherwise.

(2) Two of $w_1, \ldots, w_4$ are $\pm 1$ and two are 0: Suppose this is false. We have that $\Sigma \pm w_i = c_{02} = 0$ and $|w_i| \leq 1$ for all i. So either

(I) $w_i = 0$ for $i \leq 4$, or

(II) $w_i = 1$ for $i \leq 4$.

Suppose first (I) holds. Then we may assume $w_i = \pm 1$ for $5 \leq i \leq 8$, and then $w_i = 0$ otherwise. Consequently we have that $\sum_{i\leq 4} \pm v_i = 2$ and $\sum_{i=5}^{8} \pm v_i = 2$ and therefore $\sum_{i\geq 4} v_i^2 \geq 2$ and also $\sum_{i=5}^{8} v_i^2 \geq 2$. Now we must have $v_r \neq 0$ for $r \geq 9$, and $2^{n-2} - 3 \leq \sum_{r\geq 9} v_r^2 \leq c_{11} - 4 = 2^{n-2} - 2$. This shows that $v_r = \pm 1$ for $r \geq 9$ and , say, $\sum_{i\leq 4} v_i^2 = 3$. This is not possible since we know that $\sum_{i\leq 4} \pm v_i = 2$.

Now assume (II) holds. Then $w_r = 0$ for $r \geq 5$, and $v_r \neq 0$ for these r. Moreover $\sum_{i\leq 4} \pm v_i = 2$ and $\sum_{i\leq 4} v_i^2 \geq 2$. Consequently $2^{n-2}+1 \leq \sum_{r\geq 5} v_r^2 \leq c_{11} - 2$, a contradiction.

Now we may assume $w_3 = u_3$, $w_4 = -u_4$, moreover $w_5$ and $w_6 = \pm 1$, and $w_i = 0$ otherwise. We have $\sum_{i=1}^{4} \pm v_i = 2$ and $\pm(v_3 - v_4 + v_5 + v_6) = 2$. Hence at least three of $v_1, \ldots, v_6$ are $\neq 0$. Moreover $v_r \neq 0$ for $r \geq 7$, and we deduce $2^{n-2} - 1 \leq \sum_{r\geq 7} v_r^2 \leq c_{11} - 3$, and equality holds. This shows that $v_r = \pm 1$ for $r \geq 7$ and also that exactly three of $v_1, \ldots, v_6$ are $\pm 1$ and the others are 0. The rest follows now easily.

Now we are ready to calculate decomposition numbers.

<u>IX.6.4</u> **PROPOSITION** *Let B be any block whose Cartan matrix is of the form either*

(i) $\begin{pmatrix} 2^n & 2^{n-1} & 2^{n-1} \\ 2^{n-1} & 2^{n-2}+1 & 2^{n-2} \\ 2^{n-1} & 2^{n-2} & 2^{n-2}+1 \end{pmatrix}$ or (ii) $\begin{pmatrix} 2^n & 2^{n-1} & 2^{n-1} \\ 2^{n-1} & 2^{n-2}+1 & 2^{n-2} \\ 2^{n-1} & 2^{n-2} & 2^{n-2}+2 \end{pmatrix}$

*and* $k(B) = 2^{n-2}+3$   *and* $k(B) = 2^{n-2}+4$.

*Then the decomposition matrix of B is given by*

(i) $\begin{pmatrix} 1 & 0 & 0 \\ 1 & 1 & 0 \\ 1 & 0 & 1 \\ 1 & 1 & 1 \\ 2 & 1 & 1 \end{pmatrix}$ $(2^{n-2}-1)$-*times*   or (ii) $\begin{pmatrix} 1 & 0 & 0 \\ 1 & 1 & 0 \\ 1 & 0 & 1 \\ 1 & 1 & 1 \\ 0 & 0 & 1 \\ 2 & 1 & 1 \end{pmatrix}$ $(2^{n-2}-1)$-*times*.

*respectively.*

Proof: Let $\tilde{a}$, $\tilde{b}$ and $\tilde{c}$ be the first, second and third column of D. Put $\tilde{u} = -\tilde{a} + \tilde{b} + \tilde{c}$, $\tilde{v} = \tilde{b}$ and $\tilde{w} = \tilde{b} - \tilde{c}$, and let $\mathfrak{D}$ be the matrix with columns $\tilde{u}$, $\tilde{v}$, $\tilde{w}$. Then $\mathfrak{D}^t\mathfrak{D} = \mathfrak{C}$ where $\mathfrak{C}$ is the matrix in IX.6.1(i) or (ii). Moreover $\mathfrak{D}$ satisfies the hypothesis there, and we get the solution for $\mathfrak{D}$ and then also for D. Note that there is no ambiguity with the signs since D has entries $\geq 0$.

<u>IX.6.5</u> **PROPOSITION** *Let B be any block whose Cartan matrix is of the form*

$$\begin{pmatrix} 2^n & 2^{n-1} & 2^{n-1} \\ 2^{n-1} & 2^{n-2}+2 & 2^{n-2} \\ 2^{n-1} & 2^{n-2} & 2^{n-2}+2 \end{pmatrix}$$

*and where* $k(B) = 2^{n-2}+5$. *Then the decomposition matrix of B is given by*

$$\begin{pmatrix} 1 & 0 & 0 \\ 1 & 1 & 0 \\ 1 & 0 & 1 \\ 1 & 1 & 1 \\ 0 & 1 & 0 \\ 0 & 0 & 1 \\ 2 & 1 & 1 \end{pmatrix} \quad (2^{n-2}-1)\text{-times}.$$

Proof: Let $\tilde{a}$, $\tilde{b}$ and $\tilde{c}$ be the first, second and third column of D. Put $\tilde{u} = -\tilde{a} + \tilde{b} + \tilde{c}$, $\tilde{v} = \tilde{b}$ and $\tilde{w} = \tilde{b} - \tilde{c}$; and let $\mathfrak{D}$ be the matrix with columns $\tilde{u}$, $\tilde{v}$ and $\tilde{w}$. Then $\mathfrak{D}^t\mathfrak{D} = \mathfrak{C}$ where $\mathfrak{C}$ is the matrix in IX.6.3. Moreover, $\mathfrak{D}$ satisfies the hypothesis there, and we get a solution for $\mathfrak{D}$ and then also for D. There is no ambiguity with the signs since the entries of D are $\geq 0$.

IX.6.6 PROPOSITION *Assume that B is any block with Cartan matrix C where either*

$$(i)\ C = \begin{bmatrix} 4 & 2 & 2 \\ 2 & 2^{n-2}+1 & 1 \\ 2 & 1 & 2 \end{bmatrix} \quad \text{or} \quad (ii)\ C = \begin{bmatrix} 4 & 2 & 2 \\ 2 & 2^{n-2}+1 & 1 \\ 2 & 1 & 3 \end{bmatrix}$$

*and* $k(B) = 2^{n-2}+3$ *and* $k(B) = 2^{n-2}+4$.

*Then the decomposition matrix of B is given by*

$$(i)\begin{bmatrix} 1 & 0 & 0 \\ 1 & 1 & 0 \\ 1 & 0 & 1 \\ 1 & 1 & 1 \\ 0 & 1 & 0 \end{bmatrix} (2^{n-2}-1)\text{-times.} \qquad (ii)\begin{bmatrix} 1 & 0 & 0 \\ 1 & 1 & 0 \\ 1 & 0 & 1 \\ 1 & 1 & 1 \\ 0 & 0 & 1 \\ 0 & 1 & 0 \end{bmatrix} (2^{n-2}-1)\text{-times}$$

*respectively.*

Proof: Let $\tilde{a}$, $\tilde{b}$, $\tilde{c}$ be the columns of D and put $\tilde{u} = \tilde{a} - \tilde{c}$, $\tilde{v} = \tilde{b}$ and $\tilde{w} = \tilde{c}$. If $\mathfrak{D}$ is the matrix whose columns are $\tilde{u}$, $\tilde{v}$, $\tilde{w}$ then $\mathfrak{D}^t\mathfrak{D} = \mathfrak{C}$ where $\mathfrak{C}$ is the matrix in IX.6.1(i) or (ii) respectively. Now we deduce the result from IX.6.1.

IX.6.7 PROPOSITION *Assume B is any block with Cartan matrix C where either*

$$(i)\quad C = \begin{bmatrix} 8 & 4 & 4 \\ 4 & 2^{n-2}+2 & 2 \\ 4 & 2 & 3 \end{bmatrix} \quad \text{or} \quad (ii)\quad C = \begin{bmatrix} 8 & 4 & 4 \\ 4 & 2^{n-2}+2 & 2 \\ 4 & 2 & 4 \end{bmatrix}$$

*and* $k(B) = 2^{n-2}+4$ *and* $k(B) = 2^{n-2}+5$.

*Then the decomposition matrix of B is given by*

(i) $\begin{bmatrix} 1 & 0 & 0 \\ 1 & 1 & 0 \\ 1 & 0 & 1 \\ 1 & 1 & 1 \\ 2 & 1 & 1 \\ 0 & 1 & 0 \end{bmatrix}$ $(2^{n-2}-1)$-*times* (ii) $\begin{bmatrix} 1 & 0 & 0 \\ 1 & 1 & 0 \\ 1 & 0 & 1 \\ 1 & 1 & 1 \\ 2 & 1 & 1 \\ 0 & 0 & 1 \\ 0 & 1 & 0 \end{bmatrix}$ $(2^{n-2}-1)$-*times*

*respectively.*

The proof is the same as that of IX.6.6, using IX.6.2 and IX.6.3.

<u>IX.6.8</u> **PROPOSITION** *Suppose B is a block with Cartan matrix either*

(i) $\begin{bmatrix} 2 & 1 & 1 \\ 1 & 2^{n-2}+1 & 2^{n-2} \\ 1 & 2^{n-2} & 2^{n-2}+1 \end{bmatrix}$ *or* (ii) $\begin{bmatrix} 4 & 2 & 2 \\ 2 & 2^{n-2}+2 & 2^{n-2} \\ 2 & 2^{n-2} & 2^{n-2}+2 \end{bmatrix}$

*with* $k(B) = 2^{n-2}+3$ *with* $k(B)$ $2^{n-2}+5$.

*Then the decomposition matrix of B is given by*

(i) $\begin{bmatrix} 1 & 0 & 0 \\ 0 & 1 & 0 \\ 0 & 0 & 1 \\ 1 & 1 & 1 \\ 0 & 1 & 1 \end{bmatrix}$ $(2^{n-2}-1)$-*times* (ii) $\begin{bmatrix} 1 & 0 & 0 \\ 0 & 1 & 0 \\ 0 & 0 & 1 \\ 1 & 1 & 1 \\ 1 & 1 & 0 \\ 1 & 0 & 1 \\ 0 & 1 & 1 \end{bmatrix}$ $(2^{n-2}-1)$-*times*

*respectively.*

Proof: Let $\tilde{\mathfrak{a}}$, $\tilde{\mathfrak{b}}$, $\tilde{\mathfrak{c}}$ be the columns of D. Put $\tilde{\mathfrak{u}} = \tilde{\mathfrak{a}}$, $\tilde{\mathfrak{v}} = \tilde{\mathfrak{b}}$ and $\tilde{\mathfrak{w}} = \tilde{\mathfrak{b}} - \tilde{\mathfrak{c}}$, and let $\mathfrak{D}$ be the matrix whose columns are $\tilde{\mathfrak{u}}$, $\tilde{\mathfrak{v}}$, $\tilde{\mathfrak{w}}$. Then $\mathfrak{D}^t\mathfrak{D} = \mathfrak{C}$ where $\mathfrak{C}$ is the matrix in IX.6.1(i) or IX.6.3 respectively, and we obtain the result.

IX.6.9 PROPOSITION *Suppose that B is a block with Cartan matrix*

$$\begin{bmatrix} 2^{n-2}+1 & 2^{n-2} & 1 \\ 2^{n-2} & 2^{n-2}+2 & 2 \\ 1 & 2 & 3 \end{bmatrix}$$

*and* $k(B) = 2^{n-2}+4$. *Then the decomposition matrix of B is given by*

$$\begin{bmatrix} 1 & 0 & 0 \\ 0 & 1 & 0 \\ 0 & 0 & 1 \\ 1 & 1 & 1 \\ 0 & 1 & 1 \\ 1 & 1 & 0 \end{bmatrix} \begin{matrix} \\ \\ \\ \\ \\ (2^{n-2}-1)\text{-times.} \end{matrix}$$

Proof: Let $\tilde{a}$, $\tilde{b}$, $\tilde{c}$ be the columns of D. Define $\tilde{u} = -\tilde{a} + \tilde{b}$, $\tilde{v} = \tilde{b}$ and $\tilde{w} = \tilde{c}$, and let $\tilde{D}$ be the matrix with columns $\tilde{u}$, $\tilde{v}$, $\tilde{w}$. Then $\tilde{D}^t\tilde{D}$ is equal to the matrix $\tilde{C}$ in IX.6.2, and $\tilde{D}$ satisfies the hypothesis there. So we obtain a solution for $\tilde{D}$ and then a unique solution for D.

IX.6.10 PROPOSITION *Assume that B is a block with Cartan matrix*

$$\begin{bmatrix} 2^{n-2}+1 & 2^{n-2}-1 & 2^{n-2} \\ 2^{n-2}-1 & 2^{n-2}+1 & 2^{n-2} \\ 2^{n-2} & 2^{n-2} & 2^{n-2}+2 \end{bmatrix}, \text{ where } k(B) = 2^{n-2}+4.$$

*Then the decomposition matrix of B is of the form*

$$\begin{bmatrix} 0 & 0 & 1 \\ 0 & 1 & 1 \\ 1 & 0 & 0 \\ 1 & 0 & 1 \\ 0 & 1 & 0 \\ 1 & 1 & 1 \end{bmatrix} \begin{matrix} \\ \\ \\ \\ \\ (2^{n-2}-1)\text{-times.} \end{matrix}$$

Proof: Let $\tilde{a}$, $\tilde{b}$ and $\tilde{c}$ be the first, second and third column of D. Define $\tilde{D}$ to be the matrix with comlumns $\tilde{u} = \tilde{c} - \tilde{a}$, $\tilde{v} = \tilde{c} - \tilde{b}$ and $\tilde{w} = \tilde{c}$. Then $\tilde{D}^t\tilde{D} = \tilde{C}$ where $\tilde{C}$ is equal to the matrix in IX.6.2. Hence we obtain a solution for $\tilde{D}$ and then also for D. Since the entries of D are $\geq 0$ the signs for $\tilde{D}$ are uniquely determined.

IX.6.11 *Some generalized decomposition numbers*

We recall the following general principle from block theory: Let B be a p-block and x be a central element in some defect group of D, of order $p^k > 1$. Then the generalized decomposition matrix $D^x$ has entries which are algebraic integers in the fied $\mathbb{Q}(\zeta)$ where $\zeta$ is a primitive $p^k$-th root of unity. The matrix $D^x$ has k(B) rows and $\ell(b)$ columns where b is the Brauer correspondent of B in $C_G(x)$. Moreover, $D^x$ does not have a row of zeros only; and $(\overline{D^x})^t(D^x)$ is equal to the Cartan matrix of b.

Let now B be a tame block, we have p = 2 and take for x a central involution J of a defect group of B. Then $D^J$ has entries in $\mathbb{Z}$; moreover, the blocks b and B have the same defect group, by Brauer's First Main Theorem. In particular, the Cartan matrix, C say, of b occurs in our list. In fact, the equations $(D^J)^t(D^J)$ have already been solved. The answer is the same as in the propositions IX.6.4 to 10, except that signs may occur. However, in each row the sign is constant.

We shall use this later, in the proof of the Theorems by Brauer-Suzuki and Brauer.

We note that this method may also be used to determine the generalized decomposition matrix $D^J$ for tame blocks with two simple modules, with analogue results.

# *X Some applications*

One of the early results towards the classification of finite simple groups is the famous theorem by Brauer and Suzuki. We will give a proof, in the light of the results on tame blocks.

THEOREM *Let G be a finite group whose Syow 2-subgroup is quaternion of order $2^n$ ($n \geq 3$). Then G is not simple.*

Proof: Let $J \in G$ be an element of order 2. It suffices to find an irreducible character $\chi \neq 1_G$ of G such that $\chi(J) = \chi(1)$.

The main ingredient from character theory is a formula, due to Frobenius ([Fr], see also [F]): Suppose G is any finite group, $\{\chi_s\}$ is the set of characters of some p-block B of G, and $t_1$, $t_2 \in G$ are elements of order 2. Let

$$X_s := \frac{\chi_s(t_1)\ \chi_s(t_2)}{\chi_s(1)}.$$

(*) If $g \in G$ is an element such that for all conjugates $u_1$ of $t_1$ and $u_2$ of $t_2$, the 2-part of $u_1u_2$ is not conjugate to g then

$$\sum_s d_{si}^g X_s = 0, \text{ for all } i$$

Here $[d_{si}^g]$ is a column of the generalized decomposition matrix $D^g$.

To prove the Theorem, let p = 2 and let B be the principal 2-block of G. We take $t_1 = t_2 = :J$. The condition in (*) holds for g = J and also g = x of order 4, since, by Sylow's theorem, G does not have a subgroup of the form $C_2 \times C_2$.

The block B is tame and is an algebra of quaternion type, so by VII.3 it has at most three simple modules.

(1) We may assume that B has 3 simple modules: If B has one simple module then B must have a character $\neq 1_G$ of degree 1: By V.5.12, $k(B) = 2^{n-2} + 3$, and the Cartan number is $|\delta(B)| = 2^n$. If $\chi_i(1)$ were $> 1$ for $\chi_i \neq 1_G$ then it would follow that $2^n = \Sigma\ \chi_i(1)^2 \geq 1 + 4(2^{n-2} + 2) > 2^n$. Hence the commutator G' is a non-trivial normal subgroup of G, and we are done.

If B has two simple modules we see from the decomposition matrix, D say, that there is a non-trivial character of degree 1: There are two possibilities for D; they are given in the table at the end. In each case, there is a unique simple B-module of height zero and it corresponds to the first column, and this must be K; and D has two rows $[1 \quad 0]$. Hence we are done in this case as well.

Now there are three possible decomposition matrices, which we will list now; we take the characters of odd degree first and the characters with identical decomposition numbers at the end; and we may assume the first row and column corresponds to the trivial module.

Either or or (for $n > 3$)

$$D_1 = \begin{vmatrix} 1 & 0 & 0 \\ 1 & 1 & 0 \\ 1 & 0 & 1 \\ 1 & 1 & 1 \\ 0 & 1 & 0 \\ 0 & 0 & 1 \\ 2 & 1 & 1 \end{vmatrix} (*) \qquad D_2 = \begin{vmatrix} 1 & 0 & 0 \\ 0 & 1 & 0 \\ 0 & 0 & 1 \\ 1 & 1 & 1 \\ 1 & 1 & 0 \\ 1 & 0 & 1 \\ 0 & 1 & 1 \end{vmatrix} (*) \qquad D_3 = \begin{vmatrix} 1 & 0 & 0 \\ 1 & 1 & 1 \\ 1 & 0 & 1 \\ 1 & 1 & 0 \\ 2 & 1 & 1 \\ 0 & 0 & 1 \\ 0 & 1 & 0 \end{vmatrix} (*)$$

Here (*) means, $(2^{n-2}-1)$-times.

By IX.6.11, the generalized decomposition matrix $D^J$ has rows equal to ± the rows of D where D is one of the above matrices listed. Moreover the columns of $D^J$ (and D) are orthogonal to those of $D^x$; this is important:

(2) <u>$D^x$ has only one column, and $(\bar{D}^x)^t(D^x) = 4$:</u>

The number of columns of $D^x$ is $\ell(b)$ where b is the principal block of C(x), and the inner product is the Cartan matrix of b, see IX.6.11. From the conditions on G, we have that $\langle x \rangle$ is a Sylow 2-subgroup of C(x), and this is the defect group of b (see V.1.2) ; and the defect group of b is central. Therefore b is Morita equivalent to the group algebra $K\langle x\rangle$ (V.2.10). In particular, b has one simple module, and the Cartan number is $\dim K\langle x\rangle = 4$.

(3) <u>Four of the entries of $D^x$ are ±1 and the others are 0</u>: Let $(D^x) = [d^x_{s1}]$; then $\sum_s |d^x_{s1}|^2 = 4$. Moreover, the first row of $D^x$ corresponds to $1_G$ and so $d^x_{11} = 1$; consequently $|d^x_{s1}| \le 1$. Suppose $\chi = \chi_s$ is a character of B. Then since x and $x^{-1}$ are

conjugate, we have $\chi(x)$ is real $[\chi(x) = \chi(x^{-1}) = \overline{\chi(x)}]$. On the other hand, by the definition of $D^x$, $\chi(x) = d^x_{s1}(\dim S_x)$ where $S_x$ is the simple $b_x$-module, and therefore $d^x_{s1} \in \mathbb{Z}$.

(4) <u>The rows with characters of odd degree have non-zero entry in $D^x$</u>: It suffices to show that $\chi(1)$ is even if $(D^x)_\chi = 0$. Suppose that the entry is 0; then $\chi(x) = 0$. Now $\chi(x)$ is the trace of x on the representation affording $\chi$, and it is of the form $0 = a.1 + b(-1) + ci + d(-i)$ where $a, b, c, d \geq 0$. So $a = b$ and $c = d$, and $\chi(1) = a + b + c + d = 2a + 2c$ and is even.

We see now that $D^x$ is either (i) $[1\ -1\ -1\ 1\ 0 \ldots 0]^t$ or (ii) $[1\ 1\ 1\ -1\ 0 \ldots 0]^t$, or (iii) $[1\ 1\ -1\ 1\ 0 \ldots 0]^t$.

Now we study the generalized decomposition matrix $D^J$. The first row is $[1\ 0\ 0]$. The other rows are $\pm$ the rows of one of the matrices; we don't know in which order. We say that the first four rows of $D_1$, $D_2$, $D_3$ with respect to the order given above have "odd degree". Since the second and third column correspond to irreducible Brauer characters $\phi_2$, $\phi_3$ of $b^J$ of odd degree for $D_1$, $D_3$ and of even degree for $D_2$, we deduce $\chi(J) = \sum_i d^J_{i\chi}\, \phi_i(1)$ is odd for a row of "odd degree".

(5) <u>A row of $D^J$ of odd degree corresponds to a character $\chi$ of B of odd degree:</u> If $\chi(J) = \sum_i d^J_{i\chi}\, \phi_i(1)$ is odd, then the trace of J on a representation affording $\chi$ is of the form $a + (-1)b$, and then the degree is $a + b = a - b + 2b$.

*We assume first that* $|\delta(B)| = 8$. Let S be the submatrix of D consisting of rows 2 to 4, and E the submatrix which is formed by rows 5 to 7. Similarly let S', E' denote the submatrices of $D^J$ consisting of rows 2 to 4 and 5 to 7 respectively. Then by (4) the rows of S', E' are $\pm$ the rows of S, E for either S, E both from $D_1$ or both from $D_2$. We will not calculate the solutions for $D^J$ explicitly, to avoid cases (although this is elementary). Instead, we will exploit the orthogonality directly.

Let Y be the 3×3-matrix whose first column consists of the vector $y^t$ satisfying $D^x = [1\ \ y\ 0\ 0\ 0]^t$, and all other entries of Y are 0. We observe that in both cases for D, we have that

(6) <u>$E = S + Y$</u> .

We shall now show that this implies

(7) <u>$E' = -S' - Y$:</u> Rewriting the orthogonality condition, we have that

$$0 = D^tD^J = E_{11} + S^tS' + E^tE'$$

where $E_{11}$ is the matrix unit. Substituting $S = E - Y$ gives

$$0 = E_{11} - Y^tS' + E^t(S' + E').$$

Since $D^X$ D we obtain $0 = [1\ 0\ 0] + y^tS$ and hence $0 = E_{11} + Y^tS$. Similarly $0 = E_{11} + Y^tS'$. Therefore $0 = -2E_{11} = E^t(S' + E')$. Since E is non-singular, it remains to show that $2E_{11} = E^tY$, or equivalently $2E_{11} = Y^tE$. But $Y^tE = Y^t(S + Y) = -E_{11} + 3E_{11}$, as required.

Let $v := \underline{d}_1$ if $D^J$ is of type $D_1$ or $D_3$ and $v := \underline{d}_1 - \underline{d}_2 - \underline{d}_3$ otherwise, where $\underline{d}_i$ are the columns of $D^J$. Then obviously

(8) <u>$v_1 = 1$ and $v_i = \pm 1$ for $2 < i < 4$. Moreover, for $j > 5$, exactly one $v_j$ is $\pm 2$ and the others are 0</u>: We have $D^X v = 0$, therefore $1 + \delta_2 v_2 + \delta_3 v_3 + \delta_4 v_4 = 0$. Hence for exactly one value of i ($i \le 4$), the numbers $\delta_i$ and $v_i$ are equal. Let $\{i, j, k\} = \{2, 3, 4\}$ with $\delta_i = v_i$. By (7), we have $[v_i, v_j, v_k] = -[v_{i+3}, v_{j+3}, v_{k+3}] - [\delta_i, \delta_j, \delta_k]$; and we deduce $v_{i+3} = -2\delta_i$ and $v_{j+3} = v_{k+3} = 0$.

Using now Frobenius'formula we obtain two equations, namely

$$1 + \delta_i X_i + \delta_j X_j + \delta_k X_k = 0$$

$$1 + \sum_{m \ge 1} v_m X_m = 0$$

Adding these and substituting the values for $v_m$ gives

$$(**) \quad 2(1 + \delta_i X_i - \delta_i X_{i+3}) = 0.$$

Since $E = S + Y$ we have $\chi_{j+3}(1) = \chi_j(1) + \delta_j$; and from $E' = -S' - Y$ we see that $\chi_{j+3}(J) = -[\chi_j(J) + \delta_j]$. Substituting this into (**) and reformulating gives, setting $\delta = \delta_j$ and $\chi = \chi_j$,

$$\chi(1)[\chi(1) + \delta] + \delta[\chi(1) + \delta]\chi(J)^2 - \delta[\chi(J) + \delta]^2\chi(1) = 0$$

and therefore $[\chi(1) - \chi(J)]^2 = 0$, as required.

*The case when* $|\delta(B)| > 8$: Let Y and S be as before, and let E be the submatrix of D consisting of rows 5 to k; similarly E'. We partition E and E' as follows: Let $E_0$ and $E_0'$ be the submatrices of E and E' consisting of rows 5 to 7; denote by $E_1$ and

$E_1'$ the remaining rows. Then $E_0$ is non-singular; moreover all rows of $E_1$ are equal to the last row of $E_0$. We observe that

(6*) $\underline{S + Y = E_0}$.

Since $D^x$ D we have that $0 = [1\ 0\ 0] + y^tS$ and hence $0 = E_{11} + Y^tS$; similarly $0 = E_{11} + Y^tS'$.

(7*) $\underline{-S' - Y = E_0' + (E_0{}^t)^{-1}E_1{}^tE_1'}$: We exploit the orthogonality. Since $0 = D^tD^J$, we may write $0 = E_{11} + S^tS' + E_0{}^tE_0' + E_1{}^tE_1'$ and substituting $0 = E_{11} - Y^tS' + E_0{}^t(S' + E_0') + E_1{}^tE_1'$. As before, we know that $2E_{11} = E_0^tY$, so we have $0 = E_0^t(Y + S' + E_0') + E_1{}^tE_1'$.

(7**) $\underline{-S' - Y = E_0' + F}$ <u>where F is the matrix whose last row is the sum of all rows of $E_1'$ and the other rows are 0:</u> By construction we have that

$$E_1 = \begin{bmatrix} 0 & 0 & 1 \\ 0 & 0 & 1 \\ & \vdots & \\ 0 & 0 & 1 \end{bmatrix} E_0;$$

this implies $(E_0{}^t)^{-1}E_1{}^t = \begin{bmatrix} 0 & 0 & & 0 \\ 0 & 0 & \dots & 0 \\ 1 & 1 & & 1 \end{bmatrix}$. Substitute into (7*), and (7**) follows directly. By (6*) and the definition of $E_1$, we have :

(8) <u>The character degrees satisfy</u> $\underline{x_j(1) + \delta_j = x_{j+3}(1) \quad (2 < j < 4)}$ <u>and</u> $\underline{x_l(1) = x_7(1) \quad (l > 8)}$:

We write the i-th row of $D^J$ in the form $\epsilon_i\underline{r}_i$ where $\epsilon_i = \pm 1$, and the $\underline{r}_i$ are vectors whose entries are $\geq 0$. One of the $\underline{r}_i$ occurs with multiplicity > 1, call this $\underline{r}_i$ "multiple" vector.

(9) <u>With appropriate numbering, the character values on J satisfy the following</u>:

$$x_{j+3}(J) = -[x_j(J) + \delta_j] \quad (2 \leq j \leq 4), \text{ and}$$
$$[x_l(1)]^2 \text{ is constant, for } l \geq 8:$$

The first two identities follow directly from (7**). The rows of -S' - Y are pairwise distinct (already the second and third components distinguish them.) Hence at most one of the multiple vectors occurs in the first and second row of -S' - Y. Therefore at most two types of vectors occur in E' from row 7 onwards. We may choose the ordering such that the multiple vector occurs in rows $l \geq 8$. This proves (9).

Moreover $[-\delta_4\ 0\ 0] - \text{row}[x_4] = \text{row}[x_7] + \Sigma\ \text{row}[x_l]$, that is

$$[-\delta_4\ 0\ 0] - \epsilon_4 r_4 = \epsilon_7 r_7 + (\sum_{l \geq 8} \epsilon_l) r_8.$$

The left side is equal to $\pm r_7$. One of the second or third component of $r_4$ is 1, and distinct $r_i$ are linearly independent. So we deduce

(10) $\Sigma\ \epsilon_1 = 0$ and $\epsilon_{j+3} = -\epsilon_j$ for $2 \leq j \leq 4$.

(11) <u>We have $1 + \delta_i X_i - \delta_i X_{i+3} = 0$ for some i $(2 < i < 4)$:</u> Using Frobenius for $D^X$

$$(*) \quad 1 + \delta_2 X_2 + \delta_3 X_3 + \delta_4 X_4 = 0.$$

Consider the vector $\underset{\sim}{v}$ of the column space of $D^J$ which we define to be either the first column, if $D^J$ is of type $D_1$ or $D_3$, or $\underset{\sim}{v} = \underset{\sim}{d}_1 - \underset{\sim}{d}_2 - \underset{\sim}{d}_3$ otherwise.

Then for $2 \leq i \leq 4$, $v_i = \begin{cases} \epsilon_i & D_1 \text{ or } D_3 \\ -\epsilon_i & D_2 \end{cases}$, and for $j \geq 5$, $v_i = \pm 2$ or 0.

We have that $D^X\underset{\sim}{v} = 0$, therefore $1 + \delta_1 v_1 + \delta_2 v_2 + \delta_3 v_3 = 0$. It follows that, for exactly one i with $i \leq 4$ we have $\delta_i = v_i$. Write $\{i, j, k\} = \{2, 3, 4\}$ with $\delta_i = v_i$ (and $\delta_j = -v_j$, $\delta_k = -v_k$). Since $-S' - Y = E_0'$ it follows that

$-[v_i, v_j, v_k] - [\delta_i, \delta_j, \delta_k] = [v_{i+3}, v_{j+3}, v_{k+3}]$. We deduce $v_{i+3} = -2\delta_i$ and $v_{j+3} = v_{k+3} = 0$. From Frobenius'formula,

$$(**) \quad 1 + \delta_i X_i - \delta_j X_j - \delta_k X_k - 2\delta_i X_{i+3} + \Sigma\ v_1 X_1 = 0.$$

By (8) and (9) we have that $X_1$ is constant for $1 \geq 8$; moreover $\Sigma\ v_1 = \Sigma\ \epsilon_1$ (in (i), (iii),) or $v_1 = 0$ (in case (ii)); so in any case, $\Sigma\ v_1 X_1 = 0$. Subtracting (*) and (**) gives (11).

Now substitute $x_{i+3}(1) = x_i(1) + \delta_i$ and $x_{i+3}(J) = -[x_i(J) + \delta_i]$ and solve the equation as before.

<u>X.2</u> ***Relations for characters satisfied on elements of odd order*** Suppose B is a tame block. In [$B_3$] and [O], Brauer and Olsson determined relations satisfied by the characters of B on elements of odd order, with coefficients 0 or ±1; where the signs are undetermined. These correspond to identities relating the composition factors of reductions modulo 2.

We obtain these dirctly from the decompositon matrices, with signs uniquely determined, as follows: Express each row corresponding to a character of height > 0 as a combination of rows of height zero (the first four rows).

In the dihedral case, there is a one-one-correspondence between the signs occuring in these relations and the Morita equivalence classes of possible blocks:

Suppose now there is a group which has a tame 2-block, and one knows the character degrees of the characters in B. Then one can with this information alone determine the identities described in X.2; and therefore one may determine the Morita equxivalence class of the block, ecept possibly for socle scalars. This may be done for the blocks occring for sporadic simple groups, see [$L_2$].

We will here give another example.

X.3 *Let $n = r(r+1)/2 + 6$; then the alternating group $A_n$ has a block with dihedral defect groups. If $r \neq 0$ then B is Morita euivalent to the algebra $D(3B)_1$ that is, all these blocks are Morita equivalent to the principal 2- block of $A_7$. If $k = 0$ then B is Morita equivalent to the algebra in $D(3A)_1$, for the appropriate parameters.*

To prove this, let $B_1$ be the 2-block of the symmetric group $S_n$ with core $\lambda = (r, r-1, r-2, \ldots, 1)$. Then it is easy to see that the restriction of $B_1$ to $A_n$ is one block, B say, and that a defect group of B is dihedral of order 8.

For n = 6 and 7, the statements follow from [$E_1$] and [$E_4$]; therefore we assume $r \geq 2$. The irreducible characters $\chi^\mu$ of $B_1$ are parametrized by partitions $\mu$ of n which may be obtained by adjoining three 2-hooks to $\lambda$. If $\mu$ is a partition such that $\mu \neq \mu'$ (the conjugate partition) then the restriction of $\chi^\mu$ to $A_n$ is irreducible; this gives immediately five irreducible characters $\chi_i := \chi^{\mu_i}|_{A_n}$ for

$\mu_1 =$ 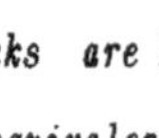 $\mu_2 =$ 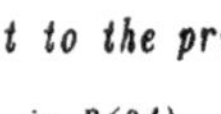 $\mu_3 =$ 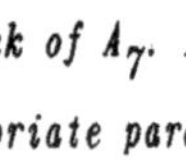

$\mu_4 =$  $\mu_5 =$ 

These must be all irreducible characters; we know k(B). (Or directily, for any $\mu$ with core $\lambda$ we have that $\mu \neq \mu'$.) By an elementary calculation, using the standard degree formulae for characters of symmetric groups [JK], one shows

$$\chi_1(1) - \chi_2(1) = \chi_3(1) - \chi_4(1) = \chi_5(1).$$

There are three possible decomposition matrices for blocks with dihedral defect groups with three simple modules. The only one where this relation holds is the one for $D(3B)_1$.

X.4 (a) In $[E_4]$, the definition of algebras of dihedral type was meant to include the condition that $\Omega$ fixes each 3-tube; this was used throughout the paper. Here we have now generalized the definition, and this hypothesis has been removed. Therefore we have obtained a number of new algebras; however, they cannot occur as blocks.

In $[E_4]$ we claimed that the algebras $D(3\mathcal{B})_1$ can only occur as blocks when a defect group has order 8. The proof given is not correct; and the question remains open whether blocks with larger defect groups in this family exist.

In some places in $[E_4]$ weakly symmetric algebras are mentioned; this should be ignored, it goes back to an earlier attempt to obtain more general results. However, for weakly symmetric algebras there are problems since $\tau$-periodicity need not be the same as $\Omega$-periodicity.

(b) The algebras $Q(2\mathcal{B})_{2,3}$ are missing in $[E_6]$; this is due to an error in distinguishing wild and tame. Also, some of the relations are not stated correctly, although the proofs are complete. This arises from omitting that certain elements in $J^3$ should be zero.

## T A B L E S

*Dihedral type*

| Ref | Quiver | Relations |
|---|---|---|
| III.1(a) |  | $XY = 0 = YX$<br>$X^m = Y^n$<br>$(m \geq n \geq 2,\ m+n > 4)$. |
| III.1(b) | | $X^2 = 0 = Y^2$<br>$XY = YX$ |
| III.1(b') | | char K = 2 and<br>$X^2 = 0,\ Y^2 = XY$<br>$XY = YX$ |
| III.1(c) | | $X^2 = 0 = Y^2$<br>$(XY)^k = (YX)^k,\ k \geq 2$. |
| III.1(c') | | char K = 2 and<br>$X^2 = (XY)^k,\ Y^2 = d(XY)^k$<br>$(XY)^k = (YX)^k,\ (XY)^kX = 0 = (YX)^kY$<br>$k \geq 2$ and $d = 0$ or $1$ |
| D(2𝒜)<br>VI.8.1 | 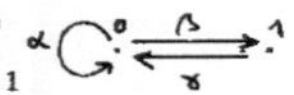 | $\gamma\beta = 0,\ \alpha^2 = c(\alpha\beta\gamma)^k$<br>$(\alpha\beta\gamma)^k = (\beta\gamma\alpha)^k$,<br>$k \geq 1$ and $c = 0$ or $1$. |

[(*) = $(2^{n-2}-1)$-times, $B_0$ = principal block; $B_1$ = some non-principal block]

| | \|Z(A)\| | Cartan matrix | blocks | decomposition matrix | Example |
|---|---|---|---|---|---|
| III.1(a) | n+m | [n+m] | - | - | |
| -"-(b*)<br>[$B_2$] | 4 | [4] | | $[1^4]$ | $KV_4$ |
| -"-(c*)<br>[$B_3$] | k+3 | [4k] | $k=2^{n-2}\geq 2$ | $\begin{bmatrix} 1^4 \\ 2 \end{bmatrix}$ (*) | $KD_{2^n}$ |
| D(2𝒜)<br>IX.4.1, [$E_8$] | k+3 | $\begin{bmatrix} 4k & 2k \\ 2k & k+1 \end{bmatrix}$ | $k=2^{n-2}$ | $\begin{bmatrix} 1 & 0 \\ 1 & 0 \\ 1 & 1 \\ 1 & 1 \\ 2 & 1 \end{bmatrix}$ (*) | $B_0(H)$<br>$L_2(q)<H$<br>$q\equiv 1 \bmod 4$ |

D(2B) 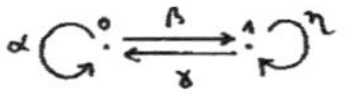
VI.8.2

$\beta\eta = 0 = \eta\gamma = \gamma\beta$
$\alpha^2 = c(\alpha\beta\gamma)^k,$
$(\alpha\beta\gamma)^k = (\beta\gamma\alpha)^k,$
$(\gamma\alpha\beta)^k = \eta^s$
$k \geq 1,\ s \geq 2$ and $c = 0$ or $1$.

D(2B) $\quad k+r+2 \quad \begin{bmatrix} 4k & 2k \\ 2k & k+r \end{bmatrix} \quad k=1,\ r=2^{n-2} \quad \begin{vmatrix} 1 & 0 \\ 1 & 0 \\ 1 & 1 \\ 1 & 1 \\ 0 & 1 \end{vmatrix}$ $\quad B_0(H$, $L_2(q)<H$, $q\equiv 3 \bmod 4$

IX.4.1, $[E_8]$

$D(3\mathcal{A})_1$
VI.9.1(i)
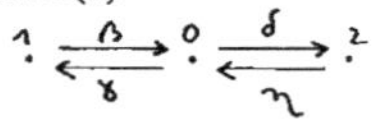

$\beta\gamma = 0 = \eta\delta$
$(\gamma\beta\delta\eta)^k = (\delta\eta\gamma\beta)^k$
$k \geq 1.$

$D(3\mathcal{A})_1 \quad k+3 \quad \begin{bmatrix} 4k & 2k & 2k \\ 2k & k+1 & k \\ 2k & k & k+1 \end{bmatrix} \quad k=2^{n-2} \quad \begin{vmatrix} 1 & 0 & 0 \\ 1 & 1 & 0 \\ 1 & 0 & 1 \\ 1 & 1 & 1 \\ 2 & 1 & 1 \end{vmatrix}$ (*) $\quad B_0(L_2(q))$, $q\equiv 1 \bmod 4$

IX.5.1
IX.6.4

$D(3\mathcal{A})_2$
VI.9.1(ii) - " -

$\beta\delta = 0 = \eta\gamma$
$(\gamma\beta)^k = (\delta\eta)^l$
$k \geq l \geq 2.$

$D(3\mathcal{A})_2 \quad \begin{bmatrix} k+1 & k & l \\ k & k+1 & 0 \\ l & 0 & l+1 \end{bmatrix}$

IX.5.4

$D(3\mathcal{B})_1$
VI.9.2(i)
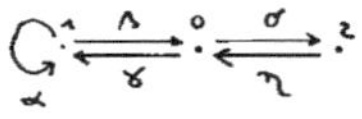

$\alpha\beta = 0 = \gamma\alpha$
$\beta\gamma = 0 = \eta\delta$
$(\gamma\beta\delta\eta)^k = (\delta\eta\gamma\beta)^k$
$\alpha^s = (\beta\delta\eta\gamma)^k,$
$k \geq 1$ and $s \geq 2.$

$D(3\mathcal{B})_1 \quad k+s+2 \quad \begin{bmatrix} 4k & 2k & 2k \\ 2k & k+s & k \\ 2k & k & k+1 \end{bmatrix} \quad s=2^{n-2},\ k=1 \quad \begin{vmatrix} 1 & 0 & 0 \\ 1 & 1 & 0 \\ 1 & 0 & 1 \\ 1 & 1 & 1 \\ 0 & 1 & 0 \end{vmatrix}$ (*) $\quad B_0(A_7)$, $n>3$ ?

IX.6.6
IX.5.1

$D(3\mathcal{B})_2$

VI.9.2(ii)

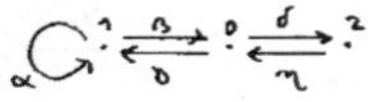

$\gamma\alpha = 0 = \alpha\beta$

$\beta\delta = 0 = \eta\gamma$

$(\gamma\beta)^k = (\delta\eta)^l$

$(\beta\gamma)^k = \alpha^s$

$k \geq 1,\ l \geq 1$ with $k+l > 2$; $s \geq 2$.

$D(3\mathcal{B})_2$ IX.5.4

$$\begin{bmatrix} k+1 & k & 1 \\ k & k+1 & 0 \\ 1 & 0 & l+1 \end{bmatrix}$$

$D(3\mathcal{D})_1$

VI.9.3(i)

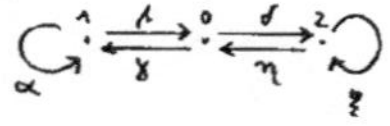

$\beta\gamma = 0 = \eta\delta$

$\alpha\beta = 0 = \gamma\alpha$

$\delta\xi = 0 = \xi\eta$

$\alpha^s = (\beta\delta\eta\gamma)^k$

$\xi^t = (\eta\gamma\beta\delta)^k$

$(\gamma\beta\delta\eta)^k = (\delta\eta\gamma\beta)^k,$

$k \geq 1;\ s, t \geq 2.$

$D(3\mathcal{D})_1$ IX.5.1 $\quad k+s+t+1$

$$\begin{bmatrix} 4k & 2k & 2k \\ 2k & k+s & k \\ 2k & k & k+t \end{bmatrix}$$

$D(3\mathcal{D})_2$

VI.9.3(ii)

- " -

$\gamma\alpha = 0 = \alpha\beta$

$\beta\delta = 0 = \eta\gamma$

$\delta\xi = 0 = \xi\eta$

$(\beta\gamma)^k = \alpha^s,\ (\eta\delta)^l = \xi^t$

$(\gamma\beta)^k = (\delta\eta)^l,$

$k, l \geq 1;\ s, t \geq 2.$

$D(3\mathcal{D})_2$ IX.5.4

$$\begin{bmatrix} k+s & k & 1 \\ k & k+s & 0 \\ 1 & 0 & l+t \end{bmatrix}$$

$D(3\mathcal{K})$

VI.9.4

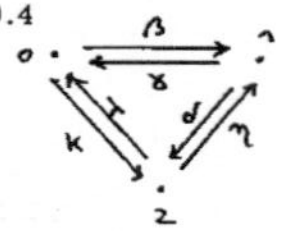

$\beta\delta = \delta\lambda = \lambda\beta = 0$

$\gamma\kappa = \kappa\eta = \eta\gamma = 0$

$(\beta\gamma)^a = (\kappa\lambda)^b,$

$(\lambda\kappa)^b = (\eta\delta)^c,$

$(\delta\eta)^c = (\gamma\beta)^a$

$a \geq b \geq c \geq 1.$

$D(3\mathcal{K})$ IX.5.2 IX.6.8 $\quad a+b+c+1$

$$\begin{bmatrix} a+b & a & b \\ a & a+c & c \\ b & c & b+c \end{bmatrix}$$

$a=b=1$, $c=2^{n-2}$

$$\begin{vmatrix} 1 & 0 & 0 \\ 0 & 1 & 0 \\ 0 & 0 & 1 \\ 1 & 1 & 1 \\ 0 & 1 & 1 \end{vmatrix} (*)$$

$B_0(L_2(q))$

$q \equiv 3 \bmod 4$

$D(3\mathcal{L})$

VI.9.5

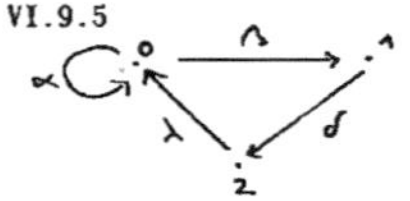


$\alpha\beta = 0 = \lambda\alpha$

$(\beta\delta\lambda)^k = \alpha^s,$

$(\delta\lambda\beta)^k\delta = 0.$

$k \geq 2,\ s \geq 2.$

$D(3\mathcal{L})$

IX.5.4

$$\begin{bmatrix} k+s & k & k \\ k & k+1 & k \\ k & k & k+1 \end{bmatrix}$$

$D(3\mathcal{Q})$

VI.9.6

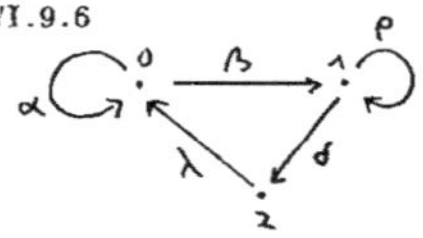


$\lambda\alpha = 0 = \alpha\beta$

$\beta\rho = 0 = \rho\delta$

$(\beta\delta\lambda)^k = \alpha^s$

$(\delta\lambda\beta)^k = \rho^t$

$s,\ t \geq 2,\ k \geq 1.$

$D(3\mathcal{Q})$

IX.5.4

$$\begin{bmatrix} k+s & k & k \\ k & k+t & k \\ k & k & k+1 \end{bmatrix}$$

$D(3\mathcal{R})$

VI.9.7

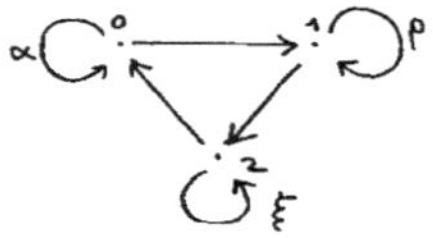


$\alpha\beta = 0 = \beta\rho = \rho\delta$

$\delta\xi = 0 = \xi\lambda = \lambda\alpha$

$\alpha^s = (\beta\delta\lambda)^k,$

$\rho^t = (\delta\lambda\beta)^k,$

$\xi^u = (\lambda\beta\delta)^k,$

$s,\ t,\ u \geq 2.$

$D(3\mathcal{R})$

IX.5.4

$$\begin{bmatrix} k+s & k & k \\ k & k+t & k \\ k & k & k+u \end{bmatrix}$$

*Semidihedral type*

| Ref. | Quiver | Relations |
|---|---|---|
| III.1(d) |  | $X^2 = (YX)^{k-1}Y$, $Y^2 = 0$<br>$(XY)^k = (YX)^k$<br>$(XY)^kX = 0.$<br>$k \geq 2$ |
| III.1(d') | | char $K = 2$ and $X^2 = (YX)^{k-1}Y - c(XY)^k$<br>$Y^2 - d(XY)^k$, $(XY)^k - (YX)^k$, $(XY)^kX = 0$<br>$c, d \in K$, not both 0. |
| $SD(2A)_1$<br>VIII.4.5 | | $\alpha^2 = c(\alpha\beta\gamma)^k$<br>$\beta\gamma\beta = (\alpha\beta\gamma)^{k-1}\alpha\beta$<br>$\gamma\beta\gamma = (\gamma\alpha\beta)^{k-1}\gamma\alpha$, $(\alpha\beta\gamma)^k\alpha = 0$<br>$k \geq 2$ and $c \in K$. |
| $SD(2A)_2$<br>VIII.4.4 | 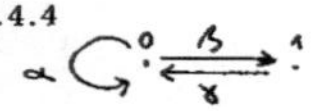 | $\gamma\beta = 0$, $(\alpha\beta\gamma)^k = (\beta\gamma\alpha)^k$<br>$\alpha^2 = (\beta\gamma\alpha)^{k-1}\beta\gamma + c(\alpha\beta\gamma)^k$<br>$k \geq 2$ and $c \in K$. |

*Semidihedral type*

| | | | | | |
|---|---|---|---|---|---|
| III.1(d*)<br>[0] | k+3 | [4k] | $k=2^{n-2}$ | $\begin{bmatrix} 1^4 \\ 2 \end{bmatrix}(*)$ | KD |
| $SD(2A)_1$<br>IX.4.1, $[E_8]$ | k+4 | $\begin{bmatrix} 4k & 2k \\ 2k & k+2 \end{bmatrix}$ | $k=2^{n-2}$ | $\begin{bmatrix} 1 & 0 \\ 1 & 0 \\ 1 & 1 \\ 1 & 1 \\ 0 & 1 \\ 2 & 1 \end{bmatrix}(*)$ | $B_1(U_3(q))$<br>$q \equiv 1 \bmod 4$ |
| $SD(2A)_2$<br>IX.4.1, $[E_8]$ | k+3 | $\begin{bmatrix} 4k & 2k \\ 2k & k+1 \end{bmatrix}$ | $k=2^{n-2}$ | as for<br>D.(2A) | $B_0(G)$<br>$L_2(q)^*$<br>$q=r^2$ odd |

| | | | | | | | |
|---|---|---|---|---|---|---|---|
| $SD(2\mathcal{B})_1$ <br> VIII.4.6 <br> 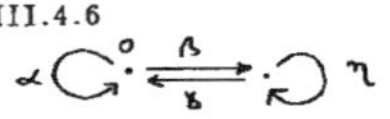 | $\gamma\beta = 0 = \eta\gamma = \beta\eta$ <br> $\alpha^2 = (\beta\gamma\alpha)^{k-1}\beta\gamma + c(\beta\gamma\alpha)^k$ <br> $\eta^t = (\gamma\alpha\beta)^k, \quad (\alpha\beta\gamma)^k = (\beta\gamma\alpha)^k$ <br> $k \geq 1,\ t \geq 2,\ c \in K.$ | $SD(2\mathcal{B})_1$ <br> IX.4.1, [$E_8$] | $k+t+2$ | $\begin{bmatrix} 4k & 2k \\ 2k & k+t \end{bmatrix}$ | $k=2$ <br> $t=2^{n-2}$ | $\left\lvert\begin{matrix} 1 & 0 \\ 1 & 0 \\ 1 & 1 \\ 1 & 1 \\ 2 & 1 \\ 0 & 1 \end{matrix}\right\rvert$ (*) | $B_1(3.M_{10})$ <br> n>4? |
| $SD(2\mathcal{B})_2$ <br> VIII.4.7 <br> -"- | $\beta\eta = (\alpha\beta\gamma)^{k-1}\alpha\beta$ <br> $\eta\gamma = (\gamma\alpha\beta)^{k-1}\gamma\alpha$ <br> $\gamma\beta = \eta^{t-1}$ <br> $\alpha^2 = c(\alpha\beta\gamma)^k$ <br> $\beta\eta^2 = 0 = \eta^2\gamma,$ <br> $k \geq 1,\ t \geq 3,\ c \in K.$ | $SD(2\mathcal{B})_2$ <br> IX.4.1, [$E_8$] | $k+t+2$ | $\begin{bmatrix} 4k & 2k \\ 2k & k+t \end{bmatrix}$ | $k=1$ <br> $t=2^{n-2}$ | as for <br> $D.(2\mathcal{B})$ | $B_1(L_3(q))$ <br> $q\equiv3 \bmod 4$ |
| $SD(2\mathcal{B})_3$ <br> VIII.4.8 <br> -"- | $\beta\eta^{s-1}\gamma = q(0)\alpha^{s+1}$ <br> $\gamma\alpha^{s-1}\beta = p(0)\eta^{s+1}$ <br> $\alpha^s\beta = 0 = \gamma\alpha^s$ <br> $\alpha^{s+2} = 0 = \eta^{s+2} = \eta^s\gamma = \beta\eta^s.$ <br> Modulo socle: $\beta\gamma \equiv \alpha^2 p(\alpha)$, <br> $\alpha\beta \equiv \beta\eta q(\eta),\ \eta\gamma \equiv \gamma\alpha p(\alpha),\ \gamma\beta \equiv \eta^2 q(\eta)$, <br> (p, q polynomials with non-zero constant term). <br> $s \geq 2$. | $SD(2\mathcal{B})_3$ <br> IX.4.1 | $s+3$ | $\begin{bmatrix} s+2 & s \\ s & s+2 \end{bmatrix}$ | $s=2^{n-2}-1$ | $\left\lvert\begin{matrix} 1 & 0 \\ 1 & 0 \\ 0 & 1 \\ 0 & 1 \\ 1 & 1 \end{matrix}\right\rvert$ (*) | ? |
| $SD(3\mathcal{A})_1$ <br> VIII.8.7 <br> 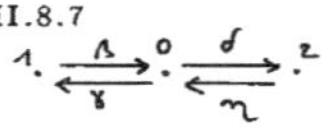 | $\beta\gamma = 0,$ <br> $\delta\eta\delta = (\gamma\beta\delta\eta)^{k-1}\gamma\beta\delta$ <br> $\eta\delta\eta = (\eta\gamma\beta\delta)^{k-1}\eta\gamma\beta,$ <br> $k \geq 1.$ | $SD(3\mathcal{A})_1$ <br> IX.5.1, IX.6.4 | $k+4$ | $\begin{bmatrix} 4k & 2k & 2k \\ 2k & k+1 & k \\ 2k & k & k+2 \end{bmatrix}$ | $k=2^{n-2}$ | $\left\lvert\begin{matrix} 1 & 0 & 0 \\ 1 & 1 & 0 \\ 1 & 0 & 1 \\ 1 & 1 & 1 \\ 0 & 0 & 1 \\ 2 & 1 & 1 \end{matrix}\right\rvert$ (*) | $B_0(U_3(q))$ <br> $q\equiv1 \bmod 4$ |

| | | | | | | |
|---|---|---|---|---|---|---|
| $SD(3\mathcal{A})_2$<br>VIII.7.9<br>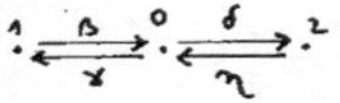 | $\gamma\beta = \delta\eta$<br>$(\beta\gamma)^{k-1}\beta\delta = 0$<br>$(\eta\delta)^{k-1}\eta\gamma = 0,$<br>$k \geq 2.$ | $SD(3\mathcal{A})_2$<br>IX.5.3 | $k+3$ | $\begin{bmatrix} k+1 & k & k \\ k & k+1 & k-1 \\ k & k-1 & k+1 \end{bmatrix}$ | | |
| $SD(3\mathcal{B})_1$<br>VIII.8.8.3<br>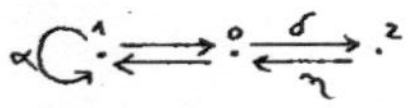 | $\alpha\beta = 0 = \gamma\alpha = \beta\gamma$<br>$\alpha^s = (\beta\delta\eta\gamma)^k$<br>$\eta\delta\eta = (\eta\gamma\beta\delta)^{k-1}\eta\gamma\beta$<br>$\delta\eta\delta = (\gamma\beta\delta\eta)^{k-1}\gamma\beta\delta$<br>$k \geq 1$ and $s \geq 2.$ | $SD(3\mathcal{B})_1$<br>IX.5.1, IX.6.6 | $k+s+3$ | $\begin{bmatrix} 4k & 2k & 2k \\ 2k & k+s & k \\ 2k & k & k+2 \end{bmatrix}$ $k=1$, $s=2^{n-2}$ | $\begin{vmatrix} 1 & 0 & 0 \\ 1 & 1 & 0 \\ 1 & 0 & 1 \\ 1 & 1 & 1 \\ 0 & 0 & 1 \\ 0 & 1 & 0 \end{vmatrix}$ (*) | $B_0(L_3(q))$<br>$q \equiv 3 \bmod 4$ |
| $SD(3\mathcal{B})_2$<br>VIII.8.8.4<br>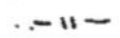 | $\eta\delta = 0$<br>$\beta\gamma = \alpha^{s-1}$<br>$\gamma\alpha = (\delta\eta\gamma\beta)^{k-1}\delta\eta\gamma$<br>$\alpha\beta = (\beta\delta\eta\gamma)^{k-1}\beta\delta\eta$<br>$k \geq 1,\ s \geq 3.$ | $SD(3\mathcal{B})_2$<br>IX.5.1, IX.6.6 | $k+s+2$ | $\begin{bmatrix} 4k & 2k & 2k \\ 2k & k+s & k \\ 2k & k & k+1 \end{bmatrix}$ $k=2$, $s=2^{n-2}$ | $\begin{vmatrix} 1 & 0 & 0 \\ 1 & 1 & 0 \\ 1 & 0 & 1 \\ 1 & 1 & 1 \\ 2 & 1 & 1 \\ 0 & 1 & 0 \end{vmatrix}$ (*) | |
| $SD(3\mathcal{C})_1$<br>VIII.5.1<br>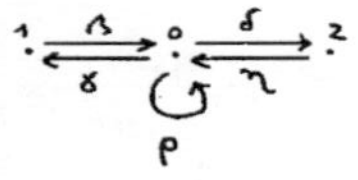 | $\beta\delta = 0 = \beta\rho = \rho\gamma$<br>$\eta\gamma = 0 = \eta\rho = \rho\delta$<br>$\rho^s = \gamma\beta = \delta\eta$<br>$\beta\gamma\beta = 0 = \eta\delta\eta,\ s \geq 3.$ | $SD(3\mathcal{C})_1$<br>IX.5.3 | $s+3$ | $\begin{bmatrix} s+1 & 1 & 1 \\ 1 & 2 & 0 \\ 1 & 0 & 2 \end{bmatrix}$ | | |

SD$(3\mathcal{C})_2$

VIII.7.10

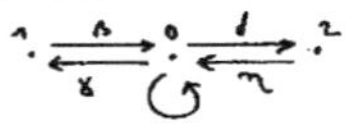

$\beta\rho = 0 = \rho\delta$

$\eta\rho = \rho\gamma = 0$

$\gamma\beta = \delta\eta$

$(\beta\gamma)^k = \rho^s$

$(\beta\gamma)^{k-1}\beta\delta = 0 = (\eta\delta)^{k-1}\eta\gamma$

$s \geq 2,\ k \geq 2.$

SD$(3\mathcal{C})_2$ $\quad k+s+2 \quad \begin{bmatrix} k+s & k & k \\ k & k+1 & k-1 \\ k & k-1 & k+1 \end{bmatrix}$ $\quad \{k, s\} = \{2, 2^{n-2}\}$

$$\begin{vmatrix} 1 & 1 & 0 \\ 0 & 1 & 0 \\ 1 & 0 & 1 \\ 0 & 0 & 1 \\ 1 & 1 & 1 \\ 1 & 0 & 0 \end{vmatrix} (*) \quad \text{and} \quad \begin{vmatrix} 0 & 0 & 1 \\ 0 & 1 & 1 \\ 1 & 0 & 0 \\ 1 & 0 & 1 \\ 0 & 1 & 0 \\ 1 & 1 & 1 \end{vmatrix} (*)$$

??

IX.5.3, IX.6.2, IX.6.10

SD$(3\mathcal{D})$

VIII.9.12

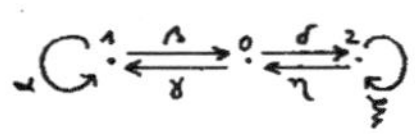

$\delta\xi = 0 = \xi\eta = \eta\delta$

$\beta\gamma = \alpha^{s-1}$

$\gamma\alpha = (\delta\eta\gamma\beta)^{k-1}\delta\eta\gamma$

$\alpha\beta = (\beta\delta\eta\gamma)^{k-1}\beta\delta\eta$

$\xi^t = (\eta\gamma\beta\delta)^k$

$k \geq 1,\ s \geq 3$ and $t \geq 2.$

SD$(3\mathcal{D})$ $\quad k+s+t+1 \quad \begin{bmatrix} 4k & 2k & 2k \\ 2k & k+s & k \\ 2k & k & k+t \end{bmatrix}$ $\quad k=1,\ t=2,\ s=2^{n-2}$

as in SD$(3\mathcal{B})_1$

$B_0(M_0)$ $n>4?$

IX.5.2, IX.6.6

SD$(3\mathcal{F})$

VIII.6.4

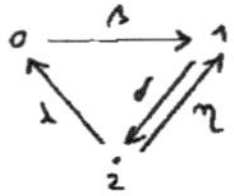

$\lambda\beta = (\eta\delta)^{k-1}\eta$

$\beta\delta\eta = 0 = \eta\delta\lambda$

$\delta\lambda\beta\delta = 0$

$k \geq 2.$

SD$(3\mathcal{F})$ $\quad k+3 \quad \begin{bmatrix} 2 & 1 & 1 \\ 1 & k+1 & k \\ 1 & k & k+1 \end{bmatrix}$

IX.5.2

SD$(3\mathcal{H})$

VIII.6.5

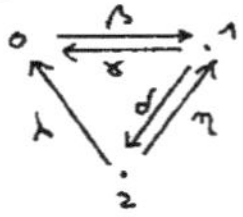

$\delta\lambda = (\gamma\beta)^{k-1}\gamma$

$\lambda\beta = (\eta\delta)^{s-1}\eta$

$\beta\delta\eta = \gamma\beta\delta = 0$

$\eta\gamma = 0$

$k,\ s \geq 2.$

SD$(3\mathcal{H})$ $\quad k+s+2 \quad \begin{bmatrix} k+1 & k & 1 \\ k & k+s & s \\ 1 & s & s+1 \end{bmatrix}$ $\quad k=2,\ s=2^{n-2}$

$$\begin{vmatrix} 1 & 0 & 0 \\ 0 & 1 & 0 \\ 0 & 0 & 1 \\ 1 & 1 & 1 \\ 0 & 1 & 1 \\ 1 & 1 & 0 \end{vmatrix} (*)$$

??

IX.5.2, IX.6.9

SD(3$\mathcal{K}$)

VIII.9.13

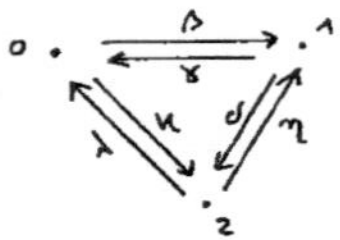


$\kappa\eta = \eta\tau = \tau\kappa = 0$

$\delta\lambda = (\tau\beta)^{a-1}\tau$

$\beta\delta = (\kappa\lambda)^{b-1}\kappa$

$\lambda\beta = ($[illegible]$)^{c-1}$[illegible]

$a \geq b \geq c \geq 2.$

SD(3$\mathcal{K}$) $\quad a+b+c+1 \quad \begin{bmatrix} a+b & a & b \\ a & a+c & c \\ b & c & b+c \end{bmatrix}$

*Quaternion type*

| Ref. | quiver | relations |
|---|---|---|
| III.1(e) |  | $X^2 = (YX)^{k-1}Y$ ,<br>$Y^2 = (XY)^{k-1}X$<br>$(XY)^k = (XY)^k$, $(XY)^kX = 0$<br>$k \geq 2$ |
| III.1(e') | | char K = 2 and<br>$X^2 = (YX)^{k-1}Y + c(XY)^k$,<br>$Y^2 = (XY)^{k-1}X + d(XY)^k$,<br>$(XY)^k = (YX)^k$,<br>$(XY)^kX = 0 = (YX)^kY$,<br>$k \geq 2$ and $c, d \in K$, $(c,d) \neq (0,0)$. |
| Q(2$\mathcal{A}$)<br>VII.7.1 | 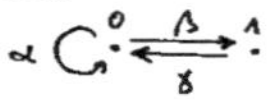 | $\gamma\beta\gamma = (\gamma\alpha\beta)^{k-1}\gamma\alpha$,<br>$\beta\gamma\beta = (\alpha\beta\gamma)^{k-1}\alpha\beta$<br>$\alpha^2 = (\beta\gamma\alpha)^{k-1}\beta\gamma + c(\beta\gamma\alpha)^k$,<br>$\alpha^2\beta = 0$<br>$k \geq 2$ and $c \in K$. |
| Q(2$\mathcal{B}$)$_1$<br>VII.7.2 | 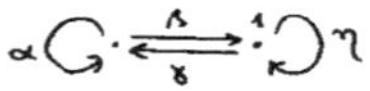 | $\gamma\beta = \eta^{s-1}$, $\beta\eta = (\alpha\beta\gamma)^{k-1}\alpha\beta$,<br>$\eta\gamma = (\gamma\alpha\beta)^{k-1}\gamma\alpha$,<br>$\alpha^2 = (\beta\gamma\alpha)^{k-1}\beta\gamma + c(\beta\gamma\alpha)^k$,<br>$\alpha^2\beta = 0$,<br>$s \geq 3$, $k \geq 2$, $a, c \in K$. |

*Quaternion type*

| | | | | | |
|---|---|---|---|---|---|
| III.1(e*)<br>[0] | k+3 | [4k] | $k=2^{n-2}$ | $\begin{bmatrix} 1^4 \\ 2 \end{bmatrix}$(*) | $KQ_{2^n}$ |
| Q(2$\mathcal{A}$)<br>IX.4.1, [$E_8$] | k+4 | $\begin{bmatrix} 4k & 2k \\ 2k & k+2 \end{bmatrix}$ | $k=2^{n-2}$ | as in SD(2$\mathcal{A}$)$_1$ | $B_0(G)$<br>$SL_2(q)<G$<br>$q \equiv 1 \bmod 4$ |
| Q(2$\mathcal{B}$)$_1$<br>IX.4.1, [$E_8$] | k+s+2 | $\begin{bmatrix} 4k & 2k \\ 2k & k+s \end{bmatrix}$ | $k=2$<br>$s=2^{n-2}$ | as in SD(2$\mathcal{B}$)$_1$ | $B_0(G)$<br>$SL_2(q)<G$<br>$q \equiv 3 \bmod 4$ |

**$Q(2\mathcal{B})_2$**

VII.7.3(i)

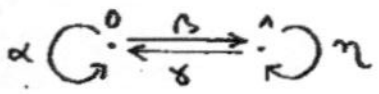


$\alpha\beta = \beta\eta, \quad \eta\gamma = \gamma\alpha$

$\beta\gamma = \alpha^2 p(\alpha),$

$\gamma\beta = \eta^2 p(\eta) + a\eta^{s-1} + c\eta^s,$

$\alpha^{s+1} = 0 = \eta^{s+1}$

$0 = \gamma\alpha^{s-1} = \alpha^{s-1}\beta,$

$p(t) \in K[t]$, $p(0) = 1$, $s > 3$,

$a, c \in K$ and $a \neq 0$.

$Q(2\mathcal{B})_2$ — IX.4.1, $[E_8]$ — $\begin{bmatrix} s+1 & s-1 \\ s-1 & s+1 \end{bmatrix}$ — $s = 2^{n-2}$ — as in $SD(2\mathcal{B})_3$ — ?

**$Q(2\mathcal{B})_3$**

VIII.7.3(ii)

–u–

$\alpha\beta = \beta\eta, \quad \eta\gamma = \gamma\alpha$

$\beta\gamma = \alpha^2 + c\alpha^3,$

$\gamma\beta = a\eta^{t-1} + d\eta^t$

$\alpha^4 = 0 = \eta^{t+1} = \gamma\alpha^2 = \alpha^2\beta$

$t \geq 3$; $a, c, d \in K$ with $a \neq 0$.

If $t = 3$ then $a \neq 1$; if $t \neq 3$ then $a = 1$

$Q(2\mathcal{B})_3$ — IX.4.1 — $\begin{bmatrix} 4 & 2 \\ 2 & t+1 \end{bmatrix}$

**$Q(3\mathcal{A})_1$**

VII.8.4

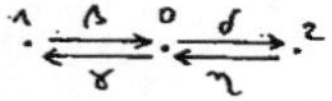


$\beta\delta\eta = (\beta\gamma)^{a-1}\beta,$

$\delta\eta\gamma = (\gamma\beta)^{a-1}\gamma$

$\eta\gamma\beta = d(\eta\delta)^{b-1}\eta,$

$\gamma\beta\delta = d(\delta\eta)^{b-1}\delta$

$\beta\delta\eta\delta = 0 = \eta\gamma\beta\gamma,$

$a \geq b \geq 2$, $0 \neq d \in K$.

If $a = b = 2$ then $d \neq 1$;

otherwise $d = 1$.

$Q(3\mathcal{A})_1$ — $k+s+2$ — IX.5.1, IX.6.5 — $\begin{bmatrix} k+s & k & s \\ k & k+1 & 1 \\ s & 1 & s+1 \end{bmatrix}$

$Q(3\mathcal{A})_2$

VII.8.3

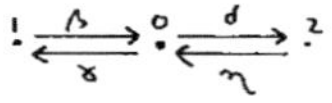


$$\beta\gamma\beta = (\beta\delta\eta\gamma)^{k-1}\beta\delta\eta,$$
$$\gamma\beta\gamma = (\delta\eta\gamma\beta)^{k-1}\delta\eta\gamma$$
$$\eta\delta\eta = (\eta\gamma\beta\delta)^{k-1}\eta\gamma\beta$$
$$\delta\eta\delta = (\gamma\beta\delta\eta)^{k-1}\gamma\beta\delta$$
$$\beta\gamma\beta\delta = 0 = \eta\delta\eta\gamma$$
$$k \geq 2$$

$Q(3\mathcal{A})_2$   $k+5$   $\begin{bmatrix} 4k & 2k & 2k \\ 2k & k+2 & k \\ 2k & k & k+2 \end{bmatrix}$   $k=2^{n-2}$

$\begin{vmatrix} 1 & 0 & 0 \\ 1 & 1 & 0 \\ 1 & 0 & 1 \\ 1 & 1 & 1 \\ 0 & 1 & 0 \\ 0 & 0 & 1 \\ 2 & 1 & 1 \end{vmatrix}$ (*)

$B_0(SL_2(q))$
$q\equiv 1 \bmod 4$

IX.5.1, IX.6.5

$Q(3\mathcal{B})$

VII.8.5

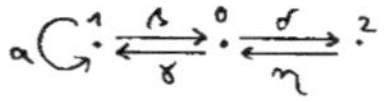


$$\beta\gamma = \alpha^{s-1}$$
$$\alpha\beta = (\beta\delta\eta\gamma)^{k-1}\beta\delta\eta,$$
$$\gamma\alpha = (\delta\eta\gamma\beta)^{k-1}\delta\eta\gamma$$
$$\eta\delta\eta = (\eta\gamma\beta\delta)^{k-1}\eta\gamma\beta$$
$$\delta\eta\delta = (\gamma\beta\delta\eta)^{k-1}\gamma\beta\delta$$
$$\alpha^2\beta = 0 = \beta\delta\eta\delta$$
$$k \geq 1,\ s \geq 3.$$

$Q(3\mathcal{B})$   $k+s+4$   $\begin{bmatrix} 4k & 2k & 2k \\ 2k & k+s & k \\ 2k & k & k+2 \end{bmatrix}$   $k=2$, $s=2^{n-2}$

$\begin{vmatrix} 1 & 0 & 0 \\ 1 & 1 & 0 \\ 1 & 0 & 1 \\ 1 & 1 & 1 \\ 2 & 1 & 1 \\ 0 & 0 & 1 \\ 0 & 1 & 0 \end{vmatrix}$ (*)

$B_0(2A_7)$
$n>4$?

IX.5.1, IX.6.7

$Q(3\mathcal{C})$

VII.8.6

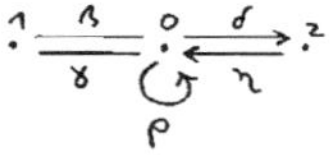


$$\beta\rho = 0 = \rho\gamma,$$
$$\eta\rho^2 = 0 = \rho^2\delta$$
$$\delta\eta - \gamma\beta = \rho^{s-1}$$
$$\eta\rho = (\eta\delta)^{k-1}\eta,$$
$$\rho\delta = (\delta\eta)^{k-1}\delta,$$
$$(\beta\gamma)^{k-1}\beta\delta = 0 = (\eta\delta)^{k-1}\eta\gamma,$$
$$k \geq 2 \text{ and } s \geq 3.$$

$Q(3\mathcal{C})$   $k+s+2$   $\begin{bmatrix} k+s & k & k \\ k & k+1 & k-1 \\ k & k-1 & k+1 \end{bmatrix}$

IX.5.3

Q(3𝒟)

VII.8.7

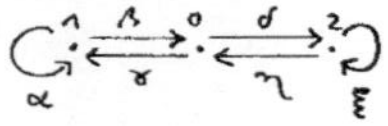


$$\begin{aligned}
\beta\gamma &= \alpha^{s-1}\\
\gamma\alpha &= (\delta\eta\gamma\beta)^{k-1}\delta\eta\gamma\\
\alpha\beta &= (\beta\delta\eta\gamma)^{k-1}\beta\delta\eta\\
\eta\delta &= \xi^{t-1},\\
\delta\xi &= (\gamma\beta\delta\eta)^{k-1}\gamma\beta\delta,\\
\xi\eta &= (\eta\gamma\beta\delta)^{k-1}\eta\gamma\beta,\\
\alpha^2\beta &= 0 = \delta\eta\delta,
\end{aligned}$$

$k \geq 1;\ s,\ t \geq 3.$

Q(3𝒟) $k+s+t+1$ $\begin{bmatrix} 4k & 2k & 2k \\ 2k & k+s & k \\ 2k & k & k+t \end{bmatrix}$

IX.5.1

Q(3𝒦)

VIII.8.8

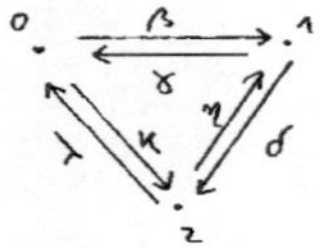


$$\begin{aligned}
\beta\delta &= (\kappa\lambda)^{a-1}\kappa\\
\eta\gamma &= (\lambda\kappa)^{a-1}\lambda\\
\delta\lambda &= (\gamma\beta)^{b-1}\gamma\\
\kappa\eta &= (\beta\gamma)^{b-1}\beta\\
\lambda\beta &= (\eta\delta)^{c-1}\eta\\
\gamma\kappa &= (\delta\eta)^{c-1}\delta\\
\gamma\beta\delta &= 0 = \delta\eta\gamma = \lambda\kappa\eta
\end{aligned}$$

$a \geq b \geq c \geq 2.$

Q(3𝒦) $a+b+c+1$ $\begin{bmatrix} a+b & a & b \\ a & a+c & c \\ b & c & b+c \end{bmatrix}$ $a=2$, $b=2$, $c=2^{n-2}$

$\begin{vmatrix} 1 & 0 & 0 \\ 0 & 1 & 0 \\ 0 & 0 & 1 \\ 1 & 1 & 1 \\ 1 & 1 & 0 \\ 1 & 0 & 1 \\ 0 & 1 & 1 \end{vmatrix}$ (*)

$B_0(SL_2(q))$

$q\equiv 1 \bmod 4$

IX.5.2, IX.6.8

<u>Bibliography</u>

[A] J.L. Alperin, Local representation theory (Cambridge studies in advanced Mathematics 11) CUP 1986

[ABG] J.L. Alperin, R. Brauer, D. Gorenstein, Finite groups with quasidihedral and wreathed Sylow 2-subgroups Trans. Amer. Math. Soc. <u>151</u>(1970), 1-262

[AB] J.L. Alperin, M. Broue, Local methods in block theory, Ann. of Math. <u>110</u>(1979) 143-157

[AF] F.W.Anderson, K.R. Fuller, Rings and Categories of Modules, Springer Garduate Text <u>13</u>(1974)

[Au] M.Auslander, Applications of morphisms determined by objects. Proc. Conf. on Representation theory Philadelhia(1976), M.Dekker 1978, 245-327

[$AR_{1,2}$] M. Auslander, I. Reiten, Representation theory of Artin algebras III: Almost split sequences, Comm. Algebra 3 (1975) 239-294 and IV, Comm. Alg. <u>5</u>(1977), 443-518

[Ba] V.A. Basev, Representations of the group $\mathbb{Z}_2$ $\mathbb{Z}_2$ in a field of characteristic 2. Dokl. Akad. Nauk. SSSR <u>141</u>(1961) 1015-1018

[Bt] R. Bautista, A characterization of finite-dimensional algebras of tame representation type, preprint 1989

[Be] D.J. Benson, Modular representation theory: New trends and methods, Springer Lecture Notes in Mathematics <u>1081</u>(1984)

[Bes] C. Bessenrodt, The Auslander-Reiten quiver of a modular group algebra revisited, preprint 1989

[BD] V.M. Bondarenko, J.A. Drozd, The representation type of finite groups Zap. Nauchn. Sem. Leningrad. Otdel. Mat. Inst. Steklov <u>57</u>(1977) 24-41. English translation: J. Soviet Math., <u>20</u>(1982), 2515-2528

[Bo] K. Bongartz, Zykellose Algebren sind nicht zugellos, Representation theory II, Springer Lecture Notes in Mathematics <u>832</u>(1980), 97-102

[BG] K. Bongartz, P. Gabriel, Covering spaces in representation theory Invent. Math. <u>65</u>(1982), 331-378

[$B_1$] R.Brauer, On groups whose order contains a prime number to the first power I and II, Amer.J.Math. <u>64</u>(1942), 401-420 and 421-440

[$B_2$] R.Brauer, Some applications of the theory of blocks of characters of finite groups III and IV, J. Algebra <u>3</u>(1966), 225-255 and <u>17</u>(1971), 489-521

[$B_3$] R. Brauer, On 2-blocks with dihedral defect groups Symposia Math. <u>13</u> (1974) 366-394

[BN] R.Brauer, C.Nesbitt, On the regular representations of algebras Proc. Nat. Acad. Sci. USA <u>23</u>(1937), 236-240

[BS] R. Brauer, M. Suzuki, On finite groups of even order whose 2-Sylow group is a quaternion group, Proc. Nat. Acad. Sci. USA <u>45</u>(1959) 1757-1759

[$Br_1$] S. Brenner, Modular representations of p-groups J. Algebra <u>15</u>(1970), 89-102

[$Br_2$] S. Brenner, Decomposition properties of some small diagrams of modules Symposia Math. Ist. Naz. Alta Mat. 13(1974), 127-141

[$Br_3$] S. Brenner, On four subspaces of a vector space J. Algebr 29(1974), 587-599

[BR] M.C.R. Butler, C.M. Ringel, Auslander-Reiten sequences with few middle terms and applications to string algebras, Comm. Alg. 15(1987), 145-179

[BuS] M.C.R. Butler, M. Shahzamanian, The construction of almost split sequences III: modules over two classes of tame local algebras Math. Ann. 247(1980) 111-122

[$C_1$] S.B. Conlon, Certain representation algebras J. Austr. Math. Soc. 5(1965) 83-99

[$C_2$] S.B. Conlon, Twisted group algebras and their representations J. Austr. Math. Soc. 4(1964), 152-173

[$CB_1$] W.W. Crawley-Boevey, On tame algebras and BOC's, Proc. L.M.S. 56(1988) 451-483

[$CB_2$] W.W. Crawley-Boevey, Functorial filtrations and the problem of an idempotent and a square zero matrix to appear in Journal of the L.M.S.

[$CB_3$] W.W. Crawley-Boevey, Functorial filtrations III: Semidihedral algebras (1988)

[CK] M.Chlebowitz, B.Kulshammer, Symmetric local algebras with 5-dimensional center to appear in Trans. A.M.S.

[CR] C.W. Curtis, I. Reiner, Methods of Representation Theory I and II, Wiley 1981 and 1987

[$Da_1$] E.C. Dade, Blocks with cyclic defect groups Ann. of Math. 84(1966), 2-48

[$Da_2$] E.C. Dade, Une extension de la theorie de Hall et Higman J. Algebra 20(1972), 570-609

[DR] V. Dlab, C.M. Ringel, Indecomposable representations of graphs and algebras, Mem. Amer. Math. Soc., 6(1976), no. 173

[Do] P.W. Donovan, Dihedral defect groups J. Algebra 56(1979) 184-206

[$DF_1$] P.W. Donovan, M. Freislich, The representation theory of finite graphs and associative algebras, Carleton Math. Lecture Notes 5(1973)

[$DF_2$] P.W. Donovan, M. Freislich, Indecomposable representations in characteristic 2 of the simple groups of order not divisible by eight, Bull. Austral Math. Soc. 15(1976) 407-419

[$DF_3$] P.W. Donovan, M. Freislich, The indecomposable representations of certain groups with dihedral Sylow subgroups Math. Ann. 238(1978) 207-216

[DS] P.Dowbor, A. Skowronski, Galois coverings of representation-infinite algebras, Comment. Math. Helv. 62(1987), 311-337

[D] Y.A. Drozd, Tame and wild matrix problems, Representation theory II, Springer Lecture Notes in Mathematics 832(1980), 242-258

[$E_1$] K.Erdmann, Principal blocks of groups with dihedral Sylow 2-subgroups, Comm

Alg. 5(1977), 665-694

[$E_2$] K.Erdmann, On 2-blocks with semidihedral defect groups, Trans. AMS. 256 (1979), 267-287

[$E_3$] K. Erdmann, Blocks whose defect groups are Klein 4-groups: a correction, J. Algebra 76(1982) 505-518

[$E_4$] K.Erdmann,,Algebras and dihedral defect groups, Proc. L.M.S 54(1987), 88-114

[$E_5$] K. Erdmann Algebras and semidihedral defect groups I, Proc. L.M.S. 57(1988) 109-150 and II, Proc. L.M.S.

[$E_6$] K. Erdmann Algebras and quaternion defect groups I and II, Math. Ann. 281(1988), 545-560 and 561-582

[$E_7$] K. Erdmann On the number of simple modules in certain tame blocks and algebras, Arch. Math. 51(1988), 34-38

[$E_8$] K. Erdmann On the local structure of tame blocks, Colloque sur les representations des groupes finis (Luminy, 1988) to appear in Asterisque

[$E_9$] K.Erdmann On the vertices of modules in the Auslander-Reiten quiver of p-groups, to appear in Math. Z.

[EM] K. Erdmann, G.O. Michler Blocks with dihedral defect groups in solvable groups Math. Z. 154(1977), 143-151

[ES] K. Erdmann, A. Skowronski On Auslander-Reiten components of blocks and self-injective biserial algebras, preprint 1989

[F] W.Feit, The Representation Theory of Finite Groups North Holland, 1982

[Fr] F.G. Frobenius, Uber die Primfaktoren der Gruppendeterminante, Sitzungsber. Preuss. Akad. Wiss. Berlin (1896), 1343-1382

[$G_1$] P.Gabriel Representations indecomposables Seminaires Bourbaki Nr.444, Springer Lecture Notes in Mathematics 431(1975), 143-169

[$G_2$] P.Gabriel Auslnder-Reiten sequences and representation-finite algebras, Representation theory I, Springer Lecture Notes in Mathematics 831(1980), 1-71

[$G_3$] P.Gabriel, The universal cover of a representation-finite algebra, Representations of Algebras, Springer Lecture Notes in Mathematics 903(1981), 68-105

[GR] P.Gabriel, C. Riedtmann, Group representations without groups, Comm. Math. Helv. 54(1979), 240-287

[GP] I.M. Gelfand, V.A. Ponomarev, Indecomposable representations of the Lorentz group, Usp. Math. Nak. 23(1968) 3-60

[Go] D. Gorenstein, Finite groups (1968) Harper and Row, New York, Evanston, London

[Go] D. Gorenstein, J.H. Walter On finite groups with dihedral Slow 2-subgroups, Ill. J. Math 6(1962), 553-593

[$Gr_1$] J.A. Green, Relative module categories for finite groups J. Pure Appl. Algebra 2(1972), 371-393

[$Gr_2$] J.A. Green, Walking around the Brauer tree, J. Austr. Math. Soc. 17(1974), 197-213

[HPR] D. Happel, U. Preiser, C.M. Ringel, Vinberg's characterization of Dynkin diagrams using subadditive functions with applications to DTr-periodic modules, in Representation theory II, Springer Lecture notes in Mathematics 832(1981), 280-294

[H] A.Heller, I.Reiner, Indecomposable representations, Ill. J. Mat. 5(1961), 314-323

[Hi] D. Higman, Indecomposable representations at characteristic p, Duke J. Math. 21(1954), 377-381

[Hu] B. Huppert, Endliche Gruppen I, Springer 1967

[J] G.J. Janusz, Indecomposable representations of groups with a cyclic Sylow subgroup Trans. A.M.S. 125 (1966), 288-295

[JK] G.D. James, A. Kerber, The representation theory of the symmetric group, Encyclopedia of Math. and Appl. 16,(1981), Addison-Wesley

[$K_1$] S. Kawata, The Green correspondence and Auslander-Reiten sequences

[$K_2$] S. Kawata, Module correspondence in Auslander-Reiten quivers for finite groups, preprint 1988

[Kh] S. Koshitani, A remark on blocks with dihedral defect groups in solvable groups, Math. Z. 179(1982), 401-406

[Ku] B. Kulshammer, Crossed products and blocks with normal defect groups, Comm. Alg. 13(1985), 410-428

[KP] B. Kulshammer, L. Puig, Extensions of nilpotent blocks, preprint 1988

[$Ku_1$] H. Kupisch, Projetive Moduln endlicher Gruppen mit zyklischer p-Sylowgruppe, J. Algebra 10(1968), 1-7

[$Ku_2$] H. Kupisch, Unzerlegbare Moduln endlicher Gruppen mit zyklischer p-Sylowgruppe, Math. Z. 108(1969), 77-104

[Kr] L. Kronecker, Algebraische Reduction der Scharen quadratischer Formen, Sitz. Konigl. Preuss. Akad. Wiss. Berlin (1896), 1375-1388; (1891), 9-17 und 33-44

[$L_1$] P. Landrock, The principal block of finite groups with dihedral Sylow 2-subgroups, J. Algebra 39(1976), 410-428

[$L_2$] P. Landrock, The non-principal 2-blocks of sporadic simple groups, Comm. Alg. 6(1978), 1865-1891

[$L_3$] P. Landrock, Finite group algebras and their modules, L.M.S. Lecture Notes 84, Cambridge University Press, 1983.

[Li] P.A. Linnell, The Auslander-Reiten quiver of a finite group, Arch. Math. 45(1985), 289-295

[MR] E. Marmolejo, C.M. Ringel, Modules of bounded length in Auslander-Reiten components, Arch. Math. 50(1988), 128-133

[MP] R. Martinez-Villa, J.A. de la Pena The universal cover of a quiver with relations, J. Pure and Appl. Algebra 30(1983) 277-292

[$M_1$] G.O. Michler, Blocks and centers of group algebras, Lecture Notes in Mathematics 246(Springer, 1972), 430-565

[$M_2$] G.O. Michler, Green correspondence between blocks with cyclic defect groups I and II, J. Algebra 39(1976), 26-51 and Proc. ICRA, Lecture Notes in Mathematics 488(Springer, 1974), 210-235

[$M_3$] G.O. Michler, Sur l'egalite des defauts d'un bloc et la correspondence de Green Proc. 1977 Antwerp Ring Theory Conf., M. Dekker, New York (1978), 59-63

[$M_4$] G.O. Michler, Ring theoretical and computational methods in representation theory, (1989), Vorlesungen aus dem Fachbereich Mathematik 18, Essen

[Mo] K. Morita, Duality of modules and its applications to the theory of rings with minimum condition, Sci. Rep. Tokyo Kyoiku Daigaku Sect. A (1958), 85-142

[Mu] W. Muller, Gruppenalgebren uber nicht-zyklischen p-Gruppen, J. Reine Ang. Math. I. 266, 10-48, II. 267(1974), 1-19

[N] T.Nakayama, On Frobenusean algebras I and II, Ann. of Math. 40(1939), 611-633 and Ann. of Math. 42(1941), 1-21

[Ok] T. Okuyama, On the Auslander-Reiten quiver of a finite group J. Algebra 110(1987), 425-430

[O] J.B. Olsson, On 2-blocks with quaternion and quasidihedral defect groups, J. Algebra 36(1975), 212-241

[$Os_1$] M. Osima, Notes on Basic Rings I and II, Math. J. of Okayama University 2(1953), 103-110 and 3(1954), 121-127

[$Os_2$] M. Osima, Some studies of Frobenius Algebras I and II, Japanese Journal of Math. XX1(1951) and Math. J. of Okayama University 3(1954)

[P] R.S. Pierce, Associative Algebras, Springer Graduate Text (1982)

[PS] Z. Pogorzaly, A. Skowronski, Selfinjective biserial standard algebras, J. Algebra, to appear.

[$Pu_1$] L. Puig, The source algebra of a nilpotent block, preprint 1981

[$Pi_2$] L. Puig, Nilpotent blocks and their source algebras Invent. Math. 93(1988), 77-116

[Re] I. Reiten, Generalized stable equivalences and group algebras J. Algebra 79(1982), 319-340

[$Ri_1$] C. Riedtmann, Algebren, Darstellungen, Uberlagerungen und zuruck, Comm. Math. Helv. 55(1980), 199-224

[$Ri_2$] C. Riedtmann, Configurations of $\mathbb{Z}D_n$, J. Algebra 82(1983), 309-327

[$Ri_3$] C. Riedtmann, Representation-finite selfinjective algebras of class $D_n$, Compositio Math. 49(1983), 231-282

[$R_1$] C.M. Ringel, The represenation type of local algebras, Lecture Notes in Mathematics 488 (Springer, 1975), 282-305

[$R_2$] C.M. Ringel, The indecomposable representations of the dihedral 2-groups, Math. Ann. 214(1975), 19-34

[$R_3$] C.M. Ringel, Tame algebras, in "Representation Theory I", Lecture Notes in Mathematics, Springer (1980), 137-287

[$R_4$] C.M. Ringel, Tame algebras and integral quadratic forms, Lecture Notes in Mathematics 1099, Springer (1984)

[S] G. Schneider, The Mathieu group $M_{11}$, Oxford M.Sc. Thesis 1979

[Sk] A. Skowronski, Selfinjective algebras of polynomial growth, Math. Ann. 285(1989), 177-199

[SW] A. Skowronski, J. Waschbusch, Representation finite biserial algebras, J. Reine Angew. Math. 345(1983), 172-181

[WW] B. Wald, J. Waschbusch, Tame biserial algebras, J. Algebra 95 (1985), 480-500

[W] P. Webb, The Auslander-Reiten quiver of a finite group, math. Z. 179(1982), 97-121

## LECTURE NOTES IN MATHEMATICS

**Edited by A. Dold, B. Eckmann and F. Takens**

**Some general remarks on the publication of monographs and seminars**

In what follows all references to monographs, are applicable also to multiauthorship volumes such as seminar notes.

§1. Lecture Notes aim to report new developments - quickly, informally, and at a high level. Monograph manuscripts should be reasonably self-contained and rounded off. Thus they may, and often will, present not only results of the author but also related work by other people. Furthermore, the manuscripts should provide sufficient motivation, examples and applications. This clearly distinguishes Lecture Notes manuscripts from journal articles which normally are very concise. Articles intended for a journal but too long to be accepted by most journals, usually do not have this "lecture notes" character. For similar reasons it is unusual for Ph.D. theses to be accepted for the Lecture Notes series.

Experience has shown that English language manuscripts achieve a much wider distribution.

§2. Manuscripts or plans for Lecture Notes volumes should be submitted (preferably in duplicate) either to one of the series editors or to Springer- Verlag, Heidelberg. These proposals are then refereed. A final decision concerning publication can only be made on the basis of the complete manuscripts, but a preliminary decision can usually be based on partial information: a fairly detailed outline describing the planned contents of each chapter, and an indication of the estimated length, a bibliography, and one or two sample chapters - or a first draft of the manuscript. The editors will try to make the preliminary decision as definite as they can on the basis of the available information. We generally advise authors not to prepare the final master copy of their manuscript (cf. §4) beforehand.

§3. Final manuscripts should contain at least 100 pages of mathematical text and should include

- a table of contents;
- an informative introduction, perhaps with some historical remarks: it should be accessible to a reader not particularly familiar with the topic treated;
- a subject index: this is almost always genuinely helpful for the reader.

§4. Lecture Notes are printed by photo-offset from the master-copy delivered in camera-ready form by the authors. Springer-Verlag provides technical instructions for the preparation of manuscripts, for typewritten manuscripts special stationery, with the prescribed typing area outlined, is available on request. Careful preparation of the manuscripts will help keep production time short and ensure satisfactory appearance of the finished book. For manuscripts typed or typeset according to our instructions, Springer-Verlag will, if necessary, contribute towards the preparation costs at a fixed rate.

The actual production of a Lecture Notes volume takes 6-8 weeks.

§5. Authors receive a total of 50 free copies of their volume, but no royalties. They are entitled to purchase further copies of their book for their personal use at a discount of 33.3 %, other Springer mathematics books at a discount of 20 % directly from Springer-Verlag.

Commitment to publish is made by letter of intent rather than by signing a formal contract. Springer-Verlag secures the copyright for each volume.

Addresses:

Professor A. Dold, Mathematisches Institut, Universität Heidelberg, Im Neuenheimer Feld 288, 6900 Heidelberg, Federal Republic of Germany

Professor B. Eckmann, Mathematik, ETH-Zentrum 8092 Zürich, Switzerland

Prof. F. Takens, Mathematisch Instituut, Rijksuniversiteit Groningen, Postbus 800, 9700 AV Groningen, The Netherlands

Springer-Verlag, Mathematics Editorial, Tiergartenstr. 17, 6900 Heidelberg, Federal Republic of Germany, Tel.: (06221) 487-410

Springer-Verlag, Mathematics Editorial, 175 Fifth Avenue, New York, New York 10010, USA, Tel.: (212) 460-1596

LECTURE NOTES

ESSENTIALS FOR THE PREPARATION
OF CAMERA-READY MANUSCRIPTS

Springer
Springer-Verlag
Berlin Heidelberg New York
London Paris Tokyo Hong Kong

The preparation of manuscripts which are to be reproduced by photo-offset require special care. Manuscripts which are submitted in technically unsuitable form will be returned to the author for retyping. There is normally no possibility of carrying out further corrections after a manuscript is given to production. Hence it is crucial that the following instructions be adhered to closely. If in doubt, please send us 1 - 2 sample pages for examination.

General. The characters must be uniformly black both within a single character and down the page. Original manuscripts are required: photocopies are acceptable only if they are sharp and without smudges.

On request, Springer-Verlag will supply special paper with the text area outlined. The standard TEXT AREA (OUTPUT SIZE if you are using a 14 point font) is 18 x 26.5 cm (7.5 x 11 inches). This will be scale-reduced to 75% in the printing process. If you are using computer typesetting, please see also the following page.

Make sure the TEXT AREA IS COMPLETELY FILLED. Set the margins so that they precisely match the outline and type right from the top to the bottom line. (Note that the page number will lie outside this area). Lines of text should not end more than three spaces inside or outside the right margin (see example on page 4).

Type on one side of the paper only.

Spacing and Headings (Monographs). Use ONE-AND-A-HALF line spacing in the text. Please leave sufficient space for the title to stand out clearly and do NOT use a new page for the beginning of subdivisons of chapters. Leave THREE LINES blank above and TWO below headings of such subdivisions.

Spacing and Headings (Proceedings). Use ONE-AND-A-HALF line spacing in the text. Do not use a new page for the beginning of subdivisons of a single paper. Leave THREE LINES blank above and TWO below headings of such subdivisions. Make sure headings of equal importance are in the same form.

The first page of each contribution should be prepared in the same way. The title should stand out clearly. We therefore recommend that the editor prepare a sample page and pass it on to the authors together with these instructions. Please take the following as an example. Begin heading 2 cm below upper edge of text area.

MATHEMATICAL STRUCTURE IN QUANTUM FIELD THEORY

John E. Robert
Mathematisches Institut, Universität Heidelberg
Im Neuenheimer Feld 288, D-6900 Heidelberg

Please leave THREE LINES blank below heading and address of the author, then continue with the actual text on the same page.

Footnotes. These should preferable be avoided. If necessary, type them in SINGLE LINE SPACING to finish exactly on the outline, and separate them from the preceding main text by a line.

Symbols. Anything which cannot be typed may be entered by hand in BLACK AND ONLY BLACK ink. (A fine-tipped rapidograph is suitable for this purpose; a good black ball-point will do, but a pencil will not). Do not draw straight lines by hand without a ruler (not even in fractions).

Literature References. These should be placed at the end of each paper or chapter, or at the end of the work, as desired. Type them with single line spacing and start each reference on a new line. Follow "Zentralblatt für Mathematik"/"Mathematical Reviews" for abbreviated titles of mathematical journals and "Bibliographic Guide for Editors and Authors (BGEA)" for chemical, biological, and physics journals. Please ensure that all references are COMPLETE and ACCURATE.

## IMPORTANT

Pagination. For typescript, number pages in the upper right-hand corner in LIGHT BLUE OR GREEN PENCIL ONLY. The printers will insert the final page numbers. For computer type, you may insert page numbers (1 cm above outer edge of text area).

It is safer to number pages AFTER the text has been typed and corrected. Page 1 (Arabic) should be THE FIRST PAGE OF THE ACTUAL TEXT. The Roman pagination (table of contents, preface, abstract, acknowledgements, brief introductions, etc.) will be done by Springer-Verlag.

If including running heads, these should be aligned with the inside edge of the text area while the page number is aligned with the outside edge noting that right-hand pages are odd-numbered. Running heads and page numbers appear on the same line. Normally, the running head on the left-hand page is the chapter heading and that on the right-hand page is the section heading. Running heads should not be included in proceedings contributions unless this is being done consistently by all authors.

Corrections. When corrections have to be made, cut the new text to fit and paste it over the old. White correction fluid may also be used.

Never make corrections or insertions in the text by hand.

If the typescript has to be marked for any reason, e.g. for provisional page numbers or to mark corrections for the typist, this can be done VERY FAINTLY with BLUE or GREEN PENCIL but NO OTHER COLOR: these colors do not appear after reproduction.

COMPUTER-TYPESETTING. Further, to the above instructions, please note with respect to your printout that
- the characters should be sharp and sufficiently black;
- it is not strictly necessary to use Springer's special typing paper. Any white paper of reasonable quality is acceptable.

If you are using a significantly different font size, you should modify the output size correspondingly, keeping length to breadth ratio 1 : 0.68, so that scaling down to 10 point font size, yields a text area of 13.5 x 20 cm (5 3/8 x 8 in), e.g.

Differential equations.: use output size 13.5 x 20 cm.

Differential equations.: use output size 16 x 23.5 cm.

Differential equations.: use output size 18 x 26.5 cm.

Interline spacing: 5.5 mm base-to-base for 14 point characters (standard format of 18 x 26.5 cm).
If in any doubt, please send us 1 - 2 sample pages for examination. We will be glad to give advice.

Vol. 1259: F. Cano Torres, Desingularization Strategies for Three-Dimensional Vector Fields. IX, 189 pages. 1987.

Vol. 1260: N.H. Pavel, Nonlinear Evolution Operators and Semigroups. VI, 285 pages. 1987.

Vol. 1261: H. Abels, Finite Presentability of S-Arithmetic Groups. Compact Presentability of Solvable Groups. VI, 178 pages. 1987.

Vol. 1262: E. Hlawka (Hrsg.), Zahlentheoretische Analysis II. Seminar, 1984–86. V, 158 Seiten. 1987.

Vol. 1263: V.L. Hansen (Ed.), Differential Geometry. Proceedings, 1985. XI, 288 pages. 1987.

Vol. 1264: Wu Wen-tsün, Rational Homotopy Type. VIII, 219 pages. 1987.

Vol. 1265: W. Van Assche, Asymptotics for Orthogonal Polynomials. VI, 201 pages. 1987.

Vol. 1266: F. Ghione, C. Peskine, E. Sernesi (Eds.), Space Curves. Proceedings, 1985. VI, 272 pages. 1987.

Vol. 1267: J. Lindenstrauss, V.D. Milman (Eds.), Geometrical Aspects of Functional Analysis. Seminar. VII, 212 pages. 1987.

Vol. 1268: S.G. Krantz (Ed.), Complex Analysis. Seminar, 1986. VII, 195 pages. 1987.

Vol. 1269: M. Shiota, Nash Manifolds. VI, 223 pages. 1987.

Vol. 1270: C. Carasso, P.-A. Raviart, D. Serre (Eds.), Nonlinear Hyperbolic Problems. Proceedings, 1986. XV, 341 pages. 1987.

Vol. 1271: A.M. Cohen, W.H. Hesselink, W.L.J. van der Kallen, J.R. Strooker (Eds.), Algebraic Groups Utrecht 1986. Proceedings. XII, 284 pages. 1987.

Vol. 1272: M.S. Livšic, L.L. Waksman, Commuting Nonselfadjoint Operators in Hilbert Space. III, 115 pages. 1987.

Vol. 1273: G.-M. Greuel, G. Trautmann (Eds.), Singularities, Representation of Algebras, and Vector Bundles. Proceedings, 1985. XIV, 383 pages. 1987.

Vol. 1274: N. C. Phillips, Equivariant K-Theory and Freeness of Group Actions on C*-Algebras. VIII, 371 pages. 1987.

Vol. 1275: C.A. Berenstein (Ed.), Complex Analysis I. Proceedings, 1985–86. XV, 331 pages. 1987.

Vol. 1276: C.A. Berenstein (Ed.), Complex Analysis II. Proceedings, 1985–86. IX, 320 pages. 1987.

Vol. 1277: C.A. Berenstein (Ed.), Complex Analysis III. Proceedings, 1985–86. X, 350 pages. 1987.

Vol. 1278: S.S. Koh (Ed.), Invariant Theory. Proceedings, 1985. V, 102 pages. 1987.

Vol. 1279: D. Ieşan, Saint-Venant's Problem. VIII, 162 Seiten. 1987.

Vol. 1280: E. Neher, Jordan Triple Systems by the Grid Approach. XII, 193 pages. 1987.

Vol. 1281: O.H. Kegel, F. Menegazzo, G. Zacher (Eds.), Group Theory. Proceedings, 1986. VII, 179 pages. 1987.

Vol. 1282: D.E. Handelman, Positive Polynomials, Convex Integral Polytopes, and a Random Walk Problem. XI, 136 pages. 1987.

Vol. 1283: S. Mardešić, J. Segal (Eds.), Geometric Topology and Shape Theory. Proceedings, 1986. V, 261 pages. 1987.

Vol. 1284: B.H. Matzat, Konstruktive Galoistheorie. X, 286 pages. 1987.

Vol. 1285: I.W. Knowles, Y. Saitō (Eds.), Differential Equations and Mathematical Physics. Proceedings, 1986. XVI, 499 pages. 1987.

Vol. 1286: H.R. Miller, D.C. Ravenel (Eds.), Algebraic Topology. Proceedings, 1986. VII, 341 pages. 1987.

Vol. 1287: E.B. Saff (Ed.), Approximation Theory, Tampa. Proceedings, 1985–1986. V, 228 pages. 1987.

Vol. 1288: Yu. L. Rodin, Generalized Analytic Functions on Riemann Surfaces. V, 128 pages, 1987.

Vol. 1289: Yu. I. Manin (Ed.), K-Theory, Arithmetic and Geometry. Seminar, 1984–1986. V, 399 pages. 1987.

Vol. 1290: G. Wüstholz (Ed.), Diophantine Approximation and Transcendence Theory. Seminar, 1985. V, 243 pages. 1987.

Vol. 1291: C. Mœglin, M.-F. Vignéras, J.-L. Waldspurger, Correspondances de Howe sur un Corps p-adique. VII, 163 pages. 1987.

Vol. 1292: J.T. Baldwin (Ed.), Classification Theory. Proceedings, 1985. VI, 500 pages. 1987.

Vol. 1293: W. Ebeling, The Monodromy Groups of Isolated Singularities of Complete Intersections. XIV, 153 pages. 1987.

Vol. 1294: M. Queffélec, Substitution Dynamical Systems – Spectral Analysis. XIII, 240 pages. 1987.

Vol. 1295: P. Lelong, P. Dolbeault, H. Skoda (Réd.), Séminaire d'Analyse P. Lelong – P. Dolbeault – H. Skoda. Seminar, 1985/1986. VII, 283 pages. 1987.

Vol. 1296: M.-P. Malliavin (Ed.), Séminaire d'Algèbre Paul Dubreil et Marie-Paule Malliavin. Proceedings, 1986. IV, 324 pages. 1987.

Vol. 1297: Zhu Y.-l., Guo B.-y. (Eds.), Numerical Methods for Partial Differential Equations. Proceedings. XI, 244 pages. 1987.

Vol. 1298: J. Aguadé, R. Kane (Eds.), Algebraic Topology, Barcelona 1986. Proceedings. X, 255 pages. 1987.

Vol. 1299: S. Watanabe, Yu.V. Prokhorov (Eds.), Probability Theory and Mathematical Statistics. Proceedings, 1986. VIII, 589 pages. 1988.

Vol. 1300: G.B. Seligman, Constructions of Lie Algebras and their Modules. VI, 190 pages. 1988.

Vol. 1301: N. Schappacher, Periods of Hecke Characters. XV, 160 pages. 1988.

Vol. 1302: M. Cwikel, J. Peetre, Y. Sagher, H. Wallin (Eds.), Function Spaces and Applications. Proceedings, 1986. VI, 445 pages. 1988.

Vol. 1303: L. Accardi, W. von Waldenfels (Eds.), Quantum Probability and Applications III. Proceedings, 1987. VI, 373 pages. 1988.

Vol. 1304: F.Q. Gouvêa, Arithmetic of p-adic Modular Forms. VIII, 121 pages. 1988.

Vol. 1305: D.S. Lubinsky, E.B. Saff, Strong Asymptotics for Extremal Polynomials Associated with Weights on ℝ. VII, 153 pages. 1988.

Vol. 1306: S.S. Chern (Ed.), Partial Differential Equations. Proceedings, 1986. VI, 294 pages. 1988.

Vol. 1307: T. Murai, A Real Variable Method for the Cauchy Transform, and Analytic Capacity. VIII, 133 pages. 1988.

Vol. 1308: P. Imkeller, Two-Parameter Martingales and Their Quadratic Variation. IV, 177 pages. 1988.

Vol. 1309: B. Fiedler, Global Bifurcation of Periodic Solutions with Symmetry. VIII, 144 pages. 1988.

Vol. 1310: O.A. Laudal, G. Pfister, Local Moduli and Singularities. V, 117 pages. 1988.

Vol. 1311: A. Holme, R. Speiser (Eds.), Algebraic Geometry, Sundance 1986. Proceedings. VI, 320 pages. 1988.

Vol. 1312: N.A. Shirokov, Analytic Functions Smooth up to the Boundary. III, 213 pages. 1988.

Vol. 1313: F. Colonius, Optimal Periodic Control. VI, 177 pages. 1988.

Vol. 1314: A. Futaki, Kähler-Einstein Metrics and Integral Invariants. IV, 140 pages. 1988.

Vol. 1315: R.A. McCoy, I. Ntantu, Topological Properties of Spaces of Continuous Functions. IV, 124 pages. 1988.

Vol. 1316: H. Korezlioglu, A.S. Ustunel (Eds.), Stochastic Analysis and Related Topics. Proceedings, 1986. V, 371 pages. 1988.

Vol. 1317: J. Lindenstrauss, V.D. Milman (Eds.), Geometric Aspects of Functional Analysis. Seminar, 1986–87. VII, 289 pages. 1988.

Vol. 1318: Y. Felix (Ed.), Algebraic Topology – Rational Homotopy. Proceedings, 1986. VIII, 245 pages. 1988.

Vol. 1319: M. Vuorinen, Conformal Geometry and Quasiregular Mappings. XIX, 209 pages. 1988.

Vol. 1320: H. Jürgensen, G. Lallement, H.J. Weinert (Eds.), Semigroups, Theory and Applications. Proceedings, 1986. X, 416 pages. 1988.

Vol. 1321: J. Azéma, P.A. Meyer, M. Yor (Eds.), Séminaire de Probabilités XXII. Proceedings. IV, 600 pages. 1988.

Vol. 1322: M. Métivier, S. Watanabe (Eds.), Stochastic Analysis. Proceedings, 1987. VII, 197 pages. 1988.

Vol. 1323: D.R. Anderson, H.J. Munkholm, Boundedly Controlled Topology. XII, 309 pages. 1988.

Vol. 1324: F. Cardoso, D.G. de Figueiredo, R. Iório, O. Lopes (Eds.), Partial Differential Equations. Proceedings, 1986. VIII, 433 pages. 1988.

Vol. 1325: A. Truman, I.M. Davies (Eds.), Stochastic Mechanics and Stochastic Processes. Proceedings, 1986. V, 220 pages. 1988.

Vol. 1326: P.S. Landweber (Ed.), Elliptic Curves and Modular Forms in Algebraic Topology. Proceedings, 1986. V, 224 pages. 1988.

Vol. 1327: W. Bruns, U. Vetter, Determinantal Rings. VII,236 pages. 1988.

Vol. 1328: J.L. Bueso, P. Jara, B. Torrecillas (Eds.), Ring Theory. Proceedings, 1986. IX, 331 pages. 1988.

Vol. 1329: M. Alfaro, J.S. Dehesa, F.J. Marcellan, J.L. Rubio de Francia, J. Vinuesa (Eds.): Orthogonal Polynomials and their Applications. Proceedings, 1986. XV, 334 pages. 1988.

Vol. 1330: A. Ambrosetti, F. Gori, R. Lucchetti (Eds.), Mathematical Economics. Montecatini Terme 1986. Seminar. VII, 137 pages. 1988.

Vol. 1331: R. Bamón, R. Labarca, J. Palis Jr. (Eds.), Dynamical Systems, Valparaiso 1986. Proceedings. VI, 250 pages. 1988.

Vol. 1332: E. Odell, H. Rosenthal (Eds.), Functional Analysis. Proceedings, 1986–87. V, 202 pages. 1988.

Vol. 1333: A.S. Kechris, D.A. Martin, J.R. Steel (Eds.), Cabal Seminar 81–85. Proceedings, 1981–85. V, 224 pages. 1988.

Vol. 1334: Yu.G. Borisovich, Yu. E. Gliklikh (Eds.), Global Analysis – Studies and Applications III. V, 331 pages. 1988.

Vol. 1335: F. Guillén, V. Navarro Aznar, P. Pascual-Gainza, F. Puerta, Hyperrésolutions cubiques et descente cohomologique. XII, 192 pages. 1988.

Vol. 1336: B. Helffer, Semi-Classical Analysis for the Schrödinger Operator and Applications. V, 107 pages. 1988.

Vol. 1337: E. Sernesi (Ed.), Theory of Moduli. Seminar, 1985. VIII, 232 pages. 1988.

Vol. 1338: A.B. Mingarelli, S.G. Halvorsen, Non-Oscillation Domains of Differential Equations with Two Parameters. XI, 109 pages. 1988.

Vol. 1339: T. Sunada (Ed.), Geometry and Analysis of Manifolds. Procedings, 1987. IX, 277 pages. 1988.

Vol. 1340: S. Hildebrandt, D.S. Kinderlehrer, M. Miranda (Eds.), Calculus of Variations and Partial Differential Equations. Proceedings, 1986. IX, 301 pages. 1988.

Vol. 1341: M. Dauge, Elliptic Boundary Value Problems on Corner Domains. VIII, 259 pages. 1988.

Vol. 1342: J.C. Alexander (Ed.), Dynamical Systems. Proceedings, 1986–87. VIII, 726 pages. 1988.

Vol. 1343: H. Ulrich, Fixed Point Theory of Parametrized Equivariant Maps. VII, 147 pages. 1988.

Vol. 1344: J. Král, J. Lukeš, J. Netuka, J. Veselý (Eds.), Potential Theory – Surveys and Problems. Proceedings, 1987. VIII, 271 pages. 1988.

Vol. 1345: X. Gomez-Mont, J. Seade, A. Verjovski (Eds.), Holomorphic Dynamics. Proceedings, 1986. VII, 321 pages. 1988.

Vol. 1346: O. Ya. Viro (Ed.), Topology and Geometry – Rohlin Seminar. XI, 581 pages. 1988.

Vol. 1347: C. Preston, Iterates of Piecewise Monotone Mappings on an Interval. V, 166 pages. 1988.

Vol. 1348: F. Borceux (Ed.), Categorical Algebra and its Applications. Proceedings, 1987. VIII, 375 pages. 1988.

Vol. 1349: E. Novak, Deterministic and Stochastic Error Bounds in Numerical Analysis. V, 113 pages. 1988.

Vol. 1350: U. Koschorke (Ed.), Differential Topology. Proceedings, 1987. VI, 269 pages. 1988.

Vol. 1351: I. Laine, S. Rickman, T. Sorvali, (Eds.), Complex Analysis, Joensuu 1987. Proceedings. XV, 378 pages. 1988.

Vol. 1352: L.L. Avramov, K.B. Tchakerian (Eds.), Algebra – Some Current Trends. Proceedings, 1986. IX, 240 Seiten. 1988.

Vol. 1353: R.S. Palais, Ch.-l. Terng, Critical Point Theory and Submanifold Geometry. X, 272 pages. 1988.

Vol. 1354: A. Gómez, F. Guerra, M.A. Jiménez, G. López (Eds.), Approximation and Optimization. Proceedings, 1987. VI, 280 pages. 1988.

Vol. 1355: J. Bokowski, B. Sturmfels, Computational Synthetic Geometry. V, 168 pages. 1989.

Vol. 1356: H. Volkmer, Multiparameter Eigenvalue Problems and Expansion Theorems. VI, 157 pages. 1988.

Vol. 1357: S. Hildebrandt, R. Leis (Eds.), Partial Differential Equations and Calculus of Variations. VI, 423 pages. 1988.

Vol. 1358: D. Mumford, The Red Book of Varieties and Schemes. V, 309 pages. 1988.

Vol. 1359: P. Eymard, J.-P. Pier (Eds.), Harmonic Analysis. Proceedings, 1987. VIII, 287 pages. 1988.

Vol. 1360: G. Anderson, C. Greengard (Eds.), Vortex Methods. Proceedings, 1987. V, 141 pages. 1988.

Vol. 1361: T. tom Dieck (Ed.), Algebraic Topology and Transformation Groups. Proceedings, 1987. VI, 298 pages. 1988.

Vol. 1362: P. Diaconis, D. Elworthy, H. Föllmer, E. Nelson, G.C. Papanicolaou, S.R.S. Varadhan. École d'Été de Probabilités de Saint-Flour XV–XVII, 1985–87. Editor: P.L. Hennequin. V, 459 pages. 1988.

Vol. 1363: P.G. Casazza, T.J. Shura. Tsirelson's Space. VIII, 204 pages. 1988.

Vol. 1364: R.R. Phelps, Convex Functions, Monotone Operators and Differentiability. IX, 115 pages. 1989.

Vol. 1365: M. Giaquinta (Ed.), Topics in Calculus of Variations. Seminar, 1987. X, 196 pages. 1989.

Vol. 1366: N. Levitt, Grassmannians and Gauss Maps in PL-Topology. V, 203 pages. 1989.

Vol. 1367: M. Knebusch, Weakly Semialgebraic Spaces. XX, 376 pages. 1989.

Vol. 1368: R. Hübl, Traces of Differential Forms and Hochschild Homology. III, 111 pages. 1989.

Vol. 1369: B. Jiang, Ch.-K. Peng, Z. Hou (Eds.), Differential Geometry and Topology. Proceedings, 1986–87. VI, 366 pages. 1989.

Vol. 1370: G. Carlsson, R.L. Cohen, H.R. Miller, D.C. Ravenel (Eds.), Algebraic Topology. Proceedings, 1986. IX, 456 pages. 1989.

Vol. 1371: S. Glaz, Commutative Coherent Rings. XI, 347 pages. 1989.

Vol. 1372: J. Azéma, P.A. Meyer, M. Yor (Eds.), Séminaire de Probabilités XXIII. Proceedings. IV, 583 pages. 1989.

Vol. 1373: G. Benkart, J.M. Osborn (Eds.), Lie Algebras, Madison 1987. Proceedings. V, 145 pages. 1989.

Vol. 1374: R.C. Kirby, The Topology of 4-Manifolds. VI, 108 pages. 1989.

Vol. 1375: K. Kawakubo (Ed.), Transformation Groups. Proceedings, 1987. VIII, 394 pages, 1989.

Vol. 1376: J. Lindenstrauss, V.D. Milman (Eds.), Geometric Aspects of Functional Analysis. Seminar (GAFA) 1987–88. VII, 288 pages. 1989.

Vol. 1377: J.F. Pierce, Singularity Theory, Rod Theory, and Symmetry-Breaking Loads. IV, 177 pages. 1989.

Vol. 1378: R.S. Rumely, Capacity Theory on Algebraic Curves. III, 437 pages. 1989.

Vol. 1379: H. Heyer (Ed.), Probability Measures on Groups IX. Proceedings, 1988. VIII, 437 pages. 1989